METHODS AND PROTOCOLS IN FOOD SCIENCE

For further volumes:
http://www.springer.com/series/16556

Methods and Protocols in Food Science series is devoted to the publication of research protocols and methodologies in all fields of food science.

Volumes and chapters will be organized by field and presented in such way that the readers will be able to reproduce the experiments in a step-by-step style. Each protocol will be characterized by a brief introductory section, followed by a short aims section, in which the precise purpose of the protocol will be clarified.

Standardized Procedures and Protocols for Starch

Edited by

Sneh Punia Bangar

Food, Nutrition and Packaging Sciences, Clemson University, Clemson, SC, USA

Humana Press

Editor
Sneh Punia Bangar
Food, Nutrition and Packaging Sciences
Clemson University
Clemson, SC, USA

ISSN 2662-950X ISSN 2662-9518 (electronic)
Methods and Protocols in Food Science
ISBN 978-1-0716-3865-1 ISBN 978-1-0716-3866-8 (eBook)
https://doi.org/10.1007/978-1-0716-3866-8

This Humana imprint is published by the registered company Springer Science+Business Media, LLC, part of Springer Nature.
The registered company address is: 1 New York Plaza, New York, NY 10004, U.S.A.

Paper in this product is recyclable.

Preface

Starch is a polysaccharide consisting of multiple glucose units and is synthesized in the amyloplasts of plant cells. Corn, wheat, rice, potato, and cassava starches are the most widely used starches in the food industry, serving as thickeners, colloid stabilizers, gelling and volume agents, adhesives, moisture retainers, texturizers, and fat substitutes. Starch plays an important role in the appearance, structure, and quality of food products. Each application of starch in the food or non-food industry requires particular functional characteristics. Native starches have unique physicochemical and functional characteristics. These properties include gelatinization temperature, retrogradation, solubility, swelling power, syneresis, and the rheological behavior of pastes and gels. The structural characteristics, shape, and botanical source of the starch influence these properties. Therefore, it is necessary to understand the specific characteristics of starch, experimentation, and protocols to meet new applications and demands in the industrial sector. This book aims to provide the protocols and methodology for understanding starch for its practical applications. Comprising 11 chapters authored by experts, this book is poised to empower researchers, food technology professionals, individuals in the food industry, and academia with a robust knowledge base on starch.

Chapter 1 highlights the significance of starch characterization and analysis and its impact on different industries. Starch characterization involves the assessment of parameters such as amylose and amylopectin content, granule size, molecular weight distribution, pasting properties, gelatinization behavior, and enzymatic digestibility. Chapter 2 introduces the research methods of starch granule morphology characteristics by combining the important methods of particle morphology characteristics research in recent years and provides a methodological reference for starch research. This chapter discusses the methods commonly used to evaluate the morphological characteristics of starch particles (such as optical microscopy, scanning electrical microscope (SEM), transmission electron microscope (TEM), polarizing microscope (PLM), confocal laser scanning microscopy (CLSM), and atomic force microscopy (AFM)). Chapter 3 focuses on an in-depth analysis of amylose and amylopectin required to understand the functional properties of starch. Characterization of amylose and amylopectin include physicochemical properties, thermal properties, rheological properties, and morphological properties. Advanced characterization techniques include different chromatographic methods, which aid in the assessment of molecular mass and chain-length distribution in both fractions. Chapter 4 discusses standardized the protocols of chromatographic techniques, sample preparation methods for different starch sources, and data interpretation of the results obtained. Chapter 5 discusses the paste properties of starch, as well as future perspectives and challenges to optimize its application in various food products. Chapter 6 examines the range of methods, protocols, and standardization of the rheological characterization of starches, an abundant complex carbohydrate biopolymer found in a variety of plants, including corn, tapioca, wheat, potatoes, peas, and rice. Chapter 7 includes an overview of the diffraction theory of X-rays by crystals, structural characterization of different starches using X-ray diffraction, and sample preparation protocols utilized in X-ray diffraction analysis for various starches. Detailed and well-explained protocol for XRD analysis of starch, along with data interpretation, will be helpful for researcher to prepare their research program regarding the structural characterization of various starches. In Chap. 8, the authors attempt to organize the starch digestibility methods under the headings of in vitro and in vivo methods. These subheadings cover the theory of the method used, materials required, sample preparation, detailed procedure, observations, inferences, and precautions, if applicable. Chapter 9 provides the reader with a basic understanding of the different physical modification methods and their influence on starch properties. Step-by-step procedures for physical modification methods have been added to simplify the understanding. In recent years, various procedures have been formulated to extract the starch with chemical means. Thus, to provide a complete overlay of chemical modification (hydrolysis, cross-linking, esterification, etherification, oxidation, and grafting, etc.) of starch, Chap. 10 has been designed, which also provides a complete overview of improved functional properties of starches after modification. Further, enzymatic modifications of native starch have been extensively explored as a green technology compared to toxic chemical approaches and energy-intensive physical approaches, resulting in modified starches with altered gelation, viscosity, and solubility characteristics. Chapter 11 provides a comprehensive discussion of protocols and methods of enzymatic modifications of starch.

Clemson, SC, USA *Sneh Punia Bangar*

Contents

Contributors

BILGE NAZLI ALTAY • *College of Engineering Technology, Packaging and Graphic Media Science, Rochester Institute of Technology, Rochester, NY, USA*

SONIKA BANYAL • *Guru Jambheshwar University of Science & Technology, Hisar, Haryana, India*

VANDANA CHAUDHARY • *College of Dairy Science and Technology, Lala Lajpat Rai University of Veterinary and Animal Sciences, Hisar, India*

WENJING CHEN • *College of Food Science, Fujian Agriculture and Forestry University, Fuzhou, China; Fujian Provincial Key Laboratory of Quality Science and Processing Technology in Special Starch, Fujian Agriculture and Forestry University, Fuzhou, China*

JEFERSON SILVA CUNHA • *Department of Food Technology (DTA), Federal University of Viçosa (UFV), University Campus, Viçosa, MG, Brazil*

AMANDA PEREIRA DA ROCHA • *Department of Food Engineering, Federal University of Ceará, Fortaleza, Ceará, Brazil*

LUCIANA GAMA DE MENDONÇA • *Department of Food Engineering, Federal University of Ceará, Fortaleza, Ceará, Brazil*

AASTHA DEWAN • *Department of Food Technology, Guru Jambheshwar University of Science and Technology, Hisar, India*

FABIO RIBEIRO DOS SANTOS • *Department of Food Technology (DTA), Federal University of Viçosa (UFV), University Campus, Viçosa, MG, Brazil*

MUKESH GAUR • *Department of Food Technology, Guru Jhambheshwar University of Science & Technology, Hisar, India*

RAKESH GEHLOT • *Centre of Food Science and Technology, CCS HAU, Hisar, Haryana, India*

ZEBIN GUO • *College of Food Science, Fujian Agriculture and Forestry University, Fuzhou, China; Fujian Provincial Key Laboratory of Quality Science and Processing Technology in Special Starch, Fujian Agriculture and Forestry University, Fuzhou, China*

RU JIA • *College of Food Science, Fujian Agriculture and Forestry University, Fuzhou, China; Fujian Provincial Key Laboratory of Quality Science and Processing Technology in Special Starch, Fujian Agriculture and Forestry University, Fuzhou, China*

PRIYANKA KAJLA • *Department of Food Technology, Guru Jambheshwar University of Science and Technology, Hisar, India*

RANJAN KAUSHIK • *Centre of Food Science and Technology, CCS HAU, Hisar, Haryana, India*

ANKIT KUMAR • *Centre of Food Science and Technology, CCS HAU, Hisar, Haryana, India*

PARVEEN KUMARI • *Department of Food Technology, Guru Jambheshwar University of Science & Technology, Hisar, India*

CHRISTOPHER L. LEWIS • *College of Engineering Technology, Manufacturing and Mechanical Engineering Technology, Rochester Institute of Technology, Rochester, NY, USA*

LU LIU • *College of Food Science, Fujian Agriculture and Forestry University, Fuzhou, China; Fujian Provincial Key Laboratory of Quality Science and Processing Technology in Special Starch, Fujian Agriculture and Forestry University, Fuzhou, China*

ANKUR LUTHRA • *Department of Food Technology, Guru Jambheshwar University of Science & Technology, Hisar, India*

MITALI MADHUMITA • *Department of Food Technology, School of Health Sciences and Technology, UPES, Dehradun, Uttarakhand, India*

PALAK MAHAJAN • *Department of Biotechnology Engineering and Food Technology, University Institute of Engineering (UIE), Chandigarh University, Mohali, Punjab, India*

SUJOSH NANDI • *Department of Agricultural and Food Engineering, Indian Institute of Technology Kharagpur, Kharagpur, West Bengal, India*

PUJA PRIYADARSHINI NAYAK • *Department of Agricultural and Food Engineering, Indian Institute of Technology Kharagpur, Kharagpur, West Bengal, India*

ANA FLÁVIA COELHO PACHECO • *Cândido Tostes Dairy Institute, Agricultural Company of Minas Gerais (EPAMIG), Juiz de Fora, MG, Brazil*

FLAVIANA COELHO PACHECO • *Department of Food Technology (DTA), Federal University of Viçosa (UFV), University Campus, Viçosa, MG, Brazil*

PAULO HENRIQUE COSTA PAIVA • *Cândido Tostes Dairy Institute, Agricultural Company of Minas Gerais (EPAMIG), Juiz de Fora, MG, Brazil*

REKHA PHOGAT • *Centre of Food Science and Technology, CCS HAU, Hisar, Haryana, India*

NEHA RANI • *Centre of Food Science and Technology, CCS HAU, Hisar, Haryana, India*

PRERNA SETHI • *Guru Jambheshwar University of Science & Technology, Hisar, Haryana, India*

KALYANI SHARMA • *Department of Biotechnology Engineering and Food Technology, University Institute of Engineering (UIE), Chandigarh University, Mohali, Punjab, India*

SHOBHIT • *Department of Food Technology, Guru Jambheshwar University of Science and Technology, Hisar, India*

SANGEETA YADAV • *Department of Food Technology, Guru Jhambheshwar University of Science & Technology, Hisar, India*

RAFAEL AUDINO ZAMBELLI • *Department of Food Engineering, Federal University of Ceará, Fortaleza, Ceará, Brazil*

BAODONG ZHENG • *College of Food Science, Fujian Agriculture and Forestry University, Fuzhou, China; Fujian Provincial Key Laboratory of Quality Science and Processing Technology in Special Starch, Fujian Agriculture and Forestry University, Fuzhou, China*

Chapter 1

Importance of Starch Characterization

Rafael Audino Zambelli, Amanda Pereira da Rocha, and Luciana Gama de Mendonça

Abstract

Starch, a complex carbohydrate widely found in plants, is crucial in numerous industries such as food, pharmaceuticals, textiles, and paper manufacturing. The characterization and analysis of starch are of utmost importance to gain insights into its structural and functional properties, enabling its optimal utilization in various applications. This abstract highlights the significance of starch characterization and analysis and its impact on different industries. Starch characterization involves the assessment of parameters such as amylose and amylopectin content, granule size, molecular weight distribution, pasting properties, gelatinization behavior, and enzymatic digestibility. These analyses provide valuable information about starch functionality and its behavior under different processing conditions. Starch characterization is vital for formulating and designing food products with desired texture, stability, and sensory attributes in the food industry. It allows for the selection of appropriate starch varieties and modification methods to achieve specific functional properties, such as thickening, gelling, and stabilization. Starch characterization and analysis are essential for understanding the structural and functional properties of starch and its suitability for various industrial applications. The accurate and comprehensive analysis provides valuable information for optimizing processing parameters, improving product quality, and promoting innovation in different industries. The continued advancement in starch characterization techniques contributes to the development of novel starch-based materials and products with enhanced performance, sustainability, and functionality.

Key words Analysis techniques, Functional properties, Industrial applications, Starch characterization, Structural properties

1 Introduction

Starch, composed of glucose monomers, is a significant component in human diets with diverse applications in the food industry and nonfood-related sectors such as paper, clothing, and construction [1]. Starches are derived from a wide range of plant sources, and this diversity plays a significant role in shaping their physical, physicochemical, and structural characteristics. The properties of starches can vary significantly depending on the specific plant from which they are obtained. Various plant species are utilized

Sneh Punia Bangar (ed.), *Standardized Procedures and Protocols for Starch*, Methods and Protocols in Food Science,
https://doi.org/10.1007/978-1-0716-3866-8_1,

for starch production, including corn, wheat, potatoes, rice, tapioca, and cassava, among others. Each of these plants possesses unique qualities that contribute to the distinctive properties of the resulting starch. Its numerous advantages include availability, affordability, ease of manufacturing, low density, high strength-to-weight ratio, inertness to active substances, biodegradability, biocompatibility, nontoxicity, and noninflammatory, making it a desirable choice for various consumer products [2].

The physical and chemical characteristics of starches, such as gelatinization temperature, swelling power, and solubility, are directly influenced by the composition and structure of the starch granules. The properties of these granules, including their size, shape, and arrangement, exhibit variations depending on the plant source [3]. Starch granules can vary in size from a few micrometers to several tens of micrometers, and their shapes can be round, oval, or irregular. These inherent characteristics are crucial in determining how starch behaves when exposed to water, heat, and other components during processing or cooking. Figure 1 shows the starch granules microstructure from different plant species in this context.

Starch granules can exist in two different compositions: simple or compound. The granules are present individually in the simple composition, while in the compound composition, they form

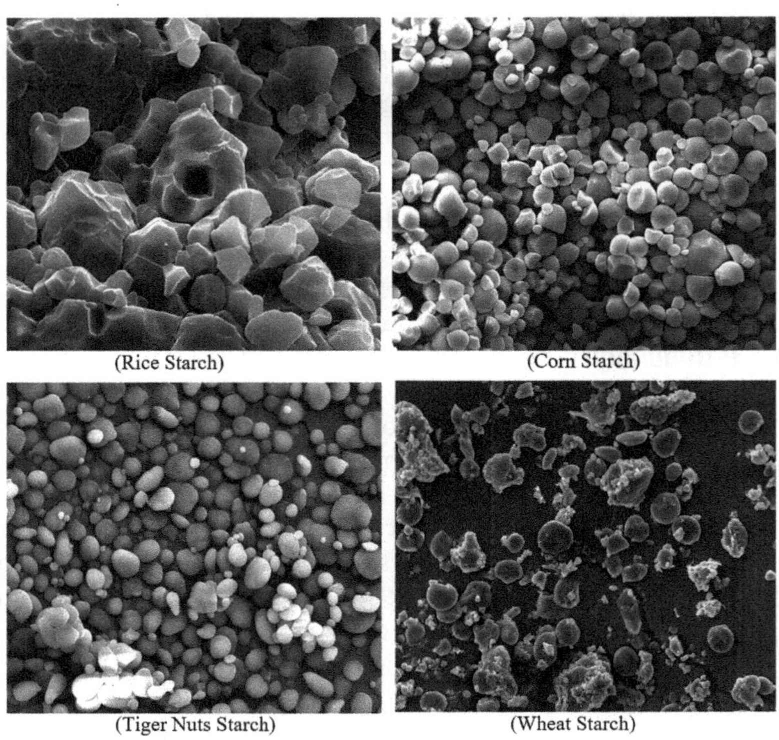

Fig. 1 Starch granules microstructure by scanning electron microscopy (SEM)

clusters. These granules exhibit a distinctive structure characterized by growth rings of amorphous and crystalline domains. Within the granule, a complex network of amylose and amylopectin molecules is responsible for the formation of these growth rings. The amylose and amylopectin are organized in alternating concentric shell-like structures, giving the granule its unique appearance. These shell-like structures have a thickness ranging from 120 to 400 nm [4].

The composition and organization of these growth rings influence the functional properties of starch granules. The crystalline regions provide the granules with structural integrity and play a role in determining their gelatinization temperature. The amorphous regions, on the other hand, contribute to the granules' solubility and swelling properties [5]. Starch has garnered significant attention in research and applications due to its exceptional film-forming ability, favorable biocompatibility, wide availability from diverse sources, low cost, renewable nature, and biodegradability. For several decades, starch has found applications in many industries, both in the food and nonfood sectors [6].

Understanding the variations in physical, physicochemical, and structural properties of starches obtained from different plant sources is essential for tailoring their use in various applications. Manufacturers can optimize their performance in food products, pharmaceuticals, textiles, paper, and many other industries by harnessing the unique characteristics of specific starches. Starch characterization is important in various fields, including food science, pharmaceuticals, materials science, and biofuels. Starch, a polysaccharide composed of glucose units, serves as a major energy source for plants and an essential component of our diet [7]. There are several compelling reasons that highlight the significance of starch characterization, which are listed below:

1. *Food science and nutrition:* Starch plays a pivotal role as an essential component in numerous food products, wherein its properties exert a profound influence on the quality, texture, and functional attributes of such foods. The process of starch characterization enables the determination of key factors, including starch composition, granule size, amylose–amylopectin ratio, gelatinization properties, retrogradation tendencies, and digestibility [8].
2. *Pharmaceutical industry:* Starch is widely employed as an excipient in pharmaceutical formulations, assuming various roles as a binder, disintegrant, and filler in tablets, capsules, and other dosage forms. The characterization process facilitates the evaluation of the physicochemical attributes of starch, including particle size, morphology, viscosity, and swelling behavior. These properties significantly impact drug release, formulation stability, and the overall performance of the pharmaceutical product [9].

3. *Materials science:* Starch has emerged as a prominent renewable and biodegradable polymer for the development of bioplastics, films, coatings, and other environmentally sustainable materials. Characterization methodologies offer valuable insights into the structural attributes, thermal behavior, mechanical strength, and barrier properties of starch-based materials [10].
4. *Biofuels and biorefineries:* Starch-based feedstocks, including corn and cassava, are frequently employed to produce biofuels, particularly ethanol. Characterization techniques are vital in assessing crucial parameters such as starch content, enzymatic digestibility, fermentation efficiency, and yield potential [11].
5. *Plant breeding and agriculture:* Starch characterization assumes a critical role within plant breeding programs focused on the development of crop varieties with desired starch properties. The analysis of starch characteristics aids plant breeders in selecting and developing cultivars featuring enhanced yield, improved nutritional composition, favorable processing properties, and increased resistance to diseases or environmental stressors [12].

2 Materials

2.1 Importance of Starch in the Food Industry

Starch is widely used in the food industry for its functional and nutritional properties. Some key contributions of starch in the food industry include the following:

2.1.1 Starch as Thickening

In general, the thickening agents used in the food industry consist of substances that can modify texture and rheological properties to improve sensory quality attributes. These substances act on the water-binding capacity, modify the structure, and change the flow properties [13]. By increasing the viscosity, they contribute to the desired thickness, smoothness, and stability of sauces, dressings, soups, and other liquid-based products. Thickeners can impart a pleasant mouthfeel, preventing unwanted separation or syneresis and ensuring a uniform texture throughout the product [14]; these thickeners are derived from diverse sources, including plants, microorganisms, and animal connective tissues.

In this context, starch is the most used hydrocolloid thickener due to the characteristics of being abundant in nature, cheap, not conferring flavor at low concentrations, and not interfering with color—the potential of potato starch as a thickener and its interaction with other components, such as pectin and starch. Potato starch is able to increase the viscosity in solution without swelling and disintegration of the starch, making it an alternative for use in the food industry, mainly because it is a clean label [15].

The starch extracted from the jackfruit seed was evaluated as a thickening agent in pepper sauce [16]. Starch had a higher amylose content, and its granules were smaller than traditional corn and potato starches. In this way, the jackfruit seed starch obtained greater thermal and mechanical shear resistance during the cooking process, suitable for use as a thickening agent. An increase in viscosity and flow resistance when adding chemically modified starch was verified. Differences in viscosity values and consistency indices (viscosity for non-Newtonian fluids) were small but significant [17].

The influence of time on the viscosity of two types of starch-based thickeners at a temperature of 20 °C was significant. The thickeners were produced with pure corn starch, maltodextrin, and nonionic polysaccharides. Viscosity was evaluated for 17 h at 30-min intervals after shearing. It was found that the thickeners showed a reduction in viscosity in the first 4 h. In addition, the thickener produced from maltodextrin was more stable compared to corn starch. This result may be related to the lower complexity of the structure of maltodextrin since it is derived from the enzymatic action of starch [18].

The application of cassava starch as a thickening agent in yogurt to avoid the phenomenon of syneresis, that is, the expulsion of liquid from the gel phase during product storage, was verified. The starches were added in the proportions of 2.5%, 5.0%, and 7.5% and caused an increase in the consistency of the yogurt, without any sensory damage, in addition to preventing the phenomenon of syneresis during 14 days of refrigerated storage of the yogurt samples [19]. Similar results were obtained in the application of thickening starches in the production of whipped yogurt. Starch inclusion was from 1.05% to 2.35%. The increase in starch concentration in yogurt formulations promoted an increase in viscosity and modulus of elasticity, as the starch granules absorbed a greater amount of water and resulted in thickening. Sensorially, the starches increased the sensory firmness and improved the cuteness of the yogurt. Higher starch inclusion values provided a cohesive texture feel [20].

2.1.2 Starch as a Binding and Gelling Agent

Starches are also recognized for their ability to act as binding and gelling agents in various industrial applications. They exhibit structure and molecular properties that provide excellent binding and gelling capabilities, making them indispensable ingredients in the food industry [13], and can bind ingredients together, providing cohesion and stability to various food formulations, improving the texture, consistency, and appearance of the final product. One of the functions of the starch binding agent can be expressed in baking, as it is used to bind the ingredients of the dough, helping to create a cohesive structure and a defined shape [21]. Its binding properties also contribute to its functionality in baked goods. In bread-making, for instance, starch can interact with gluten proteins to form a network that traps gases produced during fermentation. This binding action enhances the volume and structure of the bread, resulting in a softer texture and improved crumb quality [22].

Naturally, starch and proteins can interact with each other. Thus, mixed gel systems with different proportions of starch and protein extracted from fava beans were studied to understand the binding capacity between these two polysaccharides. It was found that the affinity for iodine was lower for samples with high protein content. In gel samples with a higher proportion of starch, there was a higher gelling capacity and viscosity. The starch granules were compacted, and the protein structures filled in the empty spaces between the starch granules. Finally, the protein content reduced the viscosity of the gels and provided them with greater softness [23].

The gelled structure formed by starch provides desirable textural attributes to food products. Starch gels can contribute to the smoothness, firmness, and stability of various food formulations. For example, in desserts such as custards, puddings, or pie fillings, starch gels can provide the desired consistency and mouthfeel. In fruit preserves or jams, starch gels can help achieve the desired gel-like texture and enhance the suspension of fruit particles [24]. In meat products such as sausages or burger patties, starch can help bind the meat particles and other ingredients, improving texture, juiciness, and overall product integrity [25]. The addition of pea protein improves the syneresis of starch solutions and stability during storage. This is probably due to the interaction of protein-starch hydrophilic groups and between the starch molecules themselves [26].

Improvement in the stability of starch gels was also identified by providing a mixture of isolated soy protein with different amounts of native corn starch and waxy corn starch [27]. The authors observed that gel strength, water-holding capacity, hydrophobic interactions, thermal stability, and gel strength increased with the inclusion of the protein. Corn starch exhibits excellent binding properties with phenolic compounds, and this behavior improves the physicochemical properties of starch, in addition to adding biological activities to phenolic compounds [28], which may have direct applications in the manufacturing of packaging films, food production, vehicles of biologically active substances, etc. The binding and gelling properties of starch can be further enhanced through modifications such as physical, chemical, or enzymatic treatments. These modifications can alter the molecular structure of starch, resulting in modified starches with specific binding and gelling properties tailored to applications [29].

2.1.3 Starch as Fat Replacement

The increasing consumer demand for low-calorie food products has led to a surge in research and development efforts by food manufacturers. This is driven by the growing awareness of the negative impact of high-energy diets on human health [30]. Starch serves as a promising fat-replacement ingredient in various food applications. Its ability to mimic the texture, mouthfeel, and functionality of fats makes it an attractive option for reducing the fat content in food formulations without compromising sensory attributes [31]. It can be used as a fat replacement in certain food products to reduce the fat content and create a lower calorie alternative and can help provide a similar texture and mouthfeel to that of fat in certain applications. It can contribute to the creamy and smooth texture desired in products such as dressings, sauces, and desserts [32].

One of the primary reasons for using starch as a fat replacement is its ability to contribute to the desired texture and creaminess in food products. Starch can create a perception of richness and smoothness in products such as sauces, dressings, and dairy alternatives. By incorporating starch, manufacturers can achieve a similar mouthfeel and sensory experience to full-fat products [33]. Modified arrowroot starch as a fat substitute in mayonnaise was studied. The starch was modified through octenyl succinic anhydride, citric acid hydrolysis, annealing, heat and moisture treatment, and acetylation. The aim was to produce different starch pastes with 30–50% fat replacement in mayonnaise formulations. Fat-reduced mayonnaise had greater emulsion stability, especially those with starches modified by octenyl succinic anhydride, annealing, and citric acid hydrolysis [34].

The effects of including waxy maize starch as a fat substitute in beef sausage formulations were studied. In general, corn starch promoted an increase in water retention and cooking loss of sausages, in addition to increasing sensory acceptance and reducing the total caloric content, reducing fat by up to 57.4% and 34.9% in the calorific value [35]. Reducing the fat content in bakery product formulations has become an important issue, especially in sugar cookies. In this sense, they developed a hybrid gel formed by mixing canola oil and candelilla oil gel to replace fat in biscuits. It was found that the inclusion of the hybrid gel promoted greater hydration and disorder in the secondary structure of starch and proteins, in addition to obtaining a 50% reduction in fat content [36]. Mungbean starches were modified by substituting functional groups with acetyl succinic as a fat substitute in cakes. It was observed that the treatment with octenyl succinic increased the viscosity of the paste and starch and increased the specific volume of the cakes without any sensory damage [37]. It is worth mentioning that various fat replacement strategies exist, and starch is just one option. Other ingredients and techniques, such as gums, fibers, emulsifiers, and innovative food processing methods, are also utilized to reduce fat content in food products [38].

Fig. 2 Amylose and amylopectin structure

2.2 Composition and Structure of Starch

Starch, a complex polysaccharide composed of two main components, namely, amylose and amylopectin, exhibits a unique structural arrangement. Its multiscale structure is complex because of the spatial alignment of amylose and amylopectin. Figure 2 shows the structures of amylose and amylopectin.

Amylose possesses a linear structure comprising glucopyranose units that are connected by α-D-(1–4) glycosidic bonds. On the other hand, amylopectin exhibits a branched structure in which the linear branches are linked by α-D-(1–4) glycosidic bonds, with periodic connections formed by α-D-(1–6) glycosidic bonds [39]. Amylose exhibits a unique property known as iodine affinity. This property allows amylose to interact with iodine molecules, forming an inclusion complex with a distinct bluish color. The formation of this complex is dependent on the size and structure of amylose. The iodine affinity of amylose can be quantified by measuring the intensity of the bluish color, which provides information about the amount of amylose present and its structural characteristics. Higher iodine affinity indicates a larger proportion of amylose and a more organized and compact structure [40, 41].

This iodine affinity test is commonly used to determine the relative amounts of amylose and amylopectin in starch samples. By measuring the intensity of the bluish color formed during the reaction with iodine, the amylose content can be estimated. Starches with high amylose content will show a more intense blue coloration, while those with high amylopectin content will have a weaker or no color change [42]. In addition to providing insights into starch composition, the iodine affinity test can also be used to study the structural variations of amylose. Different sources of starch and variations in processing conditions can lead to variations in amylose structure, affecting its interaction with iodine and resulting in differences in color intensity [39]. Amylopectin, which accounts for most starch composition, consists of external and internal chains [43]. The external chains, known as A-chains, play a significant role in the crystalline regions of starch. These A-chains are primarily responsible for the formation of tightly packed crystalline flakes within the starch granule. The crystalline regions contribute to the granule's stability and resistance to enzymatic degradation [44].

The internal chains, referred to as B-chains, form an amorphous structure within the starch granule. These B-chains are mainly present in the amorphous flakes dispersed between the crystalline regions. The amorphous flakes provide flexibility and contribute to the swelling and solubility properties of starch [45]. The differential distribution of the A- and B-chains within the granule affects the overall functionality and behavior of starch. The crystalline flakes formed by the A-chains contribute to the granule's rigidity and influence parameters such as gelatinization temperature and retrogradation. In contrast, the amorphous structure of the B-chains influences properties such as viscosity, pasting behavior, and digestibility [46].

However, it is important to point out that starch composition is variable and depends on several factors, the botanical origin being the main one. It provides different ratios between amylose and amylopectin contents, which directly impacts the physical, physicochemical, and structural properties of starch. For example, the separation between rice amylopectin and amylose occurs at a degree of polymerization (DP) of around 100, with a noticeable transition (commonly observed at DP ~ 100). Amylose is characterized by DP > 100, while amylopectin consists of chains with DP $\leq$ 100 [47].

Although high-amylose maize, barley, and potato are already accessible in the market, the development of high-amylose variants for other major crops such as wheat and rice is a more recent endeavor. These variants are expected to become commercially available in the coming years. Corn starch is known for its relatively high amylopectin content, which imparts a greater viscosity to its gel-like structures than other starches. On the other hand, potato starch contains a higher proportion of amylose, resulting in a more rigid gel network. Table 1 shows the amylose and amylopectin composition for different starch types.

The amylose and amylopectin contents of starch can vary significantly depending on factors such as the specific variety of the crop, the growing conditions it is subjected to, and the methods used for processing. It's important to acknowledge that these values are approximate and can differ between different plant varieties, cultivation practices, and processing techniques. Factors such as soil composition, climate, crop maturity stage at harvest, and post-harvest processing methods can all influence starch composition in terms of its amylose and amylopectin proportions. The amylose and amylopectin ratios in starches significantly influence their functional and physicochemical properties, including

Table 1
Composition of amylose and amylopectin for different types of starch

	Botanical origin of starch	Amylose (%)	Amylopectin (%)	References
Cereals	Wheat	25–28	72–75	[48]
	Corn	25–30	70–75	[49]
	Rice	15–30	70–85	[50]
	Oat	25–29	70–74	[51]
Tubers	Potato	20–25	75–80	[52]
	Tapioca	15–20	80–85	[7]
	Yam	20–25	75–80	[53]
	Cassava	15–25	75–85	[54]
	Tiger nuts	11.5–19.1	80.9–88.5	[55]
	Taro	18.1–22.2	75–85	[56]
Legumes	Faba	30–35	60–70	[57]
	Chickpea	30–40	60–70	[58]
	Lentil	23–32	65–75	[59]
Others	Lotus seeds	35–45	50–60	[60]
	Jackfruit	22.1–38.3	70–80	[61]

gelatinization, paste viscosity, gel stability, and solubility. It is worth noting that the amylose content can vary depending on the extraction method used for obtaining the starch. Starch extracted using water extraction methods tends to have a slightly higher amylose content. This higher amylose content enhances the film-forming ability of the starch, making it suitable for applications where film formation is desirable, such as in the production of edible films or coatings [56, 62].

The affinities of the starch components, namely, amylose and amylopectin, exhibit distinct behaviors when interacting with lipids. Amylose displays a pronounced ability to interact with lipids, forming a single-helix V-type complex [63]. Amylopectin exhibits a limited capacity to interact with lipids. This limitation can be attributed to the complex structure of its short-branched chains and the significant steric effects, which impede sufficient contact between amylopectin and lipids. Starch, being a natural polymer with a high molecular weight, possesses numerous free hydroxyl groups along its molecular chain. These hydroxyl groups facilitate the formation of hydrogen bonds between adjacent starch molecules, resulting in the development of a microcrystalline structure. This complex phenomenon gives rise to alternating regions of crystalline and noncrystalline zones within the starch particles [56].

3 Methods

3.1 Methods and Techniques for Starch Characterization

Starch, being a complex polysaccharide, requires a comprehensive set of methods and techniques to unravel its intricate characteristics. This section provides an overview of the commonly employed approaches for starch characterization, encompassing various analytical tools and methodologies. By delving into imaging techniques, spectroscopic analysis, thermal analysis, and other relevant methods, this chapter aims to provide researchers and scientists with a comprehensive toolkit for investigating the physical, chemical, and structural properties of starch. These characterization techniques offer valuable insights into starch functionality, processing, and its diverse applications in fields such as food science, pharmaceuticals, materials engineering, and more. The multiscale structure of starch granules can be analyzed and characterized using various technologies. These techniques provide valuable insights into the hierarchical organization and properties of starch at different levels [64].

In recent years, advanced imaging techniques have also emerged as powerful tools in the food structure. Techniques such as confocal microscopy, atomic force microscopy (AFM), and magnetic resonance imaging (MRI) allow for the visualization and analysis of microstructure and deformation within food samples. These imaging methods offer valuable insights into the structural changes occurring during processing and storage and their impact on the structure properties of food [65]. At the macroscopic level, imaging techniques such as scanning electron microscopy (SEM) and optical microscopy allow for the visualization of the overall size, shape, and surface morphology of starch granules. SEM provides high-resolution images, revealing details of the granule surface and any structural irregularities. Optical microscopy, on the other hand, enables the examination of granules in a larger field of view, aiding in the assessment of granule distribution and aggregation [66]. Moving to a finer scale, techniques such as transmission electron microscopy (TEM) and atomic force microscopy (AFM) provide detailed information on the internal structure and organization of starch granules. TEM allows for the visualization of the internal substructures, such as lamellae and crystalline regions, within the granule. It also enables the measurement of granule size, shape, and degree of crystallinity. On the other hand, AFM offers high-resolution imaging of the granule surface at the nanoscale, providing information about surface topography and molecular interactions [67, 68].

X-ray diffraction (XRD) is another powerful tool used to study the crystalline structure of starch granules. By analyzing the scattering patterns of X-rays passing through the granules, XRD provides information about the arrangement and packing of starch molecules within the crystalline regions. This technique can determine parameters such as crystallinity index, crystal size, and unit cell dimensions, contributing to a comprehensive understanding of the granule's structural properties [69].

In addition to imaging and diffraction techniques, spectroscopic methods Fourier-transform infrared spectroscopy (FTIR) and nuclear magnetic resonance (NMR) spectroscopy are employed to study the chemical composition and molecular interactions within starch granules. FTIR provides information about functional groups and chemical bonds present in the granule, aiding in the identification of different starch components. On the other hand, NMR spectroscopy can reveal insights into the molecular dynamics and mobility of starch molecules within the granule [70].

3.2 Characterization Methods

Characterization methods play a crucial role in understanding the properties and behavior of starch. These methods involve the analysis of various physical, chemical, and physicochemical aspects of starch samples, allowing researchers to obtain valuable insights into their structure, morphology, thermal properties, and rheological behavior. This section will discuss some commonly employed physical characterization methods for starch.

3.2.1 Differential Scanning Calorimetry

Differential scanning calorimetry (DSC) is a widely used technique for analyzing the thermal behavior of materials, including starch. It provides valuable insights into the changes in heat flow that occur during heating or cooling processes, allowing researchers to investigate various thermal transitions, such as gelatinization, melting, crystallization, and chemical reactions. By measuring the heat flow associated with these processes, DSC enables the characterization of the thermodynamic properties and stability of starch samples and is a valuable analytical technique that provides both qualitative and quantitative information about the thermal properties of solid materials. By subjecting a sample to controlled temperature changes, DSC allows for the measurement of key parameters such as melting temperature, degradation temperature, glass transition temperature, enthalpy of melting and crystallization, specific heat, latent heat, polymorphism, and material purity [71]. DSC enables the determination of melting temperatures, which indicates the point at which a solid material transitions into a liquid state. It also provides insights into the degradation temperature, which indicates the temperature at which the material starts to break down or undergo chemical decomposition. These temperature parameters are essential in understanding the stability and behavior of materials under different heating conditions [72]. The variation in heat flow is plotted as a function of temperature or time to generate a typical thermogram, as depicted in Fig. 3.

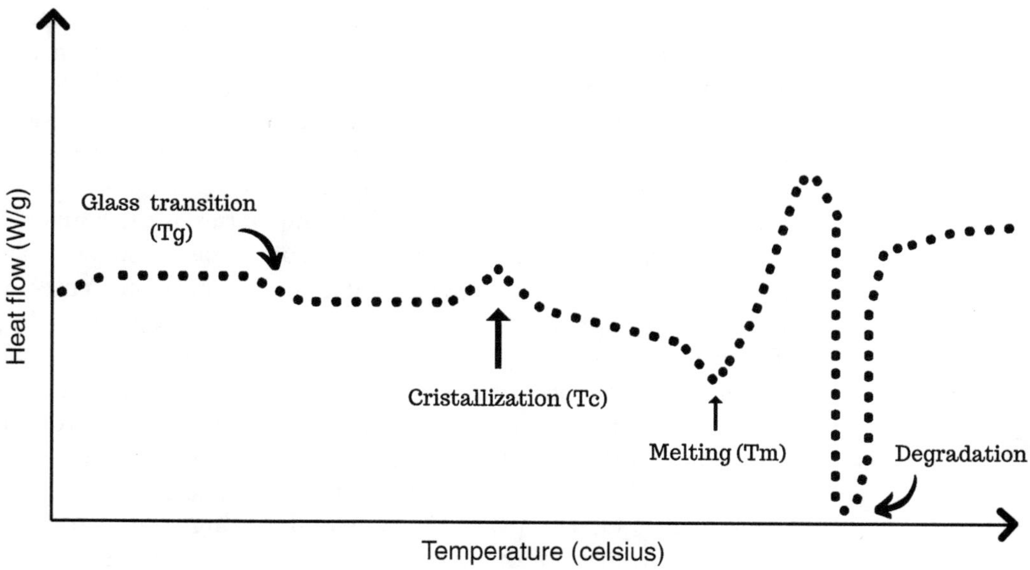

Fig. 3 Schematic representation of thermal transitions in a semicrystalline material observed in a DSC thermogram

This graphical representation provides a visual depiction of the changes in heat flow that occur during the heating or cooling process of a material. By analyzing the thermogram, researchers can gain insights into the thermal behavior, transitions, and energy changes associated with the material under investigation. The thermogram obtained through this method serves as a valuable tool for the interpretation and characterization of the material's thermal properties. It is worth noting that first-order phase transitions manifest as well-defined peaks, while second-order transitions result in variations in the heat flux curve. The initial slope of the curve corresponds to the glass transition, which represents the physical transformation that certain materials undergo when cooled below their glass transition temperature. This transition is reversible in amorphous materials, as they can shift from a rigid, glassy state to a more flexible state. During the glass transition, the molecular mobility of the material increases, leading to significant changes in its physical properties. The glass transition temperature is influenced by various factors, including the presence of plasticizers or additives, molecular weight, and chain flexibility [73].

As the temperature increases during the analysis, the material absorbs energy and undergoes microstructural rearrangement, resulting in an exothermic peak known as the crystallization temperature. This temperature marks the transition from a liquid or amorphous state to a crystalline state. During crystallization, the molecules within the material align themselves into an ordered, crystalline structure. The crystallization temperature is an important parameter for understanding the behavior and properties of materials, providing insights into the energy requirements and kinetics of the crystallization process.

The material absorbs sufficient energy at higher temperature levels to break intermolecular interactions, causing the molecules to separate. As a result, the material's viscosity decreases and melts at the melting temperature. The melting temperature is the temperature at which a solid material transitions from a crystalline state to a liquid state. At this point, the ordered arrangement of molecules in the crystalline lattice is lost, leading to a loss of structural integrity. The melting temperature is significant in assessing the thermal stability, crystalline structure, and purity of the material [74]. Finally, as more energy is imparted to the material through temperature elevation, processes such as decomposition and oxidative degradation occur, as observed during the Td period. Measuring the degradation period of a material using DSC provides valuable information about its decomposition characteristics and thermal stability. It helps in understanding how the material behaves under elevated temperature conditions and the potential for chemical breakdown or degradation [75]; therefore, it is a technique that can provide diverse information about the different types of native and modified starches.

Native starch, despite its wide availability, faces limitations in meeting the requirements of various engineering applications due to its inherent water sensitivity and subpar mechanical properties. Moreover, the high melting point of starch often surpasses its thermal degradation temperature, further constraining its potential for use in engineering applications [15]. Starch granules exhibit a complex internal structure comprising both crystalline and amorphous components. The amorphous regions lack a regular arrangement, while the crystalline regions consist of ordered chains of glucose molecules. The proportion and organization of these structures within the granules significantly impact the physicochemical properties and thermal behavior of starch [76].

The amorphous/crystalline ratio in starch granules affects their solubility, digestibility, and functional properties. Higher amorphous content generally leads to increased water absorption capacity, swelling power, and improved gel-forming ability. On the other hand, a higher proportion of crystalline regions tends to enhance the granules' resistance to enzymatic degradation and retrogradation [77]. The presence of crystalline regions contributes to the granules' melting and gelatinization temperatures, while the amorphous regions play a role in their glass transition temperature. These thermal properties are crucial for various food processing applications, such as determining the optimal heating conditions for achieving desired textural and sensory attributes in starch-based products [78, 79].

DSC is a widely used technique for investigating the thermal behavior of isolated starches. It allows for the determination of transition temperatures and corresponding enthalpies, providing valuable insights into the properties of starch. Gelatinization has been extensively studied due to its significance in starch processing for food and nonfood applications [80]. During the process of gelatinization, starch granules undergo a transformation when heated in water, resulting in the formation of a thick starch paste. Gelatinization involves the swelling of the granules and the disruption of their molecular structure, allowing them to absorb water and form a viscous paste. This phenomenon is crucial in various food preparation processes, as it contributes to desirable texture and consistency [81]. The gelatinization process of starch is characterized by a nonsharp transition, primarily due to the polydispersity of starch granules. Although gelatinization typically occurs within a temperature range of 60–80 °C, it does not occur abruptly [82]. Several factors influence the starch gelatinization process, including the starch source, amylose content, moisture content, and heating profile. These factors play a significant role in determining the extent and ease of gelatinization. For instance, high-amylose starch tends to have a lower susceptibility to gelatinization than regular starch. Moreover, higher moisture content and elevated heating temperatures typically result in more pronounced starch gelatinization [41, 82, 83].

Apart from the inherent structure of starch, various external factors can significantly influence its properties and interactions with water. These factors include temperature, pressure, water content, and small molecules such as salt, sugar, acids, and other macromolecules like hydrocolloids. These conditions can profoundly impact starch, altering its behavior, functionality, and physical characteristics when exposed to water. Understanding and controlling these external factors are essential for manipulating starch properties and optimizing its performance in different applications [84].

The determination of starch gelatinization is commonly carried out using the differential scanning calorimetry (DSC) technique. In this method, starch is subjected to controlled heating in the presence of water within a sealed container. The temperature at which gelatinization begins (onset), the peak temperature (Tp), the temperature at which gelatinization is completed (conclusion), and the associated enthalpy change (ΔH) is measured and recorded. It is worth noting that the DSC parameters can be influenced by various factors, including the rate at which heating occurs, the ratio of starch to water, and the specific experimental conditions employed. These variables can affect the observed values of To, Tp, Tc, and ΔH, potentially influencing the interpretation of the gelatinization process [85].

This information is vital for understanding the functionality and processing characteristics of starch in food formulations and the development of starch-based materials for nonfood applications. By subjecting starch samples to controlled heating or cooling cycles, DSC measures the heat flow associated with phase transitions, such as gelatinization, melting, and retrogradation. This technique provides valuable information about the temperature ranges at which these transitions occur and the corresponding enthalpy changes [86, 87]. DSC measures the heat flow and enthalpy changes that occur in a sample as a function of temperature or time, providing valuable information about phase transitions, thermal stability, and energy changes associated with the studied material [88].

The sample is typically placed in a hermetically sealed container and subjected to a controlled heating or cooling program. A reference material, which ideally does not undergo any phase transitions in the temperature range of interest, is also included for comparison. As the temperature changes, the DSC instrument measures the heat flow, which consists of a sample and reference thermocouple connected to separate temperature-controlled cells [89]. From the DSC analysis, it is possible to identify the hydrogen bonds of starches because of the elongation of the O–H bond and the vibration of the hydrogen bond and glycosidic bonds in the polysaccharide and is evidenced by peaks in known regions [90].

DSC can be used to measure the presence of exothermic or endothermic changes. In the case of starch, endothermic peaks can be observed, which are associated with its thermal degradation and subsequent carbonization at temperatures close to 300 °C [91]. This technique allows for the identification and characterization of the thermal properties of starch, providing valuable insights into its behavior and stability under different temperature conditions. Studies in the literature have reported modifications in the thermal properties of modified starches, particularly using branching enzymes and transglucosidase. In these cases, the modified starches exhibit lower gelatinization enthalpy than native starches, which can be attributed to the reduced presence of double helices in the modified starches. Consequently, the reduction in the amount of double helices in the starch structure lowers the energy required to disrupt the crystalline structure. These findings highlight the impact of enzymatic modifications on the thermal behavior of starches and provide valuable insights into the relationship between starch structure and its thermal properties [92, 93]. DSC can confirm the formation of complexes of other substances with starch since the dissociation of the complex formed during the heating promoted by the analysis results in one more endothermic peak, which was observed and confirmed by [76, 80].

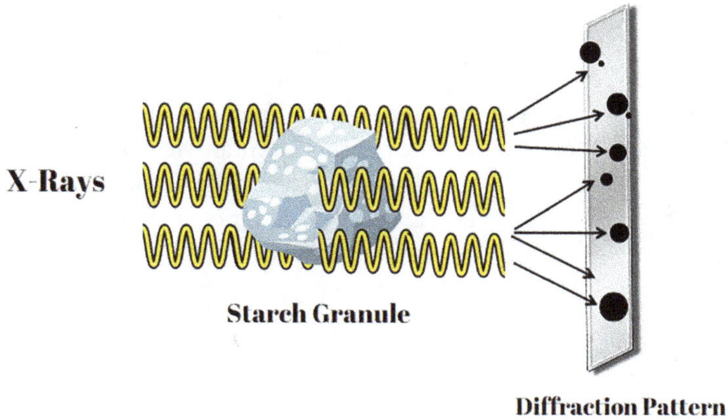

Fig. 4 Operation of X-ray analysis to obtain the diffraction pattern for starch samples

3.2.2 X-Ray Diffraction

X-rays, which are also known as X-ray radiations, are a type of high-energy electromagnetic radiation capable of penetrating various materials. They possess frequencies ranging from 30 × 1015 Hz to 30×10^{18} Hz, energies spanning from 145 eV to 124 keV, and wavelengths falling between 1×10^{-12} m and 1×10^{-9} m. As a result, X-rays have wavelengths shorter than those of ultraviolet rays but longer than those of gamma rays [62]. Figure 4 demonstrates how the X-ray analysis works to obtain the diffraction pattern for starch samples.

Initially, X-rays are propagated toward the starch sample, which, due to part of its structure being crystalline, deviates the X-rays according to the degree of crystallinity, forming a characteristic diffraction pattern for each type of starch. It is a widely used technique for analyzing the structural properties of starch. XRD involves passing X-rays through a sample and measuring the resulting diffraction pattern. This pattern provides information about the arrangement of atoms or molecules within the sample, including the crystalline structure of starch [94].

When applied to starch, XRD can reveal important details about its crystalline regions, as starch granules contain both amorphous and crystalline components. By analyzing the diffraction pattern, XRD can determine the degree of crystallinity, crystal size, and crystalline polymorphs present in the starch sample [95]. Starch typically exhibits a characteristic XRD pattern with distinct diffraction peaks corresponding to its crystalline structure. The positions and intensities of these peaks can be used to identify specific crystalline forms of starch, such as A-type, B-type, V-type, or C-type. The relative intensities of these peaks can provide insights into the organization and packing of starch molecules within the granules [96]. Starch granules display a semi-crystalline structure characterized by alternating crystalline and amorphous lamellae. Starch granules can adopt three crystalline forms depending on the degree of hydrolysis and the arrangement of amylose double helices. The A-type form produces square-shaped nanocrystals composed of left-handed double helices tightly packed in a monoclinic unit cell. The B-type form generates round nanocrystals with a high amylose content, formed by six double helices in a loosely packed hexagonal unit cell. Lastly, the C-type form consists of a B-type crystalline core structure surrounded by A-type crystalline starch granule crystals [65].

The arrangement of helices and the presence of water molecules between them facilitate the formation of starch crystallites through hydrogen bonding. This phenomenon enables crystal unit assembly, contributing to starch structure development at a long-range scale [40]. XRD analysis is particularly valuable for studying the effects of various factors on the crystalline structure of starch. For example, changes in processing conditions, such as temperature, moisture content, or additives, can alter the degree of crystallinity and the crystalline polymorphs present in starch. XRD can be used to assess these structural changes and understand their impact on the functional properties of starch, such as digestibility, gelatinization behavior, and textural characteristics [97, 98].

This analysis can provide information on crystallinity and pattern of botanical origin; it is a nondestructive and robust analysis. However, it is limited to essentially crystalline materials, is time-consuming, requires large amounts of sample, and may have low diffracted X-ray intensity depending on the material to be analyzed [99]. It also provides valuable information about the crystalline and amorphous regions within starch. Larger peaks in the patterns correspond to the crystalline regions, which are primarily associated with the presence of amylopectin. On the other hand, smaller peaks represent the amorphous regions characterized by amylose [100].

X-ray diffraction provides a means to investigate alterations in the crystalline fraction of starch caused by hydrothermal effects, such as cooking. These effects disrupt the type A polymorphs and induce changes in the molecular-level hydrogen bonding. Consequently, the previously disordered starch chains undergo reassembly into helical structures, leading to the emergence of B-type polymorphs [101]. Therefore, it is widely employed as a primary method for examining the crystalline nature and interatomic distances of specimens. In a particular investigation, native arrowhead starch samples exhibited a crystalline pattern consistent with the A-type structure, characterized by diffraction peaks at reflection angles of 15°, 17°, 18°, and 23° [102].

Starch derived from cereal grains typically exhibits diffraction peaks at diffraction angles of approximately 15°, 17°, 18°, and 23°. These diffraction patterns confirm that the starch possesses the typical characteristics of A-type cereal starch. This finding provides valuable insights into the structural properties of cereal grain starch and further contributes to our understanding of its composition and behavior [103]. The gelatinization process of starch can cause the double helical structure to break down, resulting in the loss of characteristic peaks associated with starch [104]. XRD analysis revealed that the formation of a starch-polyphenol complex follows a similar pattern to the formation of a starch-lipid complex, forming a V-type lens structure [105]. This observation suggests that the interaction between starch and polyphenols is similar to that between starch and lipids. The V-type lens structure formed in both complexes indicates a specific arrangement of the molecules, which is critical for their functional properties and potential applications [106]. It is possible to visualize the phenomenon of retrogradation of starches through XRD analysis since the starch molecules reassemble to form recrystallization through hydrogen bonds. The formation of crystals at peak $2\theta = 17°$ in potato starch is due to recrystallized amylopectin [107].

3.2.3 Scanning Electron Microscopy

Starch has a unique hierarchical structure that affects its digestion and subsequent nutritional impact. At the microscopic level, starch granules are composed of layers known as lamellae, consisting of alternating amylose and amylopectin regions [108]. SEM is a powerful technique used to examine the microstructure and surface characteristics of starch. It allows for detailed observations of starch granules at high magnification, providing valuable insights into their morphology, size, shape, and surface features [109]. When applied to the starch analysis, SEM can provide the following information:

1. *Granule size and shape:* SEM enables the measurement and characterization of starch granules in size and shape. Starch granules can vary significantly in size, ranging from a few micrometers to several tens of micrometers. SEM images allow researchers to determine the size distribution of starch granules in a sample. The shape of starch granules can also vary, such as spherical, ellipsoidal, oval, or irregular, and SEM can capture these variations [110].
2. *Surface characteristics:* SEM provides a detailed view of the surface features of starch granules. Starch granules can have a smooth or rough surface, and SEM images can reveal the presence of surface irregularities, pores, cracks, or even fine structural details. These surface characteristics can influence the functional properties of starch, such as its water absorption capacity, gelatinization behavior, and interactions with other food components during processing [111].
3. *Structural organization:* SEM allows for the examination of starch granules at a microscale level, providing insights into their internal structure and organization. It can reveal the

presence of lamellae within starch granules, which are composed of alternating amylose and amylopectin regions. The organization and arrangement of these lamellae can affect the digestibility and functional properties of starch [15].

4. *Starch modifications:* SEM can also be used to study starch modifications, such as physical, chemical, or enzymatic treatments. Researchers can assess the impact of various treatments on the granule structure and surface morphology by comparing SEM images of modified starches with unmodified ones. This information is crucial for understanding the effects of modifications on starch functionality, such as improved stability, enhanced thickening, or gelling properties, or altered digestibility [112].

From SEM image analysis, several structural parameters of starch can be defined. For example, in the work developed by [113], corn starch granules showed a polygonal shape in their native state. However, from the process of gelatinization and retrogradation, the starch molecules showed crosslinking to form a dense and uniform network structure. In addition, SEM images allowed an understanding of the interaction between starch and phenolic compounds, where the complexity of the structure of the phenolic compound causes damage to the surface structure of the starch through the appearance of holes and cracks. From these results, it is possible to mold the starch for several applications and understand the interaction phenomena. SEM images provided the identification of slight corrosion and variation of shapes and sizes in the structure of taro starch modified with octenyl succinic anhydride; in addition, ball milling caused the development of smaller starch fragments that presented a rougher surface than native taro starch [114]. Figure 5 shows an example of a micrograph obtained from a scanning electron microscope.

The digestion and breakdown of starch are strongly dependent on its structure [115]. A porous honeycomb structure with a "cell wall" shape in potato starch through scanning electron microscopy was observed [116]. Furthermore, they verified the inclusion of cations and their effect on the microstructure of potato starch. The size and shape of the cavities varied according to the inclusion of cations, promoting marked changes in the potato starch microstructure. The morphology of potato and pea starch granules was visualized through SEM and observed that the potato starch granules were large and round, while the pea starch presented a similar shape to a bean. The identified variation in granule sizes by SEM was consistent with the particle size distribution of the granules [117]. The gelatinization process of potato starch was followed using SEM. After heating the starch solution for 20 min, the gel still has a granular structure, and the borders between the granules can be associated with the borders between the ruptured and intact

Fig. 5 Micrograph of Cassava starch obtained from scanning electron microscope

granules, indicating that the swelling process of the granules is not yet complete. After 45 min of heating, it was possible to observe this event [118].

3.2.4 Fourier-Transform Infrared Spectroscopy

FTIR is a widely used analytical technique for the characterization of starch. It provides valuable information about the molecular structure, functional groups, and chemical composition of starch samples. FTIR spectroscopy works by measuring the absorption and transmission of infrared radiation by the sample, allowing the identification and quantification of different chemical bonds and molecular groups [119]. FTIR spectroscopy is a nondestructive and relatively rapid technique, making it suitable for routine analysis of starch samples. It allows for qualitative and quantitative characterization of starch composition, molecular structure, and modifications. The information obtained from FTIR spectroscopy contributes to our understanding of starch functionality, processing behavior, and its impact on food quality and nutrition [120]. Figure 6 shows the main FTIR analysis procedures.

The sample under investigation needs to be prepared for analysis. This may involve various methods depending on the nature of the sample, such as grinding a solid sample into a fine powder, diluting a liquid sample, or spreading a thin film of a solid or liquid onto a suitable IR-transparent substrate, such as a salt plate or KBr pellet [121]. The prepared sample is placed in the holder or cell and inserted into the FTIR instrument. The instrument generates an

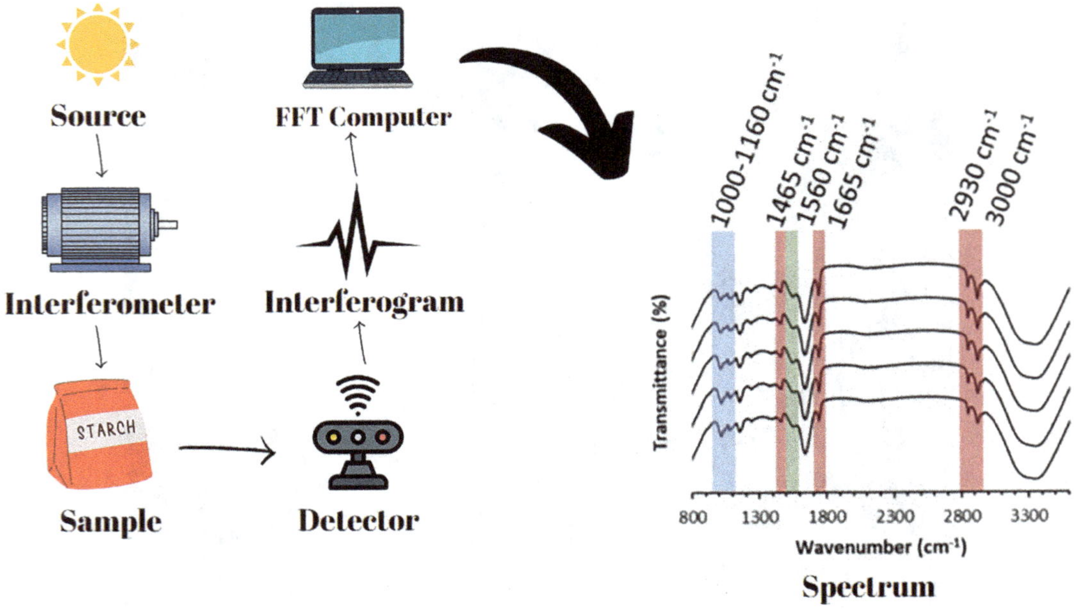

Fig. 6 FTIR analysis procedures

infrared beam, which is directed onto the sample. The sample interacts with the beam, absorbing specific frequencies of infrared light based on its molecular composition [122].

The instrument detects the transmitted or reflected light from the sample and measures the intensity of the light as a function of the infrared wavelength. The resulting data called an interferogram is obtained by performing a Fourier transform on the detected signal. This data represents the infrared spectrum of the sample [123]. The acquired interferogram is processed and converted into a spectrum, representing the intensity of the absorbed infrared radiation as a function of wavelength or wavenumber. The spectrum is typically displayed as a plot of transmittance, absorbance, or a related quantity [124]. FTIR spectroscopy applied to starch analysis provides valuable insights into the following aspects.

1. *Identification of functional groups:* FTIR spectra of starch can identify various functional groups in the starch molecule. The peaks observed in the spectrum correspond to specific vibrations of the chemical bonds within the starch structure. For example, peaks in the 3000–3700 cm^{-1} range indicate the presence of hydroxyl (OH) groups, while peaks in the 1000–1200 cm^{-1} range indicate the presence of glycosidic linkages between glucose units [125].
2. *Characterization of molecular structure:* FTIR can provide information about the molecular structure of starch, including the relative proportion of amylose and amylopectin. The presence of specific absorption bands or peak ratios can indicate the

degree of branching, crystallinity, and the arrangement of glucose units within the starch molecule [126].

3. *Assessment of starch modifications:* FTIR spectroscopy can be used to detect and characterize modifications in starch, such as chemical treatments or enzymatic reactions. By comparing the FTIR spectra of modified starch samples with unmodified ones, researchers can identify changes in peak intensities or shifts, indicating modifications in functional groups or molecular structure [127].
4. *Quantification of starch composition:* FTIR spectroscopy can be combined with chemometric methods to quantify the composition of starch samples. By establishing calibration models using reference standards, it is possible to determine the amylose content or the ratio of amylose to amylopectin in each starch sample. This information is valuable for understanding the functional properties and digestibility of starch [128].
5. *Analysis of starch–water interactions:* FTIR spectroscopy can provide insights into the interactions between starch and water molecules. By measuring the changes in specific absorption bands related to starch–water interactions, researchers can assess the water absorption capacity, gelatinization behavior, and structural changes of starch during hydration or cooking processes [79].

The interactions of wheat starch with sage seed gum through the FTIR technique were studied [129]. According to the spectra formed, the interactions between starch and gum increased as the amount of gum increased. In this way, the connection between water and COO- groups resulted in blocking interactions between amylose and gum chains. Starch structural changes were captured in the bands at 1047 and 1022 cm^{-1}. Bands formed in the spectral region of 800–1200 cm^{-1} reflect changes in the conformation of polymers and in the hydration of processed starches, while bands in the regions of 1047 and 1022 cm^{-1} are associated with the ordered and amorphous structures of starch. Furthermore, when evaluating the gelatinization process using the FTIR technique, the authors verified an increase in the intensity of the peaks in the region of 1022 cm^{-1} when the preheating time of the starch solutions was high to characterize the loss of the ordered structure [60].

During the gelatinization process, FTIR detects a disruption in the molecular order by quantifying short-range glucan structures; thus, the absorbance bands at 1022 cm^{-1} are sensitive to change in amorphous structures, while the bands at 944 and 1047 cm^{-1} provide information on the characteristics of crystalline samples because of intramolecular hydrogen bonding between hydroxyl groups [73, 130]. Thirteen native starches were evaluated by FTIR. The main differences were observed in the intensity of two regions of the spectrum: the OH vibration, between 3650 and 3000 cm^{-1}, and the adsorption region at 1200–800 cm^{-1}. It was verified that the vibration band of the glucan ring is overlapped by vibration bands and stretching and bending C–OH and vibration in the glycosidic bond C–O–C, which suggests that type A starches have a greater amount of ordered short-range double helices than B-type starches. The most important structural differences between the different starch species are observed in 1500 to 400 cm^{-1} regions, as they provide information on changes in the polymeric structure and starch conformation [107, 131].

The starch retrogradation process was investigated using the FTIR technique based on the ratio variation between the 1047/1022 cm^{-1} spectra. As storage time increased, there was a positive variation in the 1047/1022 cm^{-1} ratio, with a good linear correlation with the early phase of retrogradation (0–7 days) and the late phase (7–35 days). Increasing the retrogradation time promoted a reduction in the distance between the hydrogen bonds and the double helices in the starch structure [132]. The reduction of fat content in sugar cookies by gelatinized corn starch hydrogel was studied. It was verified that the inclusion of the hybrid gel promoted greater hydration and disorder in the secondary structure of the starch through the FTIR analysis [36].

3.3 Starch Characterization in Industrial Applications

3.3.1 Papermaking and Textiles

In the papermaking and textiles industries, starch is widely used for its functional properties. Characterizing starch in these applications is important to optimize its performance. By characterizing starch properties in papermaking and textiles, manufacturers can optimize processes, improve product quality, and enhance the overall performance of starch-based materials in these industries [132].

Starch is commonly used as a key component in the paper coating and sizing processes in papermaking. The size of starch particles directly affects their behavior and functionality in the papermaking system. Small starch particles are desirable because they have a higher surface area and are less likely to interfere with the bonding of paper fibers. This promotes improved starch retention onto the paper fibers during the manufacturing process. Moreover, small particle size facilitates better dispersibility and uniform distribution of starch in the paper coating formulation, leading to enhanced paper properties such as smoothness, printability, and opacity [103, 133]. Starch is utilized in the textile industry for various purposes, including fabric sizing, finishing, and printing. Starch is applied to improve the weaving process in fabric sizing by providing temporary stiffness and stability to yarns. Larger starch particles are typically preferred in this application to form a cohesive and protective film on the yarns, enhancing their tensile strength and preventing breakage during weaving [134]. On the other hand, when starch is used for fabric finishing or printing, smaller particles are generally favored. Smaller particles ensure better penetration and adhesion of the starch to the fabric fibers, resulting in improved fabric characteristics such as wrinkle resistance, softness, and dimensional stability [135].

Below is an overview of how starch is characterized in the papermaking and textiles industries:

1. *Particle size distribution:* Starch particle size distribution is crucial in both industries. For papermaking, starch particles need to be small to avoid interference with the paper fibers and improve retention. In textiles, larger particles may cause uneven distribution and affect fabric properties [136].
2. *Rheological behavior:* Understanding the rheological properties of starch is vital in papermaking and textiles. The viscosity and flow behavior of starch solutions influence processes such as coating, sizing, and printing. Rheological measurements help determine starch's suitability and adjust its concentration and viscosity for desired application performance [137].
3. *Gelatinization temperature:* Gelatinization temperature, measured using techniques like differential scanning calorimetry, is important in starch sizing for papermaking and textile applications. Knowing the gelatinization temperature helps control the starch's behavior during processing and ensures proper film formation and binding properties [69].
4. *Adhesive and binding strength:* Starch is used as an adhesive in papermaking and textiles. Evaluating the adhesive and binding strength of starch-based formulations helps assess their effectiveness in bonding fibers, coatings, and fabrics, ensuring optimal product quality and performance [138].
5. *Retrogradation and film properties:* In textiles, starch-based coatings and finishes contribute to fabric properties such as stiffness, wrinkle resistance, and water repellency. Studying starch retrogradation behavior and film properties helps develop coatings that maintain their functionality over time and withstand various textile processing conditions [139].
6. *Application techniques:* Starch characterization also involves evaluating the suitability of application techniques such as spraying, padding, or printing for specific papermaking or textile processes. Understanding how starch interacts with different application methods ensures efficient utilization and desired results [140].
7. *Compatibility with other additives:* Starch is often combined with other additives in papermaking and textiles to enhance performance. Characterization includes assessing starch compatibility with additives such as fillers, binders, and dyes to ensure proper formulation and compatibility for successful integration [141].

3.3.2 Bioplastics and packaging Materials

Bioplastics are a type of plastic derived from renewable sources such as plants, and they offer potential advantages over traditional petroleum-based plastics. One common type of bioplastic is starch-based bioplastics, which are derived from starch-rich crops such as corn, wheat, or potatoes. These bioplastics are often used in packaging materials [142]. Regarding packaging materials, starch-based bioplastics can be used for various applications, including food packaging, disposable cutlery, bags, and films. They can provide a renewable and biodegradable alternative to conventional plastic packaging, reducing the environmental impact associated with waste disposal [143].

The demand for biodegradable and sustainable packaging materials is growing, and starch-based bioplastics are gaining attention in the market. However, regulations and certifications regarding biodegradability and composability vary across different regions, so it's essential to ensure that the specific starch-based bioplastic product meets the required standards [144]. Starch-based bioplastics are typically made by extracting starch from plant sources and then processing it to create a biopolymer. The starch is usually modified to improve its properties and make it more suitable for use as a plastic material. Some common modifications include blending the starch with other biodegradable polymers or adding plasticizers to enhance flexibility [145]. Starch, as a renewable and biodegradable polymer, has found various applications in the field of bioplastics and packaging materials. Its unique properties make it a suitable alternative to conventional plastics derived from fossil fuels [146].

Starch is widely recognized as a promising material for developing biodegradable food packaging regardless of its source. Edible starches derived from various sources such as corn, potato, sweet potato, wheat, and cassava are commonly used to create starch-based biodegradable food packaging [147]. Starch-based bioplastics have properties that can be tailored depending on the desired application. They are generally transparent or translucent, which makes them suitable for packaging materials where visibility of the contents is important. However, their brittleness and lower mechanical strength can limit their use in applications that require high durability or impact resistance [148]. The use of these edible starches as a basis for packaging materials offers several advantages. First, these starches are abundantly available from renewable plant sources, making them a sustainable choice. Additionally, their biodegradability ensures that the packaging can naturally break down in the environment, reducing waste accumulation and environmental pollution [147].

These bioplastics are commonly used for producing films and coatings. These films can be used in food packaging, where their biodegradability and low toxicity make them a desirable alternative to traditional plastics. Starch-based coatings can also be applied to paper or cardboard products to improve their moisture resistance and enhance their barrier properties [148]. The process of developing starch-based biodegradable food packaging typically involves blending the starch with other natural ingredients and additives to improve its functional properties. Depending on the desired characteristics of the packaging material, these additives may include plasticizers, reinforcing agents, antioxidants, or antimicrobial agents [149].

The resulting starch-based packaging exhibits desirable attributes such as good mechanical strength, barrier properties, and moisture resistance, ensuring the protection and preservation of food products. Furthermore, these materials are often edible or potentially consumed, providing the additional benefit of reducing plastic waste and offering an alternative to nonedible packaging [58]. Ongoing research and innovation in starch-based biodegradable food packaging aim to further enhance its performance and versatility. Scientists are exploring methods to improve the water resistance, flexibility, and shelf stability of these materials, as well as investigating novel approaches like incorporating nanotechnology to achieve desired properties. However, starch-based bioplastics do have some limitations. They are generally more brittle and have lower mechanical strength than traditional plastics. They also have higher water sensitivity, leading to problems such as swelling and loss of mechanical properties when exposed to moisture [150]. To overcome these limitations, researchers are actively working on improving the properties of starch-based bioplastics through various techniques, including blending with other polymers, or incorporating additives [151].

It's worth noting that while starch-based bioplastics offer advantages in terms of biodegradability and renewable sourcing, the overall environmental impact of bioplastics depends on various factors, including the cultivation of feedstock crops, processing methods, and end-of-life disposal. Proper waste management systems, such as industrial composting facilities, are necessary to ensure the full biodegradability of starch-based bioplastics and minimize their impact on the environment [152]. Starch-based bioplastics can also be modified by incorporating plasticizers, which help increase their flexibility and reduce brittleness. Common plasticizers used include glycerol and sorbitol. The addition of plasticizers improves the processability and mechanical properties of starch-based bioplastics, making them more suitable for various applications [153].

3.4 Starch Characterization in Nutritional Studies

Starch characterization plays a crucial role in nutritional studies as it provides valuable insights into the properties and behavior of starch in various food sources. Understanding the characteristics of starch is essential for assessing its impact on human health, including its digestibility, glycemic response, and overall nutritional value [154].

Nutritional studies often focus on evaluating the composition and structure of starch, such as the ratio of amylose to amylopectin, degree of branching, and molecular weight distribution. These factors influence the functional properties and digestion kinetics of starch. Methods commonly employed in starch characterization include spectroscopic techniques such as infrared spectroscopy (IR) and nuclear magnetic resonance (NMR), as well as chromatographic techniques like high-performance liquid chromatography (HPLC). These techniques allow researchers to analyze the chemical bonds, molecular structure, and physicochemical properties of starch samples [155]. In nutritional studies, a significant focus is placed on characterizing the digestibility of starch. Starch can be classified into different types, including rapidly digestible starch (RDS), slowly digestible starch (SDS), and resistant starch (RS), each with distinct rates of digestion and varying effects on postprandial blood glucose levels. Assessing the digestibility of starch allows for the determination of important parameters such as the glycemic index (GI) and glycemic load, which play a crucial role in managing blood sugar levels and reducing the risk of metabolic disorders, including diabetes [156].

Starch characterization also aids in evaluating the effects of food processing on starch properties. Techniques such as gelatinization and retrogradation can modify the structure and digestibility of starch. These changes influence the release of glucose during digestion and the subsequent physiological response [157]. Furthermore, starch characterization contributes to the development of functional foods and dietary strategies. By understanding the physicochemical properties of starch, researchers can modify its behavior in food formulations, leading to products with improved texture, stability, and nutritional profiles. For instance, the incorporation of resistant starch in food products can enhance their dietary fiber content and promote gut health [158].

Starch characterization in nutritional studies involves a comprehensive analysis of various properties and attributes of starch to gain a deeper understanding of its impact on human health and nutrition. Here are some key aspects of starch characterization in nutritional studies.

1. *Composition and structure:* Starch is composed of two main components, amylose, and amylopectin, with different ratios depending on the source. Characterizing the composition and structure of starch involves determining the amylose/amylopectin ratio, degree of branching, and molecular weight distribution. This information provides insights into the functional properties and digestibility of starch [159]. Characterizing starch composition and structure is crucial for understanding its functional properties and how it behaves during digestion. It helps assess starch digestibility, glycemic response, and

potential health implications, such as its impact on blood sugar levels and satiety. By studying the composition and structure of starch, researchers can make informed decisions about food formulation, processing techniques, and dietary recommendations for different population groups [160].

2. *Digestibility:* The digestibility of starch is a critical parameter in nutritional studies. Different types of starch exhibit varying rates of digestion and affect postprandial blood glucose levels differently. RDS is quickly broken down into glucose, causing a rapid increase in blood sugar levels. SDS undergoes slower digestion, resulting in a more gradual release of glucose. RS is not fully digested in the small intestine and reaches the large intestine, where it acts as a substrate for beneficial gut bacteria [161].

Starch digestibility can be assessed through various methods in nutritional studies.

In vitro methods: These methods simulate the digestive process outside of the human body. They involve subjecting starch samples to enzymatic digestion with amylases and measuring the release of glucose. In vitro, methods provide valuable information on starch digestibility and its potential impact on glycemic response.

In vivo methods: In vivo studies involve human subjects consuming specific foods or test meals containing starch. Blood samples are collected at regular intervals to measure glucose and insulin responses. These studies provide direct insights into the postprandial glycemic and insulinemic effects of starch in real-life scenarios.

1. *Physicochemical properties:* Starch characterization involves evaluating the physicochemical properties of starch, including granular morphology, size, and structural organization. Techniques such as microscopy, X-ray diffraction, and thermal analysis provide insights into starch crystallinity, gelatinization behavior, and retrogradation tendencies. These properties influence the functionality and processing characteristics of starch in food applications [162].
2. *Impact of processing:* Food processing techniques, such as cooking, milling, and extrusion, can modify the structure and properties of starch. Characterizing the changes that occur during processing helps understand how these alterations affect starch digestibility, functionality, and nutritional outcomes. For example, processing can induce gelatinization, which disrupts the starch granule structure and influences digestibility [71].
3. *Functional foods and product development:* Starch characterization in nutritional studies contributes to the development of functional foods and the optimization of product formulations. Understanding the physicochemical properties of starch allows researchers to modify its behavior and create food products

with improved texture, stability, and nutritional profiles. For instance, incorporating resistant starch into food products increases dietary fiber content and promotes gut health [163].

By conducting comprehensive starch characterization in nutritional studies, researchers can gain valuable insights into the properties and behavior of starch. This knowledge enhances our understanding of starch's role in human nutrition, aids in formulating healthier food products, and supports evidence-based dietary recommendations for improved health and well-being.

3.4.1 Digestibility of Starch and Glycemic Index

Starch, a complex carbohydrate, is the primary source of glycemic energy in various dietary staples such as cereals, legumes, roots, and tubers. It plays a crucial role in human nutrition, providing a significant energy source for bodily functions and physical activities [164]. Starch digestibility refers to the extent to which starch is broken down and converted into glucose during digestion. It is crucial in understanding how different types of starch impact blood sugar levels, nutritional value, and overall health [165]. The digestibility of starch is influenced by various factors, including its structural properties, processing methods, and interactions with other diet components. Upon ingestion, starch undergoes a process of enzymatic digestion in the human body. Salivary amylase initiates the breakdown of starch into smaller polysaccharides in the mouth, while pancreatic amylase continues the process in the small intestine, converting these polysaccharides into glucose molecules [166]. The process of starch digestion begins in the mouth and continues predominantly in the gastrointestinal tract. Starch present in food undergoes step-by-step hydrolysis into glucose, primarily facilitated by the enzymes α-amylase and pancreatin. This enzymatic breakdown increases blood sugar levels as glucose is released [167]. The kinetics of glucose release during starch digestion vary depending on several factors. One important factor is the structure and composition of the starch. Starch with a high amylopectin content tends to be more rapidly digestible, rapidly increasing blood glucose levels. In contrast, starch with a higher amylose content or specific physical properties may exhibit slower digestion and a more gradual release of glucose [112].

The structural characteristics of starch play a significant role in its digestibility. Starch with a higher amylopectin content tends to be more rapidly digestible than starch with higher amylose content. The branching in amylopectin allows for more efficient enzymatic breakdown, resulting in a quicker release of glucose during digestion. On the other hand, amylose has a more linear structure, making it less accessible to digestive enzymes and leading to slower digestion [168]. Rapidly digestible starch is quickly broken down into glucose during digestion, rapidly increasing blood sugar levels. Slowly digestible starch is digested more slowly, resulting in a gradual release of glucose into the bloodstream. On the other hand, resistant starch cannot be fully digested in the small intestine and reaches the large intestine intact, which serves as a substrate for beneficial gut bacteria. This classification provides a valuable framework for understanding the varying effects of different types of starch on postprandial blood glucose levels and overall health [15]. Researchers continue to study the implications of these starch types in relation to nutrition, metabolism, and disease prevention.

Understanding starch digestibility is essential for various reasons. It affects the glycemic response of foods, influencing postprandial blood sugar levels. Foods with high, rapidly digestible starch content can lead to a rapid spike in blood glucose, which may have implications for individuals with diabetes or those seeking to manage their blood sugar levels. On the other hand, slowly digestible starch and resistant starch are associated with a more gradual glucose release and potential health benefits such as improved glycemic control and enhanced gut health [169].

The high glycemic index (GI) of native starches poses challenges regarding digestibility and nutrition. To address this issue, there is a need for modifications in starchy foods to lower their GI scores. Implementing such modifications can effectively improve the digestibility and nutritional value of these foods, ultimately leading to a more desirable dietary experience. Individuals can better manage their blood sugar levels and overall health by developing starchy foods with reduced GI, making these modifications a valuable solution in combating the drawbacks associated with native starches [170].

4 Notes

1. Through comprehensive analysis and examination of various starch parameters, such as granule size, amylose/amylopectin ratio, and gelatinization behavior, researchers and industry professionals can gain valuable insights into starch functionality and tailor its use in a wide range of fields, including food science, pharmaceuticals, and materials engineering. By characterizing starch at a molecular and structural level, scientists can elucidate the relationships between starch composition and its functional properties, enabling the development of improved processing techniques and innovative applications. This knowledge opens doors to optimizing the texture, stability, digestibility, and nutritional attributes of starch-based products, leading to enhanced consumer experiences, and improved overall quality.
2. Furthermore, starch characterization techniques serve as valuable tools in quality control and assurance, facilitating the

identification of starch variants, impurities, and potential adulterations. This ensures product consistency and authenticity while supporting regulatory compliance within relevant industries. In the food industry, starch characterization allows for the optimization of product formulations, texture, and stability. By understanding how different starch properties affect viscosity, gel formation, and retrogradation, manufacturers can develop improved food products with enhanced sensory attributes and extended shelf life. Additionally, starch characterization aids in controlling the nutritional profile and digestibility of starch-based foods, leading to healthier and more balanced dietary options.

3. Moreover, starch characterization plays a vital role in drug formulation and delivery systems in the pharmaceutical industry. By analyzing the characteristics of starch particles, researchers can tailor drug encapsulation and release mechanisms, ensuring optimal drug stability, solubility, and bioavailability. This knowledge is crucial for developing efficient and effective pharmaceutical products. In materials engineering, starch characterization enables the production of eco-friendly and sustainable materials. Understanding starch properties such as molecular weight, amylose content, and gelatinization behavior helps researchers design biodegradable films, coatings, and packaging materials. These starch-based alternatives offer advantages over traditional petroleum-based materials, reducing environmental impact and promoting a circular economy.
4. Furthermore, starch characterization serves as a quality control tool across industries. Manufacturers can ensure product consistency, purity, and compliance with regulatory standards by identifying starch variants, impurities, and adulterations. This enhances consumer confidence and safeguards the industry reputation.

Acknowledgments

This work was supported by Central Analítica of the Federal University of Ceará.

References

1. Luchese CL, Benelli P, Spada JC, Tessaro IC (2018) Impact of the starch source on the physicochemical properties and biodegradability of different starch-based films. J Appl Polym Sci 135:465–472
2. Zhong C, Xiong Y, Lu H, Luo S, Wu J, Ye J, Liu C (2022) Preparation and characterization of rice starch citrates by superheated steam: a new strategy of producing resistant starch. LWT Food Sci Technol 154:112–119
3. Alcázar-Alay SC, Meireles MAM (2015) Physicochemical properties, modifications, and applications of starches from different

botanical sources. Food Sci Technol 35: 215–236

4. Sundarram A, Murthy TPK (2014) α-amylase production and applications: A review. J Appl Environ Microbiol 2:166–175
5. Sun C, Fang Y (2021) In: Fang Y, Zhang H, Nishinari K (eds) Replacement of fat or starch, vol 1. Food Hydrocolloids: Functionalities and Applications, pp 409–444
6. Lu Z, Liu Y, Lee YEJ, Chan A, Lee PR, Yang H (2023) Effect of starch addition on the physicochemical properties, molecular interactions, structures, and in vitro digestibility of the plant-based egg analogues. Food Chem 403:118–126
7. Wang S, Guo P (2020) In: Wang S (ed) Botanical sources of starch, vol 1. Starch Structure, Functionality, and Application in Foods, pp 9–27
8. Schimer M, Höchstötter A, Jekle M, Arendt E, Becker T (2013) Physicochemical and morphological characterization of different starches with variable amylose/amylopectin ratio. Food Hydrocoll 32:52–63
9. Manek RV, Builders PF, Kolling WM, Emeje M, Kunle OO (2012) Physicochemical and binder properties of starch obtained from *Cyperus esculentus.* AAPS PharmSciTech 13: 379–388
10. Onyeaka H, Obileke K, Makaka G, Nwokolo N (2022) Current research and applications of starch-based biodegradable films for food packaging. Polymers 14:1126–1134
11. Bušić A, Marđetko N, Kundas S, Morzak G, Belskaya H, Šantek MI, Komes D, Novak S, Šantek B (2018) Bioethanol production from renewable raw materials and its separation and purification: A review. Food Technol Biotecnol 56:289–311
12. Liu S, Du C, Feng J, Jia Y, Hao Z, Xie Y, Wang C, Ma D (2023) Characterization of starch physicochemical properties and grain transcriptome reveal the mechanism for resistant starch accumulation. Agronomy 13: 1482–1488
13. Himashree P, Sengar AS, Sunil CK (2022) Food thickening agents: sources, chemistry, properties, and applications – a review. Int J Gastron Food Sci 27:468–477
14. Wu B (2015) Structural design approaches for creation of reduced fat products. Doctoral dissertations. 296 p
15. Wang X, Fang J, Cheng L, Gu Z, Hong Y (2023) Interaction of starch and non-starch polysaccharides in raw potato flour and their effects on thickening stability. Int J Biol Macromol 242:124–133
16. Rengsutthi K, Charoenrein S (2011) Physicochemical properties of jackfruit seed starch (*Artocarpus heterophyllus*) and its application as a thickener and stabilizer in chilli sauce. LWT Food Sci Technol 44:1309–1313
17. Block AE, Bolhuis DP, Arnaudov LN, Velikov KP, Stieger M (2023) Influence of thickeners (microfibrillated cellulose, starch, xanthan gum) on rheological, tribological and sensory properties of low-fat mayonnaises. Food Hydrocoll 136:242–250
18. Dewar RJ, Joyce MJ (2006) Time-dependent rheology of starch thickeners and the clinical implications for dysphagia therapy. Dysphagia 21:264–269
19. Agyemang PN, Akonor PT, Tortoe C, Johnsona PNT, Manu-Aduening J (2020) Effect of the use of starches of three new Ghanaian cassava varieties as a thickener on the physicochemical, rheological, and sensory properties of yogurt. Scientific African 9:521–529
20. Wong SS, Wicklund R, Bridges J, Whaley J, Koh YB (2020) Starch swelling behavior and texture development in stirred yogurt. Food Hydrocoll 98:274–282
21. Mason WR (2009) Starch use in foods. Starch Chem Technol 3:745–795
22. Zambelli RA, Galvão AMMT, Mendonça LG, Leão MVS, Carneiro SV, Lima ACS, Melo CAL (2018) Effect of different levels of acetic, citric and lactic acid in the cassava starch modification on physical, rheological, thermal and microstructural properties. Food Sci Technol Res 24:747–754
23. Nilsson K, Johansson M, Sandstrom C, Rohnisch HE, Hedenqvist MS, Langton M (2023) Pasting and gelation of faba bean starch-protein mixtures. Food Hydrocoll 138:494–502
24. Bitencourt BS, Guedes JS, Saliba ASMC, Sartori AGO, Torres LCR, Amaral JEPG, Alencar SM, Maniglia BC, Augusto PED (2023) Mineral bioaccessibility in 3D printed gels based on milk/starch/-carrageenan for dysphagic people. Food Res Int 170:113–121
25. Pathraje D, Carlin J, Der T, Wanasundara JPD, Shand PJ (2023) Generating multifunctional pulse ingredients for processed meat products, vol 12. Scientific Evaluation of Infrared-Treated Lentils, pp 1722–1730
26. Ribotta PD, Colombo A, Rosell CM (2012) Enzymatic modifications of pea protein and its application in protein–cassava and corn starch gels. Food Hydrocoll 27:185–190
27. Zhao C, Miao Z, Qi Q, Zheng Q, Mao Y, Chu Z, Zhang H, Xu X, Zheng M, Liu J (2023) Interactions of soy protein isolate

with common and waxy corn starches and their effects on acid-induced cold gelation properties of complexes. Food Chemistry: X 18:671–679

28. Mavdc A, Dg B, La A (2021) Antioxidant/pro-oxidant and antiproliferative activities of phenolic-rich foods and extracts: A cell-based point of view. Adv Food Nutr Res 98:253–280
29. Punia S (2020) Barley starch modifications: physical, chemical, and enzymatic—A review. Int J Biol Macromol 144:578–585
30. Sun C, Fang Y, Replacement of fat or starch (2021) In: Fang Y, Zhang H, Nishinari K (eds) Food hydrocolloids: functionalities and applications, vol 1, pp 409–444
31. Agyei-Amponsah J, Macakova L, DeKock HL, Emmambux MN (2021) Effect of substituting sunflower oil with starch-based fat replacers on sensory profile, tribology, and rheology of reduced-fat mayonnaise-type emulsions. Starch 73:200–210
32. Cao C, Wang C, Yuan D, King B, Sun F, Liu Q (2021) Effects of acetylated cassava starch on the physical and rheological properties of multicomponent protein emulsions. Int J Biol Macromol 183:1459–1474
33. Tsyganov MS, Ezhokova GO, Kharitonova MA, Nikitina EV (2022) Cassava starch as an effective texture corrector of fat-free dairy products based on symbiotic starter culture. Int J Food Sci 10:870–883
34. Park JJ, Olawuyi IF, Lee WY (2020) Characteristics of low-fat mayonnaise using different modified arrowroot starches as fat replacer. Int J Biol Macromol 153:215–223
35. Mohammadi M, Oghabi F (2012) Development of low-fat and low-calorie beef sausage using modified starch as fat replacement agent. J Sci Food Agric 92:1291–1296
36. Barragán-Martínez LP, Román-Guerrero A, Vernon-Carter EJ, Alvarez-Ramirez J (2022) Impact of fat replacement by a hybrid gel (canola oil/candelilla wax oleogel and gelatinized corn starch hydrogel) on dough viscoelasticity, color, texture, structure, and starch digestibility of sugar-snap cookies. Int J Gastron Food Sci 29:563–570
37. Punia S, Siroha AK, Sandhu KS, Kaur M (2019) Rheological and pasting behavior of OSA modified mungbean starches and its utilization in cake formulation as fat replacer. Int J Biol Macromol 128:230–236
38. Bourouis I, Pang Z, Liu X (2023) Recent advances on uses of protein and/or polysaccharide as fat replacers: textural and tribological perspectives: A review. J Agric Food Res 11:519–527
39. Seung D (2020) Amylose in starch: towards an understanding of biosynthesis, structure, and function. New Phytol 228:1490–1504
40. Pérez S, Bertoft E (2010) The molecular structures of starch components and their contribution to the architecture of starch granules: A comprehensive review. Starch 62: 389–420
41. Wang S, Wang S, Guo P, Liu L, Wang S (2017) Multiscale structural changes of wheat and yam starches during cooking and their effect on in vitro enzymatic digestibility. J Agric Food Chem 65:156–166
42. Brust H, Orzechowski S, Fettke J (2020) Starch and glycogen analyses: methods and techniques. Biomol Ther 10:1020–1031
43. Chen C, Fu W, Chang Q, Zheng B, Zhang Y, Zheng H (2019) Moisture distribution model describes the effect of water content on the structural properties of lotus seed resistant starch. Food Chem 286:449–458
44. Huang J, Wang Z, Fan L, Ma S (2022) A review of wheat starch analyses: methods, techniques, structure, and function. Int J Biol Macromol 203:130–142
45. Miller K, Reichert CL, Schmid M, Loeffler M (2022) Physical, chemical and biochemical modification approaches of potato (Peel) constituents for bio-based food packaging concepts: A review. Food Secur 11:2927–2939
46. Ren Y, Yuan TZ, Chigwedere CM, Ai Y (2021) A current review of structure, functional properties, and industrial applications of pulse starches for value-added utilization. Compr Rev Food Sci Food Saf 20:3061–3092
47. Zhao Y, Henry RJ, Gilbert RG (2021) Starch structure-property relations in australian wild Rices compared to domesticated Rices. Carbohydr Polym 271:118–125
48. Huang PV, Maeda T, Morita N (2006) Waxy and high-amylose wheat starches and flours—characteristics, functionality, and application. Trends Food Sci Technol 17:448–456
49. Mani R, Bhattachary M (1998) Properties of injection moulded starch/synthetic polymer blends—III. Effect of amylopectin to amylose ratio in starch. Eur Polym J 34:1467–1475
50. Patindol JA, Siebenmorgen TJ, Wang YJ (2015) Impact of environmental factors on rice starch structure: A review. Starch 67:42–54
51. Shamekh S, Forssell P, Poutanen K (1994) Solubility pattern and recrystallization behavior of oat starch. Starch 46:129–133

52. Basiak W, Lenart A, Debeaufort F (2017) Effect of starch type on the physico-chemical properties of edible films. Int J Biol Macromol 98:348–356
53. Zhang Y, Rempel C, Liu Q (2014) Thermoplastic starch processing and characteristics—A review. Crit Rev Food Sci Nutr 54:1353–1370
54. Rolland-Sabaté A, Sánchez T, Buléon A, Colonna P, Jaillais B, Ceballos H, Dufour D (2012) Structural characterization of novel cassava starches with low and high-amylose contents in comparison with other commercial sources. Food Hydrocoll 27:161–174
55. Yu Y, Lu X, Zhang T, Zhao CZ, Guan S, Pu Y, Gao F (2022) Tiger nut (*Cyperus esculentus L.*): nutrition, processing, function and applications. Foods 11:601–610
56. Jane J, Shen L, Chen J, Lim S, Kasemsuwan T, Nip WK (1992) Physical and chemical studies of Taro starches and flours. Cereal Chem 69: 528–535
57. Bashir K, Aggarwal M (2016) Effects of gamma irradiation on the physicochemical, thermal, and functional properties of chickpea flour. LWT Food Sci Technol 69:614–622
58. Hoover R, Ratnayake WS (2002) Starch characteristics of black bean, chickpea, lentil, navy bean, and pinto bean cultivars grown in Canada. Food Chem 78:489–498
59. Chen C, Li G, Zhu F (2021) A novel starch from lotus (*Nelumbo nucifera*) seeds: composition, structure, properties, and modifications. Food Hydrocoll 120:899–908
60. Zhang Y, Li B, Xu F, He S, Zhang Y, Sun L, Zhu K, Li S, Wu G, Tan L (2021) Jackfruit starch: composition, structure, functional properties, modifications, and applications. Trends Food Sci Technol 107:268–283
61. Xu H, Zhou J, Liu X, Yu J, Copeland L, Wang S (2021) Methods for characterizing the structure of starch in relation to its applications: a comprehensive review. Crit Rev Food Sci Nutr 38:49–60
62. Hernando I, Llorca E, Quiles A (2018) Emerging techniques for microstructural analysis. In: El-Bakry MMAR, Sanchez A, Mehta BM (eds) Microstructure of dairy products. Wiley, p 720
63. Chakraborty I, Pallen S, Shetty Y, Roy N, Mazumder N (2020) Advanced microscopy techniques for revealing molecular structure of starch granules. Biophys Rev 12:105–122
64. Gallant DJ, Bouchet B, Baldwin PM (1997) Microscopy of starch: evidence of a new level of granule organization. Carbohydr Polym 32:177–191
65. Falsafi SR, Rostamabadi H, Assadpour E, Jafari SM (2020) Morphology and microstructural analysis of bioactive-loaded micro/nanocarriers via microscopy techniques; CLSM/SEM/TEM/AFM. Adv Colloid Interf Sci 280:166–175
66. Blazek J, Gilbert EP (2011) Application of small-angle X-ray and neutron scattering techniques to the characterisation of starch structure: A review. Carbohydr Polym 85: 281–293
67. Warren FJ, Gidley MJ, Flanagan BM (2016) Infrared spectroscopy as a tool to characterise starch ordained structure. A joint FTIR-ATR, NMR, XRD and DSC study. Carbohydr Polym 139:35–42
68. Leyva-Porras C, Cruz-Alcantar P, Espinosa-Solís V, Martínez-Guerra E, Piñón-Balderrama CI, Martínez IC, Saavedra-Leos M (2020) Application of differential scanning calorimetry (DSC) and modulated differential scanning calorimetry (MDSC) in food and drug industries. Polymers 12:5–12
69. Tomassetti S, Aquilanti A, Muciaccia PF, Coccia G, Mankel C, Koenders EAB, Di Nicola G (2022) A review on thermophysical properties and thermal stability of sugar alcohols as phase change materials. J Energy Storage 55:456–462
70. Zhang Y, Xu P, Zhu J, Yan S, Zhang J, Li L (2023) The emergence of considerable room temperature magnetocaloric performances in the transition metal high-entropy alloys. Mater Today Phys 23:101–110
71. Liu X, Huang S, Chao C, Yu J, Copeland L, Wang S (2022) Changes of starch during thermal processing of foods: current status and future directions. Trends Food Sci Technol 119:320–337
72. Nikolaidis AK, Achilias DS (2018) Thermal degradation kinetics and viscoelastic behavior of poly(methyl methacrylate)/Organomodified montmorillonite nanocomposites prepared via In situ. Bulk Radical Polymerization Polymers 10:491–500
73. Kang X, Gao W, Cui B, El-Aty AMA (2023) Structure and genetic regulation of starch formation in sorghum (*Sorghum bicolor* (L.) Moench) endosperm: A review. Int J Biol Macromol 239:124–135
74. Nagy R, Máthé E, Csapó J, Sipos P (2021) Modifying effects of physical processes on starch and dietary fiber content of foodstuffs. PRO 9:17–24
75. Donmez D, Pinho L, Patel B, Desam P, Campanella OH (2021) Characterization of starch–water interactions and their effects on

two key functional properties: starch gelatinization and retrogradation. Curr Opin Food Sci 39:103–109

76. Zambelli RA, Pinto LIF, Junior ECS, de Lima ACV, Goiana ML, Mendonça LG (2018) Effect of black sesame (*Sesamum indicum*) and flaxseed flour (*Linum usitatissimum*) on bread quality using surface response methodology. J Eng Process Manag 10:32–40
77. Biliaderis CG, Page CM, Maurice TJ, Juliano BO (1986) Characterization of rice starches: A polymeric approach to phase transitions of granular starch. J Agric Food Chem 34:6–14
78. Wang Z, Ma S, Sun B, Wang F, Huang J, Wang X, Bao Q (2021) Effects of thermal properties and behavior of wheat starch and gluten on their interaction: a review. Int J Biol Macromol 177:474–484
79. Liu Y, Yu J, Copeland L, Wang S, Wang S (2019) Gelatinization behavior of starch: reflecting beyond the endotherm measured by differential scanning calorimetry. Food Chem 284:53–59
80. Ratnayake WS, Jackson DS (2007) A new insight into the gelatinization process of native starches. Carbohydr Polym 67:511–529
81. Biliaderis CG (2009) Structural transitions and related physical properties of starch. Starch Chem Technol 3:293–372
82. Zeng FK, Liu H, Liu G (2014) Physicochemical properties of starch extracted from *colocasia esculenta* (L.) schott (*Bun-long taro*) grown in Hunan, China. Starch 66:142–148
83. Karim AA, Norziah MH, Seow CC (2000) Methods for the study of starch retrogradation. Food Chem 71:9–36
84. Mamat H, Hill SE (2018) Effect of ingredients on the mass loss, pasting properties and thermal profile of semi-sweet biscuit dough. J Therm Anal Calorim 134:1413–1428
85. Demetzos C (2008) Differential scanning calorimetry (DSC): A tool to study the thermal behavior of lipid bilayers and liposomal stability. J Liposome Res 18:159–173
86. Rahaman A, Kumari A, Zeng XA, Farooq MA, Siddique R, Khalifa I, Siddeeg A, Ali M, Manzoor MF (2021) Ultrasound based modification and structural-functional analysis of corn and cassava starch. Ultrason Sonochem 80:795–812
87. Kalita D, Kaushik N, Mahanta CL (2014) Physicochemical, morphological, thermal and IR spectral changes in the properties of waxy rice starch modified with vinyl acetate. J Food Sci Technol 51:2790–2796
88. Li Q, Liu F, Zhang L, Nelson BJ, Zhang S, Ma C, Tao X, Cheng J, Zhang X (2012) *In situ* construction of potato starch-based carbon nanofiber/activated carbon hybrid structure for high-performance electrical double layer capacitor. J Power Sources 207:199–204
89. Guo L, Tao H, Cui B, Janaswamy S (2019) The effects of sequential enzyme modifications on structural and physicochemical properties of sweet potato starch granules. Food Chem 277:504–514
90. Hassanein WS, Ispirli H, Dertli E, Yilmaz MT (2023) Structural characterization of potato starch modified by a 4,6-α-glucanotransferase B from lactobacillus reuteri E81. Int J Biol Macromol 242:124–130
91. Purohit SR, Jayachandran LE, Raj AS, Nayak D, Rao PS (2019) X-Ray diffraction for food quality evaluation. In: Zhong J, Wang X (eds) Evaluation technologies for food quality, vol 1. Woodhead Publishing Series in Food Science, Technology and Nutrition, pp 579–594
92. Lopez-Rubio A, Flanagan BM, Gilbert EP, Gidley MJ (2008) A novel approach for calculating starch crystallinity and its correlation with double helix content: A combined XRD and NMR study. Biopolymers 89:761–768
93. Rodriguez-García ME, Hernandez-Landaverde MA, Delgado JM, Ramirez-Gutierrez CF, Ramirez-Cardona M, Millian-Malo BM, Londoño-Restrepo SM (2021) Curr Opin Food Sci 37:107–111
94. Pei-Ling L, Xiao-Song H, Qun S (2010) Effect of high hydrostatic pressure on starches: a review. Starch 62:615–628
95. Wang S, Li C, Copeland L, Niu Q, Wang S (2015) Starch retrogradation: a comprehensive review. Food Sci Food Saf 14:568–585
96. Schafranski K, Ito VC, Lacerda LG (2021) Impacts and potential applications: a review of the modification of starches by heat-moisture treatment (HMT). Food Hydrocoll 117:106–115
97. Pozo C, Rodríguez-Llamazares S, Bouza R, Barral L, Castaño J, Muller N, Restrepo I (2018) Study of the structural order of native starch granules using combined FTIR and XRD analysis. J Polym Res 25:1–8
98. Li N, Cai Z, Guo Y, Xu T, Qiao D, Zhang B, Zhao S, Huang Q, Niu M, Jia C, Lin L, Lin Q (2019) Hierarchical structure and slowly digestible features of rice starch following microwave cooking with storage. Food Chem 295:475–483
99. Sangwongchai W, Tananuwong K, Krusong K, Natee S, Thitisaksakul M (2023)

Starch chemical composition and molecular structure in relation to physicochemical characteristics and resistant starch content of four Thai commercial rice cultivars differing in pasting properties. Polymers 15:574–582

100. Abedi E, Pourmohammadi K, Jahromi M, Niakousari M, Torri L (2019) The effect of ultrasonic probe size for effective ultrasound-assisted Pregelatinized starch. Food Bioprocess Technol 12:1852–1862

101. Deng N, Bian X, Luo S, Liu C, Hu X (2023) The starch-polyphenol inclusion complex: preparation, characterization, and digestion. Food Biosci 53:655–672

102. Liu Y, Zheng R, Li J (2022) High latent heat phase change materials (PCMs) with low melting temperature for thermal management and storage of electronic devices and power batteries: critical review. Renew Sust Energ Rev 168:783–799

103. Wang Y, Jean-Phillipe R, Saulnier L, Kansou K (2022) How does starch structure impact amylolysis? review of current strategies for starch digestibility study. Food Secur 11: 1223–1230

104. Miranda JAT, Carvalho LMJ, Vieira ACM, Castro IM (2019) Scanning electron microscopy and crystallinity of starches granules from cowpea, black and carioca beans in raw and cooked forms. Food Sci Technol 39:718–724

105. Fuentes C, Kang I, Lee J, Song D, Sjoo M, Choi J, Lee S, Nilsson L (2019) Fractionation and characterization of starch granules using field-flow fractionation (FFF) and differential scanning calorimetry (DSC). Anal Bioanal Chem 411:3665–3674

106. Żołek-Tryznowska Z, Bednarczyk E, Tryznowski M, Kobiela T (2023) A comparative investigation of the surface properties of corn-starch-microfibrillated cellulose composite films. Materials 16:3320–3332

107. Guleria P, Yadav BS (2022) Effect of chemical treatments on the functional, morphological, and rheological properties of starch isolated from pigeon pea (*Cajanus cajan*). Curr Res Food Sci 5:1750–1759

108. Chen N, Gao HX, He Q, Yu ZL, Zeng WC (2018) Influence of structure complexity of phenolic compounds on their binding with maize starch. Food Strcut 33:286–299

109. Liu C, An F, He H, He D, Wang Y, Song H (2018) Pickering emulsions stabilized by compound modified areca taro (*Colocasia esculenta* (L.) Schott) starch with ball milling and OSA. Colloids Surf A Physicochem Eng Asp 556:185–194

110. García-Alonso A, Goñi I (2000) Effect of processing on potato starch: In vitro availability and glycaemic index. Food Nahrung 44: 19–22

111. Zhuang Y, Wang Y, Yang H (2023) Characterizing digestibility of potato starch with cations by SEM, X-ray, LF-NMR, FTIR. Food Chem 424:396–404

112. Martens BMJ, Gerrits WJJ, Bruininx EMAM, Schols HA (2018) Amylopectin structure and crystallinity explains variation in digestion kinetics of starches across botanic sources in an in vitro pig model. J Anim Sci Biotechnol 9:91–99

113. Cieśla K, Sartowska B, Królak E (2015) SEM studies of the structure of the gels prepared from untreated and radiation modified potato starch. Radiat Phys Chem 106:289–302

114. Johnson JB, Walsh K, Naiker M (2020) Application of infrared spectroscopy for the prediction of nutritional content and quality assessment of faba bean (*Vicia faba* L.). Legum Sci 2:40–55

115. Kizil R, Irudayaraj J, Seetharaman K (2002) Characterization of irradiated starches by using FT-Raman and FTIR spectroscopy. J Agric Food Chem 50:3912–3918

116. Tkachenko Y, Niedielski P (2022) FTIR as a method for qualitative assessment of solid samples in geochemical research: a review. Molecules 27:8846–8855

117. Fadlelmoula A, Pinho D, Carvalho VH, Catarino SO, Minas G (2022) Fourier transform infrared (FTIR) spectroscopy to analyse human blood over the last 20 years: A review towards lab-on-a-chip devices. Micromachines 13:187–199

118. Lilo T, Morais CLM, Shenton C, Ray A, Gurusinge N (2022) Revising Fourier-transform infrared (FT-IR) and Raman spectroscopy towards brain cancer detection. Photodiagn Photodyn Ther 38:785–796

119. Han Y, Han L, Yao Y, Li Y, Liu X (2018) Key factors in FTIR spectroscopic analysis of DNA: the sampling technique, pretreatment temperature and sample concentration. Anal Methods 10:2436–2443

120. Hong T, Yin JY, Nie SP, Xie MY (2021) Applications of infrared spectroscopy in polysaccharide structural analysis: progress, challenge, and perspective. Food Chem 12:100–121

121. Yu H, Cheng L, Yin J, Yan S, Liu K, Zhang F, Xu B, Li L (2013) Structure and physicochemical properties of starches in lotus (*Nelumbo nucifera Gaertn.*) rhizome. Food Sci Nutr 1:273–283

122. Hosseinpourpia R, Echart AS, Adamopoulos S, Gabiliondo N, Eceiza A (2018) Modification of pea starch and dextrin polymers with isocyanate functional groups. Polymers 10:939–960
123. Cozzolino D, Degner S, Eglinton J (2014) A review on the role of vibrational spectroscopy as an analytical method to measure starch biochemical and biophysical properties in cereals and starchy foods 3:605–621
124. Rooyen JV, Simsek S, Oyeyinka SA, Manley H (2023) Wheat starch structure–function relationship in breadmaking: a review. Compr Rev Food Sci Food Saf 22:2292–2309
125. Dankar I, Haddarah A, Omar FEL, Pujolà M, Sepulcre F (2018) Characterization of food additive-potato starch complexes by FTIR and X-ray diffraction. Food Chem 260:7–12
126. Adewale P, Yancheshmeh MS, Lam E (2022) Starch modification for non-food, industrial applications: market intelligence and critical review. Carbohydr Polym 291:590–599
127. Dal AEB, Hubbe MA (2021) Hydrophobic copolymers added with starch at the size press of a paper machine: a review of findings and likely mechanisms. Bioresources 16:2138–2180
128. Aldalbahi A, El-Naggar ME, El-Newehy MH, Rahaman M, Hatshan MR, Khattab TA (2021) Effects of technical textiles and synthetic nanofibers on environmental pollution. Polymers 13:155–164
129. Zhu Z, Shen S (2014) Effect of amphoteric grafting branch on the adhesion of starch to textile fibers. J Adhes Sci Technol 28:1695–1710
130. Zhou T, Zhang L, Zhao R, Liu Q, Liu W, Hu H (2022) Effects of particle size distribution of potato starch granules on rheological properties of model dough underwent multiple freezing-thawing cycles. Food Res Int 156:111–118
131. Makroo HA, Naqash S, Saxena J, Sharma S, Majid D, Dar BN (2021) Recovery and characteristics of starches from unconventional sources and their potential applications: a review. Appl Food Res 1:100–112
132. Gadhave RVI, Gadhave CR (2022) Adhesives for the paper packaging industry: An overview. Open J Polymer Chem 12:38–49
133. Petrova L, Grishin R, Zimmurov A, Kozlova O, Odintsova O, Sangeeva E (2023) Improvement of textile materials processing techniques by applying aqueous dispersions of polymers. Coatings 13:462–470
134. Wei Q, Zheng H, Zhu M (2021) Starch-based surface-sizing agents in paper industry. An overview. Paper Biomater 6:54–61
135. Balea A, Fuente E, Monte MC, Merayo N, Campano C, Negro C, Blanco A (2020) Industrial application of nanocelluloses in papermaking: a review of challenges, technical solutions, and market perspectives. Molecules 25:526–534
136. Abe MM, Martins JR, Sanvezzo PB, Macedo JV, Branciforti MC, Halley P, Botaro VR, Brienzo M (2021) Advantages and disadvantages of bioplastics production from starch and lignocellulosic components. Polymers 13:2484–2492
137. Atiwesh G, Mikhael A, Parrish CC, Banoub J, Le TAT (2021) Environmental impact of bioplastic use: a review. Heliyon 7:918–930
138. Cruz RSM, Krauter V, Krauter S, Agriopolou S, Weinrich R, Herbes C, Scholten PBV, Uysal-Unalan I, Sogut E, Kopacic S, Lahti J, Rutkaite R, Varzakas T (2022) Bioplastics for food packaging: environmental impact, trends and regulatory aspects. Food Secur 11:3087–3099
139. Amaraweera SM, Gunathilake C, Guanawardene OHP, Fernando NML, Wanninayaka DB, Dassanayake RS, Rajapaksha SM, Manamperi A, Fernando CAN, Kulatunga AK, Manipura (2021) A. development of starch-based materials using current modification techniques and their applications: a review. Molecules 28:6880–6890
140. Marichelvam MK, Jawaid M, Asim M (2019) Corn and rice starch-based bioplastics as alternative packaging materials. Fibers 7:32–40
141. Oliveira MA, Gonzaga MLC, Bastos MSR, Magalhães HCR, Benevides SD, Furtado RF, Zambelli RA, Garruti DS (2020) Packaging with cashew gum/gelatin/essential oil for bread: release potential of the citral. Food Packag Shelf Life 23:431–440
142. Tan SX, Andriyana A, Ong HC, Lim S, Pang YL, Ngoh GC (2022) A comprehensive review on the emerging roles of nanofillers and plasticizers towards sustainable starch-based bioplastic fabrication. Polymers 14:664–670
143. Yuvaraj D, Iyyappan J, Gnanasekaran R, Ishwarya G, Harshini RP, Dhithya V, Chandran M, Kanishka V, Gomathi K (2021) Advances in bio food packaging—An overview. Heliyon 7:998–1005
144. Díaz-Montes E (2022) Polysaccharide-based biodegradable films: An alternative in food packaging. Polyssacharides 3:761–775

145. Galvão AMMT, Araújo AWO, Carneiro SV, Zambelli RA, Bastos MSR (2018) Coating development with modified starch and tomato powder for application in frozen dough. Food Packag Shelf Life 16:194–203
146. Méité N, Konan LK, Tognonvi MT, Oyetola S (2022) Effect of metakaolin content on mechanical and water barrier properties of cassava starch films. S Afr J Chem Eng 40: 186–194
147. Galvão AMMT, Zambelli RA, Araújo AWO, Bastos MSR (2018) Edible coating based on modified corn starch/tomato powder: effect on the quality of dough bread. LWT 89:518–524
148. Jayarathoa S, Andersson M, Andersson R (2022) Recent advances in starch-based blends and composites for bioplastics applications. Polymers 14:4557–4564
149. Moshood TD, Nawanir G, Mahmud F, Mohamad F, Ahmad MH, Abdul GA (2022) Sustainability of biodegradable plastics: new problem or solution to solve the global plastic pollution? Curr Res Green Sustain Chem 5: 273–280
150. Vieira AGA, da Silva MA, Santos LO, Beppu MM (2011) Natural-based plasticizers and biopolymer films: a review. Eur Polym J 47: 254–263
151. Hu P, Zhao H, Duan Z, Linlin Z, Wu D (2004) Starch digestibility and the estimated glycemic score of different types of rice differing in amylose contents. J Cereal Sci 40:231–237
152. Magallanes-Cruz PA, Flores-Silva PC, Bello-Perez LA (2017) Starch structure influences its digestibility: a review. Concise Rev Hypoth Food Sci 82:2016–2023
153. Tuaño APP, Barcellano ECG, Rodriguez MS (2021) Resistant starch levels and in vitro starch digestibility of selected cooked Philippine brown and milled Rices varying in apparent amylose content and glycemic index. Food Chem Molecul Sci 2:10–19
154. Patel H, Royall PG, Gaisford S, Williams GR, Edwards CH, Warren FJ, Flanagan BM, Ellis PR, Butterworth PJ (2017) Structural and enzyme kinetic studies of retrograded starch: inhibition of α-amylase and consequences for intestinal digestion of starch. Carbohydr Polym 164:154–161
155. Sanz T, Salvador A, Fiszman SM (2008) Resistant starch (RS) in battered fried products: functionality and high-fibre benefit. Food Hydrocoll 22:543–549
156. Khatefov EB, Goldstein VG, Krivandin AV, Wasserman LA (2023) Main characteristics of processed grain starch products and physicochemical features of the starches from maize (*Zea mays L.*) with different genotypes. Polymers 15:1976–1984
157. Lal MK, Singh B, Sharma S, Singh MP, Kumar A (2021) Glycemic index of starchy crops and factors affecting its digestibility: a review. Trends Food Sci Technol 111:741–755
158. Zhu J, Bai Y, Gilbert G (2023) Effects of the molecular structure of starch in foods on human health. Food Secur 12:2263–2270
159. Breyton AE, Goux A, Lambert-Porcheron S, Meynier A, Sothier M, Van Den Berghe L, Brack O, Disse E, Laville M, Vinoy S, Nazare JA (2021) Starch digestibility modulation significantly improves glycemic variability in type 2 diabetic subjects: a pilot study. Nutr Metab Cardiovasc Dis 4:237–246
160. Pino-Hernández E, Fasolin LH, Ballesteros LF, Pinto CA, Saraiva JA, Abrunhosa L, Teixeira JA (2023) Structural and physicochemical properties of starch from rejected chestnut: hydrothermal and high-pressure processing dependence. Molecules 28:700–712
161. Palanisamy A, Parimalavalli R (2022) Resistant starch: A functional ingredient in dairy products. J Food Process Preserv 46:171–182
162. Bai Y, Li X, Ji H, Wang Y, Zheng D, Wang Y, Jin Z (2021) A review of the design and architecture of starch-based dietary foods. Engineering 7:663–673
163. Li C, Hu Y, Li S, Yi X, Shao S, Yu W, Li E (2023) Biological factors controlling starch digestibility in the human digestive system. Food Sci Human Wellness 12:351–358
164. Yang Z, Zhang Y, Wu Y, Ouyang J (2023) Factors influencing the starch digestibility of starchy foods: a review. Food Chem 406:135–143
165. Bello-Pérez LA, Flores-Silva PC, Sifuentes-Nieves I, Agama-Acevedo E (2021) Controlling starch digestibility and glycaemic response in maize-based foods. J Cereal Sci 99:222–229
166. Lehmann U, Robin F (2007) Slowly digestible starch—its structure and health implications: a review. Trends Food Sci Technol 18: 346–355
167. Zhang B, Huang Q, Luo FX, Fu X (2012) Structural characterizations and digestibility of debranched high-amylose maize starch complexed with lauric acid. Food Hydrocoll 28:174–181
168. Lin L, Cai C, Gilbert RG, Li E, Wang J, Wei C (2016) Relationships between amylopectin

molecular structures and functional properties of different-sized fractions of normal and high-amylose maize starches. Food Hydrocoll 52:359–368

169. Zhang Z, Tian X, Wang P, Jiang H, Li W (2019) Compositional, morphological, and physicochemical properties of starches from red adzuki bean, chickpea, faba bean, and baiyue bean grown in China. Food Sci Nutr 7:2485–2494

170. Bangar SP, Ashogbon AO, Singh A, Chaudhary V, Whiteside WS (2022) Enzymatic modification of starch: a green approach for starch applications. Carbohydr Polym 85:265–272

Chapter 2

Morphological Characterization of Starches

Lu Liu, Ru Jia, Wenjing Chen, Baodong Zheng, and Zebin Guo

Abstract

Starch particles prepared by different sources and processing methods have different forms, and the characterization of starch particles plays a crucial role in studying their interaction with other substances and exploring the application of starch materials. At present, there are many techniques to detect the morphological characteristics of starch, and the methods and procedures used in different laboratories may cause differences in the detection results of the same starch. Here, we examine the methods commonly used to evaluate the morphological characteristics of starch particles (such as optical microscopy, scanning electrical microscope, transmission electron microscope, polarizing microscope, confocal laser scanning microscopy, and atomic force microscopy) were described and the new detection methods in recent years were evaluated.

Key words Morphological characterization, Optical microscopy, SEM, TEM, PLM, CLSM, AFM

1 Introduction

The size of starch particles is generally between 0.1 and 200 μm, and the structure of starch particles is compact and diverse, which affects the performance of starch [1]. At present, the research on the morphological characteristics of starch mainly includes the particle morphology of natural starch [2], modified starch [3], the dynamic changes of starch in plants [4], and the changes of starch particle morphology during processing/digestion [5]. Through different detection techniques, the morphology of starch particles can be analyzed from multiple dimensions.

Common methods for detecting the morphology of starch particles include microscope, scanning electron microscope (SEM), transmission electron microscope (TEM), polarizing microscope (PLM), confocal laser scanning microscopy (CLSM), atomic force microscope (AFM), etc. Among them, the starch distribution and particle profile in iodine-stained plant tissue fluid

Sneh Punia Bangar (ed.), *Standardized Procedures and Protocols for Starch*, Methods and Protocols in Food Science,
https://doi.org/10.1007/978-1-0716-3866-8_2,

can be observed through a low-power microscope. For some suspension particles that are inconvenient to observe with SEM and other electron microscopes or require a complicated preparation process, such as Pickering emulsion [6], optical microscopy is a convenient method, but its magnification and observation effect are slightly inferior to electron microscopy.

In electron microscopy, an electron beam is used instead of light to form an enlarged image of samples, as electrons increase the resolution by a factor of 1000. When the electron beam shines on the sample, the interaction produces various signals that can be captured to get an image [7].

Of the tests mentioned above, the electron microscope includes two typical electron microscopes: SEM and TEM. Comparing the two tests, SEM uses a fine electron probe to scan the surface of the sample, producing an image with a three-dimensional appearance, while in TEM, electrons are transmitted through a sample, showing a two-dimensional image of the inside of a cell. SEM also has a good contrast, if the observer has a certain angle with the electron beam to observe the SEM image of the sample, there will be a sense that the eyes are looking along the axis of the electron beam, and it seems to be shining on the sample from the electron collector placed on the side [8]. Therefore, it contributes to making the rough surface show extremely clear and strong stereo acuity, so it is a powerful tool for studying solid sample surface morphology. Thus, it is widely used in the observation of starch particle morphology. As for TEM, it has been used to characterize the ultrastructure and specific observations of certain biomolecules in edible films [9]. TEM, also with tungsten wires, allows micron and nanometer resolution [7]. In the study of starch, TEM is often used to detect the appearance structure of starch particles, starch paste, and starch derivatives, which is of great importance to the analysis of the morphology of starch particles. Although the two instruments operate in completely different ways, they both use accelerating electrons and electromagnetic lenses to generate images, and both operate under high vacuum conditions.

In addition, since starch has birefringence characteristics, PLM can be used as a means of starch characterization. Under PLM, the density and refractive index differences between the crystalline and amorphous structures within starch particles, that is, anisotropy, result in polarization crossing and the appearance of polarization cross [10]. Due to gelatinization in the starch modification process, the birefringent light of the modified starch will be weakened compared with the raw starch, so PLM is a common method to study the structural changes of modified starch.

CLSM is a high-resolution microscopic imaging technique. It is a method of providing detailed morphology by visualizing clear images of the structure of matter, as well as a technique suitable for obtaining 2D and 3D images of particles under mild conditions.

This method can obtain an optical cross-section across a thick section of the sample [11]. The samples used for CSLM observation do not require complex preparation and are not limited to the observation of raw starch particles, which can be monitored in their original environment as a residue of the gelatinization process and particles in the starch paste [12]. In this way, starch particles from their plant source to their final application can be visualized with a microscopy technique. CSLM allows the staining of specific compounds, so the distribution of these compounds through the kernel can be more pronounced than with other microscopy techniques.

Different from the above detection methods, AFM obtains information such as the morphology, adhesion, and elasticity of the sample surface by contacting the sample surface with the tip of the needle. AFM can detect the morphology, properties, modifications, and interactions of different starch sources at the micro- and nano-structural levels. and does not require complex sample preparation conditions. The structural information obtained by AFM technology provides a qualitative and quantitative basis for the study of some physicochemical properties of different starch systems [13].

Therefore, this chapter introduces the research methods of starch granule morphology characteristics of lotus seed by combining the important methods of particle morphology characteristics research in recent years and provides a methodological reference for starch research.

2 Materials

2.1 Optical Microscopy

Optical microscopy (also known as light microscopy) is a basic microscopic observation device, which is mainly composed of eyepiece (one or two), objective lens, stage, aperture, light source (lighting system), and so on. The resolution of conventional optical microscopes is limited by the proximity to the wavelength of visible light (390–760 nm), so, given the wavelength and refractive index of the sample, optical microscopes cannot resolve samples below ~0.2 μm [14]. If the sample is transparent or has high light transmittance, light passes through the sample and is collected by the objective lens onto the eyepiece. But if the sample is translucent or opaque, light will be reflected from the surface to the eyepiece [15]. Optical microscopy is often used to describe the size and shape of a sample [16]. Although it is an ancient microscopic observation method, it requires only simple sample preparation and can be viewed by the naked eye, and the rapid photographic characteristics make it still indispensable in the study of infrastructure. In the study of starch morphology, optical microscopy is often used in Pickering emulsions prepared from starch [6]. The following materials are required for the observation of solid starch particles or liquid starch solutions (suspensions):

1. Pure water (or ethanol solution)
2. Slide, cover slide
3. Dropper

2.2 Scanning Electron Microscopy

SEM is a kind of electron microscope that uses a high-energy focused electron beam to scan the surface of a sample and obtain sample information. Therefore, it uses the incident electron beam as the illumination source, and its energy can be from less than 100 eV to more than 30 keV [17]. SEM is often used to characterize the microstructure and morphology, and it typically uses tungsten wires to produce electron beams with a resolution of about 1 μm, depending on the sample [18]. The basic components of SEM are shown in Fig. 1. The electron beam scans the surface of the sample and uses the information generated by the interaction of electrons and substances to conduct imaging in combination with the principle of electron optics. In SEM, the electrons accelerated from the gun pass through several electron lenses to the anode, forming an image of the electron source. In most SEM, the electron beam is deflected twice inside the objective lens and scans the sample in a predetermined manner. The primary electron produces secondary electron, light, X-ray, infrared, ultraviolet, and semiconductor effects on the specimen surface. Any one of these signals can be measured. Freeze-drying is one of the most common methods for preparing samples [19]. The sample preparation of SEM is simple, the best specimen preparation is the least preparation. Some samples can be directly observed in the electron microscope without preparation, so it is closer to the natural state of the substance, and the results can be obtained quickly, so it is widely used in the observation of starch particle morphology:

1. Tweezers: Used to pick up samples, etc.
2. Rubber suction bulb: Used to blow off floating powder
3. Conductive double-sided tape
4. Aluminum sample stub

2.3 Transmission Electron Microscope

TEM is an electro-optical instrument used to observe the submicroscopic or ultrastructure of an object. The basic components of TEM are shown in Fig. 1. It is a high-resolution, high-magnification electronic optical instrument that uses an electromagnetic lens with an electron beam of extremely short wavelength as an illumination source to focus imaging to show the surface or internal characteristics of the object. Its electron energy is between 60 and 150 keV (usually about 100 keV) or 200 keV–3 MeV (medium- or high-voltage electron microscopy). TEM consists of three systems,

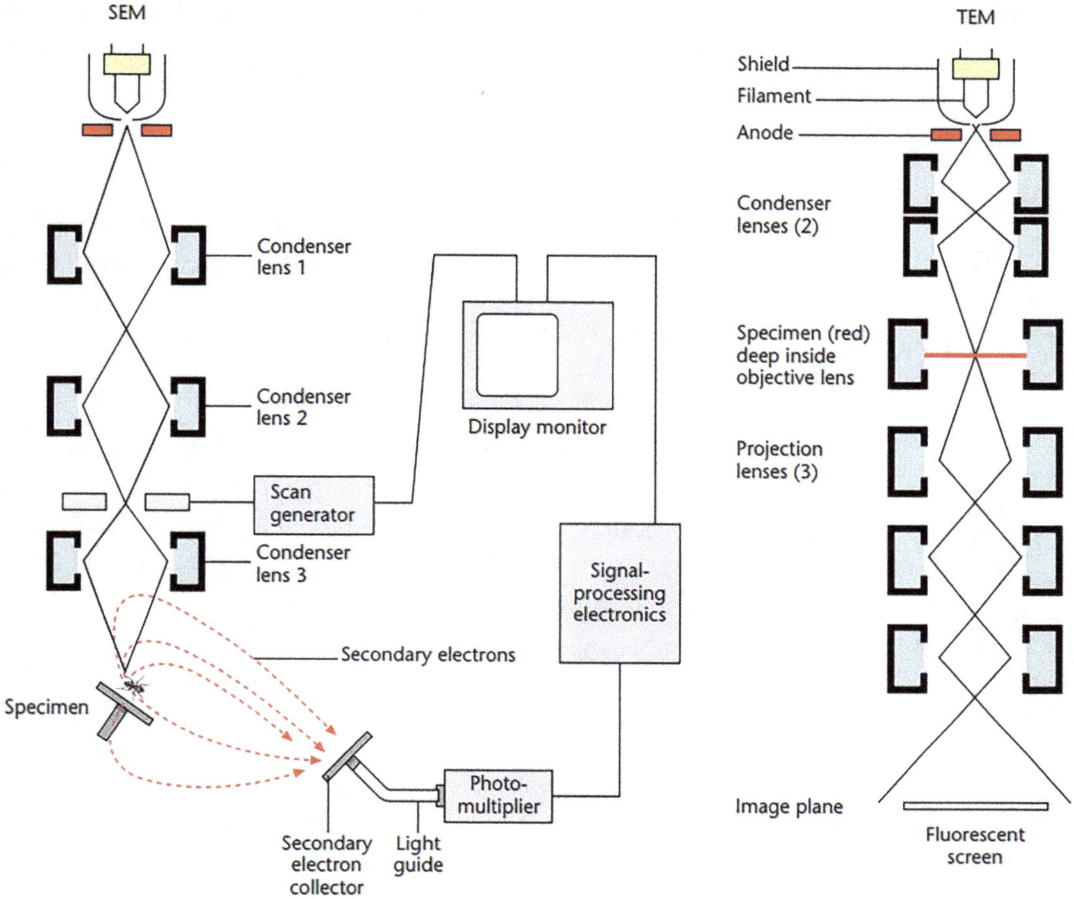

Fig. 1 Basic components making up scanning electron microscopy and transmission electron microscopy

namely, an optical system, a vacuum system, and an electronic line control system [20]. Its greatest feature is high resolution and magnification, but because electrons are easily scattered during detection, the penetration is extremely low, so the sample to be tested is required to be thin enough [21]. If the sample is thin enough and at least partially translucent to electrons (bacteria, viruses, isolated cell wall fragments, and macromolecules), the sample can be observed directly. Negative or positive staining can be used to increase the contrast of micrographs of such samples. But starch is larger and needs to be embedded in resin to obtain thin slices that can be stained with heavy metal salt (structure) and examine the sections in the path of the electron beam [22]. As with light microscopy, the sample must be preserved chemically, and the chemical fixative used must be able to quickly penetrate small pieces of tissue and be placed in a supporting matrix that prevents the tissue from collapsing when the knife cuts through it [23].

1. 3% Glutaraldehyde: Add 30 mL of 50% glutaraldehyde solution to a 500-mL volumetric bottle and then add ultrapure water and volume to 500 mL. Store at 4 °C.
2. 0.1 M sodium cacodylate buffer (pH = 7.2): Weigh 10.7 g of sodium dimethyl arsenate into a 500-mL volumetric bottle, add 400 mL of ultrapure water, and adjust the pH to 7.2 with HCl. Finally, set the volume to 500 mL. Store at 4 °C.
3. 3% Agar aqueous solution: Put 3.0 g of agar powder into a beaker, add 100 mL of ultrapure water, stir well, high-pressure disinfected, refrigerate at 4 °C, and set aside. Reheat and boil before use.
4. Saturated 2,4-dinitrophenylhydrazine.
5. 15% Acetic acid solution: Weigh 15.0 g of glacial acetic acid into a beaker and add 85 mL of ultrapure water.
6. 1% aqueous periodic acid solution: Weigh 1.0 g of periodic acid into a beaker and add 100 mL of ultrapure water. Store at 4 °C.
7. Saturated aqueous thiosemicarbazide solution.
8. 1% aqueous silver nitrate solution: Weigh 1.0 g of silver nitrate into the beaker, add 100 mL of ultrapure water, and generally add an appropriate amount of nitric acid to adjust the pH value and stability;
9. An ethanol series (30–100%): Remove 30 mL of anhydrous ethanol into a beaker and add 70 mL of ultrapure water to get 30% ethanol. The formulation of 50%, 70%, and 90% ethanol is the same as above.
10. 10% H_2O_2.
11. Rubber suction bulb.

2.4 Polarizing Microscope

PLM is a microscopic device used to detect the anisotropy and birefringence of substances. PLM involves inserting a polarizer and analyzer into the optical system of an optical microscope, using single-wavelength light as the light source to detect certain ordered structures and their refractive optical properties within the organism [24]. Any substance with birefringence properties can be clearly distinguished under a polarizing microscope. Starch particles are a type of spherical crystal system with optical anisotropy. Their birefringence will be demonstrated under a polarized light microscope, and due to the presence of polycrystalline structures (amorphous and crystalline) inside, a polarized cross will appear at the navel of the particles. In the crystalline and amorphous regions, the arrangement of starch molecular chains is not the same: in the former part, they are arranged in an orderly manner, but in the latter part, they behave in a disorderly manner. In terms of refractive index and density, these two structures differ in the amorphous region, leading to the occurrence of anisotropy. Once polarized light passes through starch particles, a Maltese cross will occur [25].

1. 50% glycerol solution: Measure 50 mL of glycerol, add distilled water, and bring to volume 100 mL. Shake well with oscillation. Store at 4 °C.
2. Polarization microscope (*see* **Note 1**).
3. Micropipette.
4. Glass slide and cover glass.

2.5 Confocal Laser Scanning Microscopy

Confocal laser scanning microscopy (CLSM) is a morphological feature determination tool associated with laser-induced fluorescence imaging. It can provide extremely thin focused images through true deep optical slicing methods and collect the microscopic morphology of substances (including microorganisms) in a three-dimensional environment [11]. In a nutshell, the basic principle of CLSM is to filter the fluorescence signal by laser-excited point scanning and spatial emission from the focus back to the confocal focus [26]. According to Fig. 2, laser excitation light is reflected by a dichroic mirror to the objective lens and refracted to the specimen on the focal plane. At this time, the fluorescence light emitted by the specimen passes through the objective lens again, and the fluorescence carrying image information has a long wavelength; thus, it passes directly to the dichroic mirror and reaches the photodetector through the detection pinhole, which becomes an electrical signal and is sent to the computer. In this process, only the

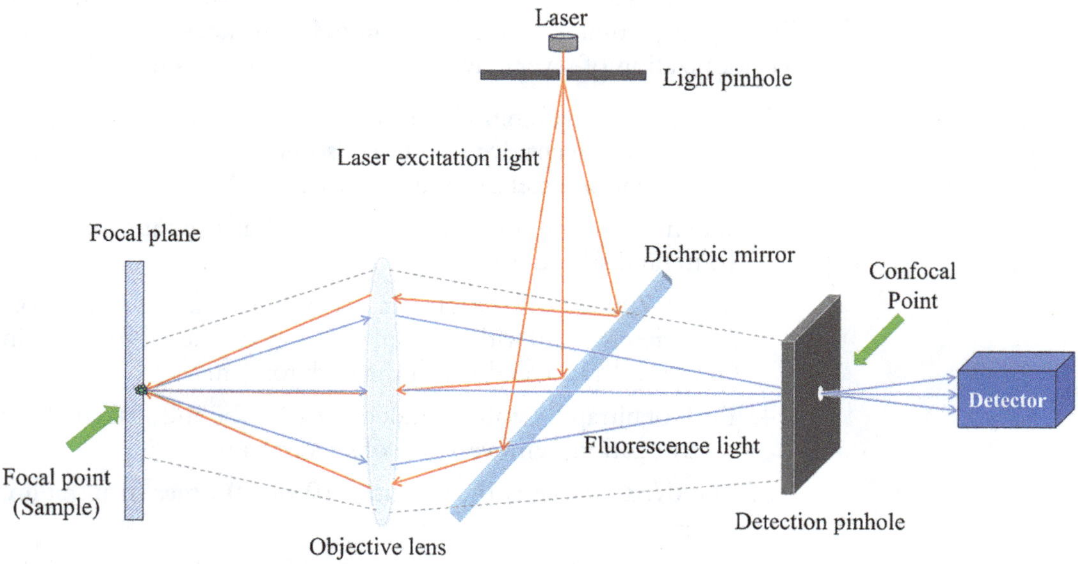

Fig. 2 Basic principles of confocal laser scanning microscopy

light emitted by points in the focal plane can pass through the detection pinhole, and the light emitted by points outside the focal plane is out of focus at the detection pinhole, and most of it cannot pass through the central hole. Therefore, the viewing target points on the focal plane appear bright, while the nonviewing points appear black as the background. In addition, the fluorescence light generated by laser irradiation, together with a small amount of reflected laser, is recollected by the objective lens and sent to a dichroic mirror.

The observation of starch morphological characteristics through CLSM generally involves fluorescent staining of starch (or substances bound to starch) with fluorescent staining agents. Each fluorescent molecule is characterized by two characteristic spectra, including an excitation spectrum of the wavelength at which the fluorescent dye is excited and an emission spectrum of the wavelength emitted by the excited fluorescent dye. The dyeing effect of different dyes is shown in Table 1 and Fig. 3. Compared with traditional wide-field microscopes, image resolution using CLSM is generally 15% higher in the *x–y* direction and 30% higher in the *z*-direction [11]. Moreover, noninvasive confocal optical sectioning technology can be used for the examination and three-dimensional reconstruction of living specimens [27]. At the same time, it also has some limitations, including the limited excitation wavelength used by the laser, the low efficiency of point-by-point scanning, and even the phototoxicity and photobleaching of the sample that may be triggered by some laser powers [11].

Images generated after CLSM detection need to be processed by image analysis software (Imaris, Amira, Matlab, ImageJ, etc.) to extract standardized geometric descriptors, such as microbial biological volume (mm^3), sample thickness (mm), or co-localization of several dyes (such as live/dead staining) [11].

1. 2 mg/mL of safranin O solution was prepared with ultrapure water (fluorescent dye selection also needs to be based on the equipment parameters used, *see* **Note 2**).
2. Starch and ultrapure water were mixed at the ratio of 1 mg: 10 μL and stirred evenly.
3. About 200 mg of the sample was weighed in an Eppendorf tube, then stained with safranin O (1 mg sample + 1 μL safranin O) at a 1:1 ratio, and left in the dark for 1 h.
4. 1 mL of ultrapure water is added to each tube and shaken. After centrifugation, remove 1 mL of supernatant.
5. The dyed sample is then diluted 10 or 20 times in ultrapure water.
6. A 200 μL stained sample drop is added to the hole with the slide.

Table 1
Dye used in the study of starch by confocal laser scanning microscopy

Dye	λ_{ex} [a] (nm)	λ_{em} [a] (nm)	Fluorescence color	Application	References
Safranin O	520	585	Red	The degradation rate of biodegradable starch microspheres was measured.	[28]
Rhodamine B	543	600–635	Red	The morphology and structure of giant starch granules in orchid pseudobulbs were studied.	[29, 30]
Green tea polyphenol	488	515	Green	The morphological characteristics of granules of starch-tea polyphenol complex were studied.	[5, 31]
Merbromin	488	500–634	Green	The effects of protease treatment on the surface and interior of starch granules were studied.	[32, 33]
Fluorescein isothiocyanate isomer (FITC)	488	500–525	Green	The distribution of polysaccharides on the surface of starch particles was observed.	[30]
Chromoionophore XI or 5-hexadecanoylaminofluorescein	488	480–640	Green	The characteristics of starch granules from several plant sources were studied.	[3]
APTS and sodium cyanoborohydride	488	500–535	Yellow	The internal structure and morphology of high amylose granules in barley mutant Amo1 were studied.	[34]
Ag^{+}	488	–	–	Localization of reaction sites in modified starch granules.	[35]

[a] λ_{ex} excitation wavelength, λ_{em} emission wavelength

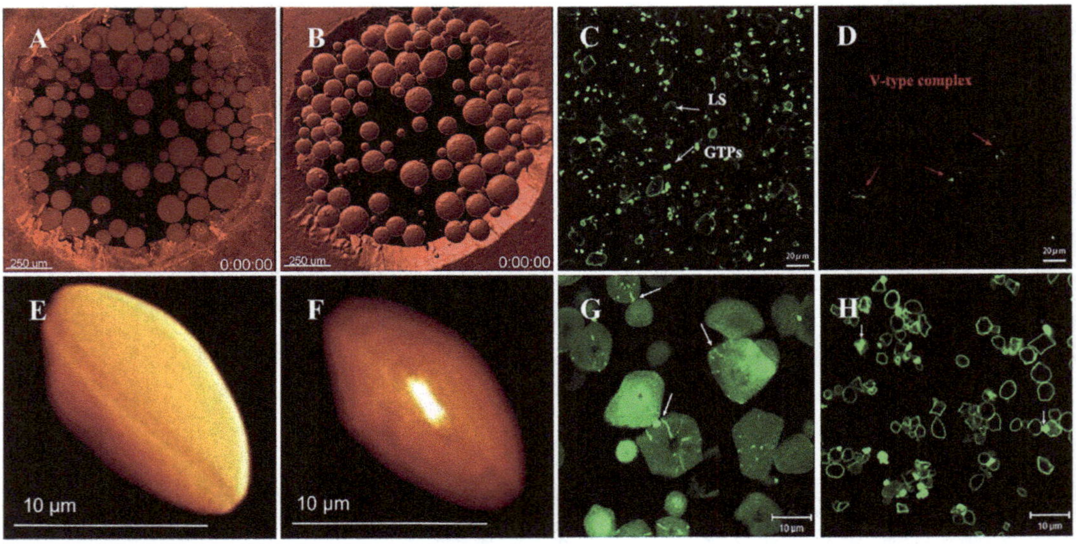

Fig. 3 Confocal laser scanning microscopy. Safranin O stain: (**a**) Degradable starch microspheres, (**b**) with surfaces added using the IMARIS software; green tea polyphenol stain: (**c**) mixture of lotus starch and tea polyphenols, (**d**) lotus starch-tea polyphenol complex; APTS and sodium cyanoborohydride stain: (**e**) surface image of equatorial groove in a Midas starch granule, (**f**) optical slice of equatorial groove in a Midas starch granule; merbromin stain: (**g**) corn starch and (**h**) rice starch

2.6 Atomic Force Microscopy

AFM utilizes weak interaction between the probe and the sample for imaging, so it can not only conductive samples but also non-conductive samples. AFM has atomic-level spatial resolution and can characterize the surface morphology of samples at the nano-scale. Moreover, AFM testing is simple in sample preparation and causes little damage to the sample [36]. AFM testing is not limited by environmental conditions and can image the surface microstructure of samples in atmospheric and solution environments [37]. In addition, AFM can also determine the mechanical properties of samples [38]. Given the advantages of AFM itself, its application scope is becoming increasingly widespread. Currently, it has been successfully applied in fields such as chemistry, physics, biology, and materials science, especially in the research of biological samples such as proteins [39], nucleic acids [40], cells [41], and bacteria [42], demonstrating enormous advantages. In recent years, it has also been frequently used to observe the apparent morphology of starch [36]. AFM mainly consists of a laser diode, a sharp tip, a cantilever, a piezo scanner tube, a photodetector, and a feedback control system [43]. When the probe scans the surface of the sample, the interaction between the probe and the sample surface (Van der Waals force) causes the cantilever to bend, and the laser beam emitted from the laser at the end of the cantilever is reflected on the photodetector. When the probe scans along the topography of the sample surface, the angle of the laser reflected on the photodetector changes according to Hooke's law. The intensity variations of the laser spot in different quadrants generate electrical signals, which are used by the feedback system to control the elongation and contraction of the piezoceramic scanning tube, resulting in an atomic force image of the surface topography of the sample [44]. According to the difference in the force domain range of the interaction force between the probe tip and the surface molecules of the sample to be tested during atomic force microscopy imaging, the working modes of atomic force microscopy can be divided into contact mode, noncontact mode, and tapping mode. Generally, the tapping mode is used for scanning starch [45]. Tapping mode is a working mode that falls between contact mode and noncontact mode. In this mode, the tip contacts the sample surface intermittently and periodically. Before scanning the surface of the sample, the probe vibrates at a certain amplitude. When the probe contacts the sample to be tested, the amplitude of the cantilever will also change due to the fluctuation of the surface morphology of the sample. The feedback system adjusts the distance between the probe and the sample through changes in the amplitude of the cantilever to maintain a constant amplitude of the cantilever and finally converts this signal to obtain a morphology feature map of the sample. In tapping mode, due to the intermittent contact between the needle tip and the surface of the sample to be tested, the force on the sample is much smaller compared to the contact mode. Therefore, the advantage of tapping mode is to avoid damage to the sample caused by prolonged contact between the needle tip and the sample surface. Meanwhile, the image resolution obtained by tapping mode can be comparable to that of contact mode. Therefore, the tapping mode is more suitable for measuring samples that are relatively soft and fragile and have significant surface fluctuations.

1. Atomic force microscope
2. Magnetic heating stirrer
3. Pipette
4. Mica sheet

3 Methods

3.1 Optical Microscopy

3.1.1 Starch Powder

1. A small amount of starch particles was placed on a slide and moistened with 1–2 drops of 50% ethanol-aqueous solution.
2. Put on the cover glass, using a microscope to find samples by adjusting the coarse and fine foci spiral, adjusting the light and dark degree of the visible range by adjusting the aperture, and observing the microscopic morphology of the particles in the best observation environment.

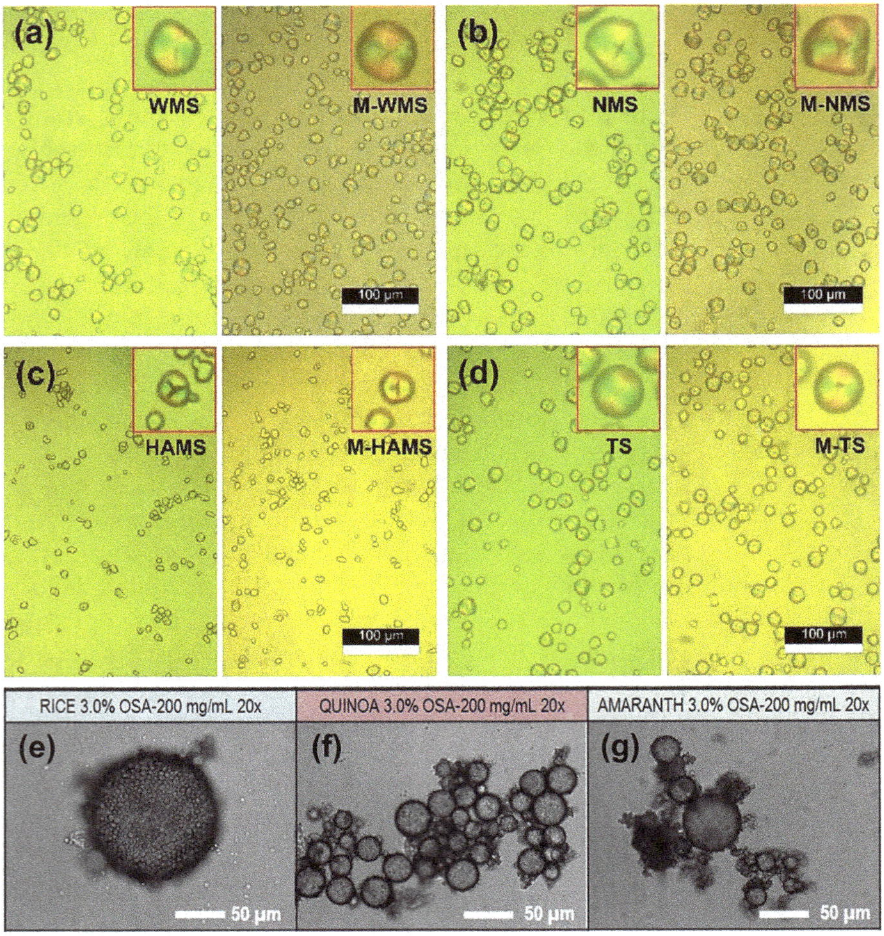

Fig. 4 Microscopic images of original starch and esterified starch (M–). (**a**) WMS (waxy maize starch), (**b**) NMS (normal maize starch), (**c**) HAMS (high-amylose maize starch), and (**d**) TS (tapioca starch). Optical micrographs of 3.0% OSA-modified starch granule stabilized emulsions (200 mg starch/mL oil): (**e**) rice starch, (**f**) quinoa starch, and (**g**) amaranth starch

3.1.2 Liquid Starch Sample

1. Take a uniform starch solution (suspension), and then follow the steps in Subheading "3.1.1 Starch Power", **step 2**.

The starch sample and Pickering emulsion prepared by starch observed by optical microscope are shown in Fig. 4.

3.2 Scanning Electron Microscopy

3.2.1 Powder Samples

1. Paste a layer of conductive double-sided tape on the aluminum sample stub, take a small amount of sample powder, and gently tap or shake the sample spoon with tweezers or toothpicks to make the sample evenly fall on the conductive tape [46]. After the sample is glued firmly, use the rubber suction bulb to blow away or knock off the sample that is not stuck on the surface (*see* **Note 3**).
2. The sample stub was placed in the sample chamber of the ion coater, and the sample was sputtered with gold at different

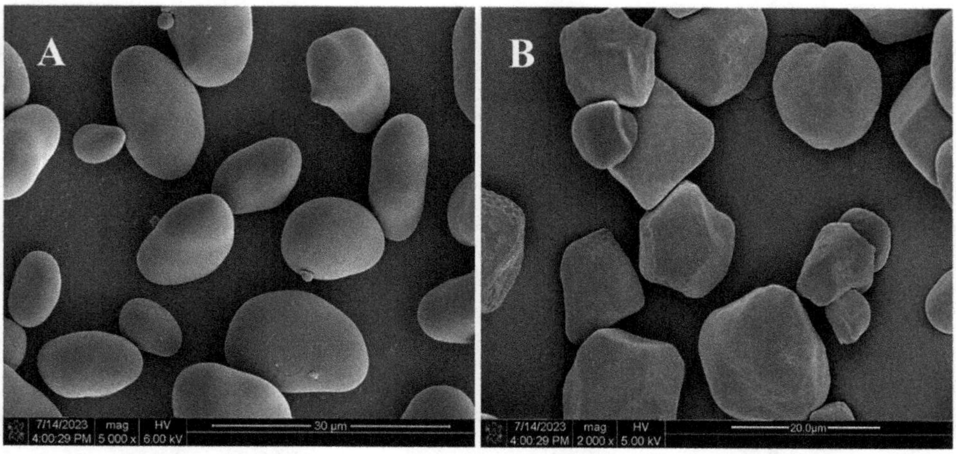

Fig. 5 Scanning electron microscopy. (**a**) Lotus seed starch. (**b**) Corn starch

currents and times as required [47]. The thickness of the gold film was about 15 nm, and the sputtering time was 2 min (*see* **Note 4**).

3. The prepared samples are placed in a SEM sample chamber and observed in a high-vacuum mode with an acceleration voltage of 6 keV at 1000–20,000 times the resolution (*see* **Note 5**). Figure 5 shows the lotus seed starch and corn starch particles photographed by scanning electron microscopy;
4. Enter the test software page, and set the sample size according to the actual size of the sample table. After the "on" key is pressurized, the sample is observed at low magnification first, and the sample area can be found by dragging the mouse. Switch the angle of view to the condition of high magnification, adjust the multiple by rotating "magnification" and "Stigmator X/Y" to eliminate astigmatism, adjust the focal length by "Focus," and adjust the light and dark and contrast by "Brightness" and "Contrast" buttons, until observers get the image that meets their requirements. Capture the image when it is clear.

3.2.2 Block Samples

1. For block samples, there is no need to do too much pretreatment, but the sample size needs to match the SEM sample stub, and the height should be 5–10 mm.
2. Paste a layer of conductive tape on the sample stub, and fix the cross-section center of the sample on the conductive tape.
3. Place the sample table in the sample chamber of the ion sputtering instrument, and spray gold on the sample at different currents and times according to needs. The thickness of the gold film is about 50 nm (*see* **Note 4**).
4. *See* Subheading 3.2.1, **step 3**.
5. *See* Subheading 3.2.1, **step 4**.

3.3 Transmission Electron Microscope

1. Appropriate samples are taken out in test tubes, 3% glutaraldehyde is added, and fixed at room temperature in 0.1 M sodium cacodylate buffer for 2 h and fixed at 4 °C for 6 h.
2. The fixed sample is preencapsulated in a 3% agar aqueous solution, cut into 1-mm^3 cubic samples, and fixed again for 1 h.
3. After fixing, wash them with buffer and ultrapure water twice for 20 min each time.
4. To block the aldehydes left over from the previous step, the sample cubes should be immersed in a saturated 2,4-dinitrophenylhydrazine and soaked in 15% acetic acid solution for 1 h at room temperature.
5. Wash four times in ultrapure water for 15 min at a time.
6. The sample blocks are oxidized in 1% periodic acid aqueous solution for 45 min, washed according to **step 5**, and then soaked in a saturated thiosemicarbazide aqueous solution for 24 h.
7. Repeat **step 5** after washing, and place the sample cubes in 1% silver nitrate aqueous solution for 3 days in the dark.
8. After rinsing the stained cubes with ultrapure water, they are dehydrated with ethanol of different gradients (30%, 50%, 70%, 90%, and 100%). Dehydrate 30–90% ethanol for 15 min and 100% ethanol for 20 min.
9. The sample cubes are embedded in epoxy resin.
10. The oxidation reaction is controlled in parallel by substituting 10% H_2O_2 for periodic acid.
11. The sample cubes are embedded in the ultrathin slicer and cut into 90–100 nm slices. After completion, the samples are pulled out with a carrier net (*see* **Note 6**).
12. Use a needle to lift the front end of the single-rod metal ring, place the sample face down in the sample hole, and press down the metal ring. Use the rubber suction bulb to blow away any dust that may have stuck to the copper wire.
13. Align with the dowel pin and position the empty position, slowly push the single line rod, push until there is resistance, and push forward a little until the red light is on and the vacuum is started. When the red light goes out, turn the handle counterclockwise until the single-row rod enters the cabin body and closes the door.
14. Observe and shoot at 60 kV [48]. Figure 6 [4, 49] shows the starch granules and starch nanoparticles observed by transmission electron microscopy, respectively.
15. Focus by continuously adjusting the spot with the trackball and Intensity buttons (*see* **Note 7**). Find the position of interest,

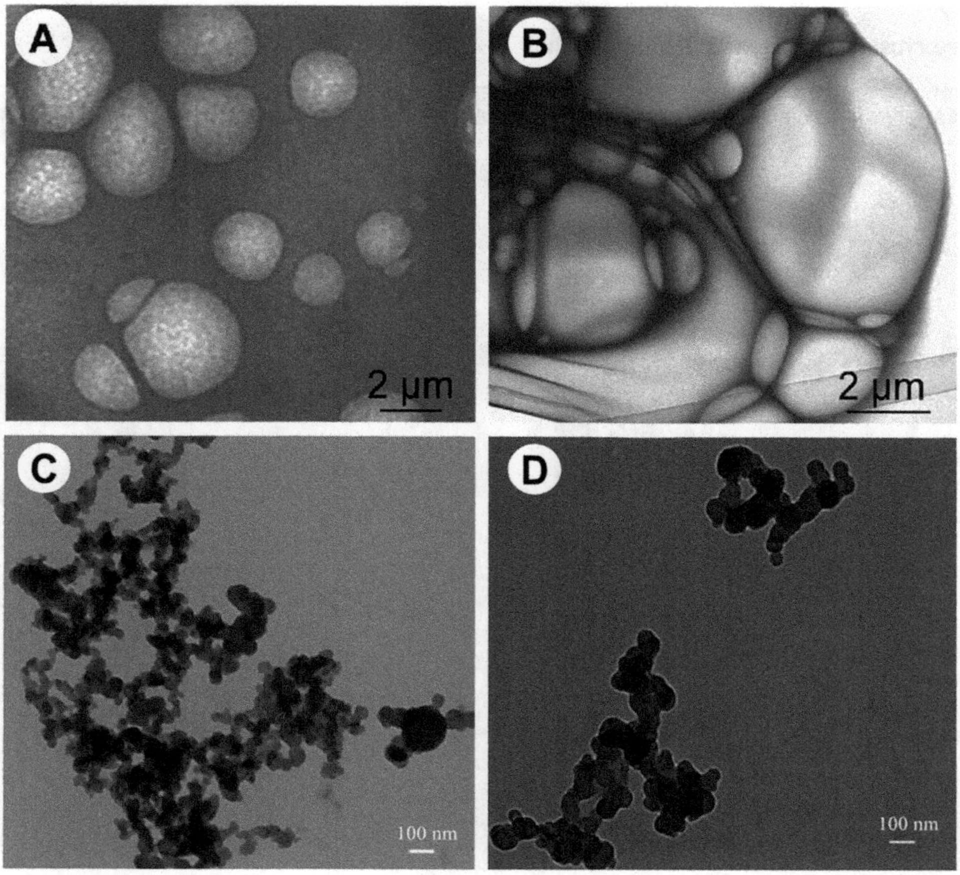

Fig. 6 Transmission electron microscopy. (**a**) Small starch granules in spruce needles. (**b**) Large starch granules in spruce needles. Starch nanoparticles at 90 °C, 20% moisture (**c**) and 30% moisture (**d**)

and adjust the focus with the "Focus" button; the upper knob is the focus knob, and the lower button can adjust the step (the longer the step, the greater the focus degree). Find the sample location and click preview, press the "R1" key to switch the screen to the camera software, adjust the focal length, and capture the target image.

16. Regions of a large mass in the image cause electrons to deflect to the point where they can be effectively removed from the optical axis of the microscope. In these regions, the electrons are subtracted to make them show darker when they are finally projected onto the screen. Regions with lower mass scatter electrons to a lesser extent and show brighter on the screen.

3.4 Polarizing Microscope

1. Preparation of starch suspension: Disperse 2 g of starch in 100 mL of 50% glycerol to obtain a concentration of 2% (w/v) [50].

Table 2
Maltese cross properties of different starches

The variety of starch	Picture	The position of the Maltese cross	References
Corn starch	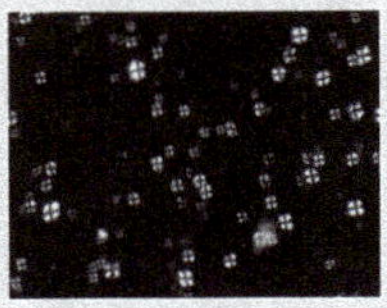	Center	[25]
Wheat starch	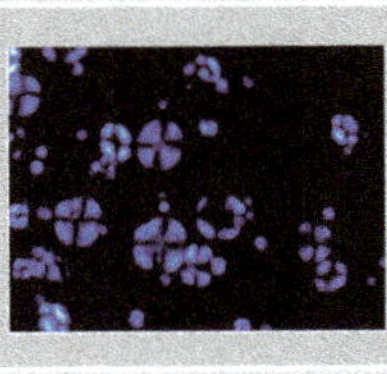	Center	[52]
Rice starch	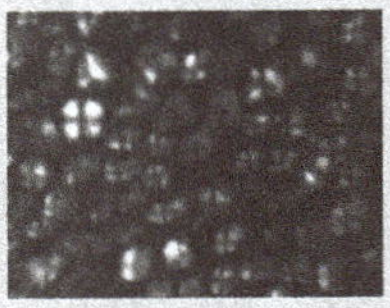	Center	[51]
Potato starch	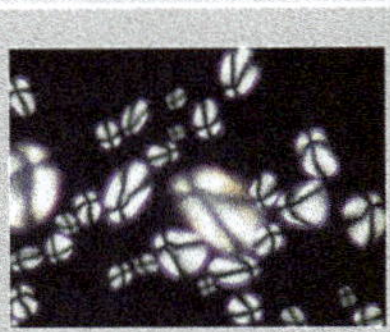	Leaning toward one end of the particle	[53]

2. Preparation of specimen: Use a micropipette to pipette 20 μL of the starch suspension and drop it onto a clean glass slide; after natural drying, cover it with a clean cover glass slide [51].
3. Place the specimen on a polarizing microscope stage, observe at the required magnification, and capture a polarized cross-image of starch particles.
4. The type of starch can be determined by observing the position of the Maltese cross in the starch image. Different starches have different positions of the Maltese cross, such as the polarized cross of corn starch in the center, while the Maltese cross of potato starch leans toward one end of the particles (Table 2). In addition, the crystal structure of starch can be determined by the Maltese Cross. Due to the presence of polycrystalline structures, a Maltese cross may appear on the particles. In the crystalline and amorphous regions, the arrangement of starch molecular chains is different: In the crystalline region, they are arranged in order, but in the amorphous region, their properties are disordered. Therefore, the degree of crystallization of starch can be determined by the number of Maltese crosses in the image.

3.5 Confocal Laser Scanning Microscopy

1. CLSM imaging was performed by the inverted microscope, and a 60× Oil Immersion Objective was used for an objective lens (the objective multiple was adjusted according to the actual situation).
2. Safranin O was excited at 561 nm by a DPSS laser with 0.700% power, and the emission spectrum was measured at 570–620 nm by a GaAsp detector.
3. All samples' laser gain and intensity remained constant, and the TPMT detector was turned on to record the transmitted light (*see* **Note 8**). All measurements were taken at room temperature.
4. For more CLSM considerations, *see* **Note 9**.

Many fluorescence dyes are used in confocal laser scanning microscopy to study the morphological characteristics of starch, and the corresponding excitation wavelength and emission wavelength of each stain are different. Table 1 lists the fluorescent stain used by eight laboratories including the above detection methods to observe starch particles through laser confocal and the corresponding excitation light wavelength. The observed starch particles after staining with some of the fluorescent stain are shown in Fig. 3.

3.6 Atomic Force Microscopy

1. Preparation of starch suspension: Disperse 10 mg of starch in 50 mL of distilled water to obtain a 0.02% (w/v) starch solution [54].
2. Heat and stir in a water bath using a magnetic heating stirrer at 90 °C for 30 min.
3. Use a pipette to absorb starch solution 20 μL and quickly drip onto a mica sheet.
4. Dry with a hairdryer and dry at room temperature for 24 h, and then wait for observation with Atomic force microscopy.
5. Open the Atomic force microscopy and fix the sample on the sample table so that the probe can scan any optional representative area. According to the characteristics of the instrument, initialize the software and instrument, and select the tapping mode. The scanning frequency is 300–400 kHz, and the tip radius is between 5 and 10 nm [54] (*see* **Notes 10–12**).
6. Scan the image and save it. The atomic force microscopy scanning diagram of some starch is shown in Fig. 7 [2].
7. Analyze the blocklet structure size, surface roughness, and cross-sectional characteristics of the sample particle surface using the NanoScope Analysis software (*see* **Note 13**).

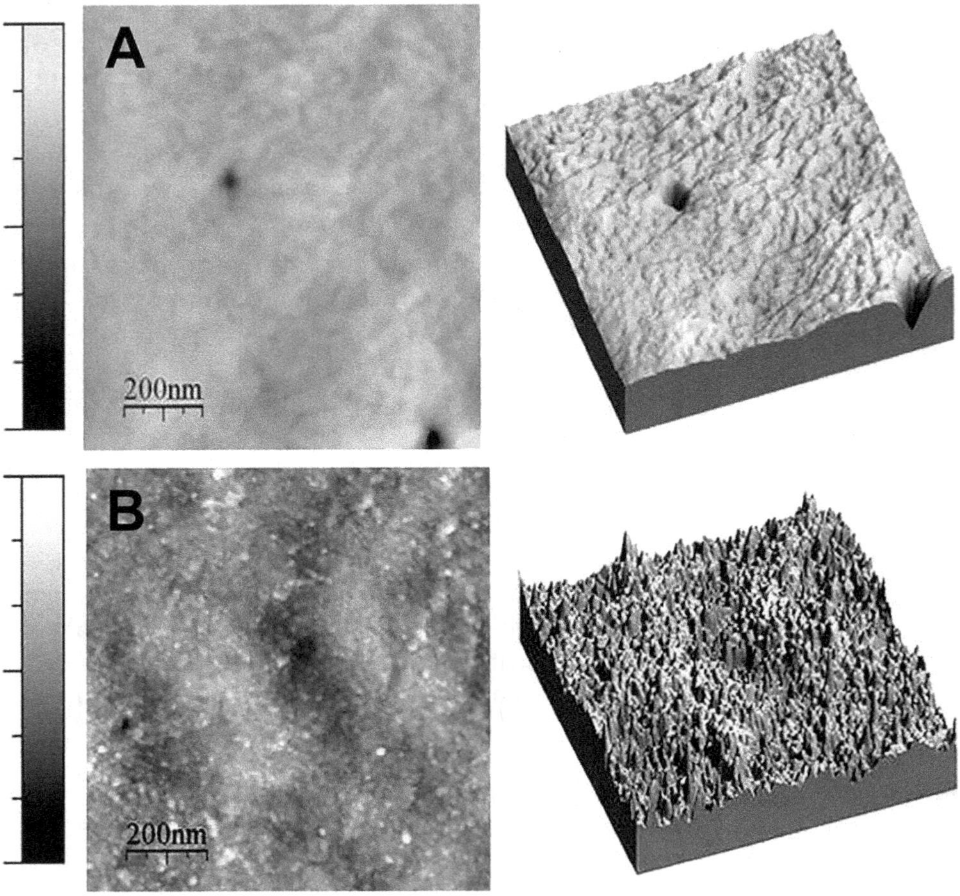

Fig. 7 Atomic force microscopy images of the surface of native corn starch granules (**a**) (height difference 82.2 nm, scan size 1 × 1 μm) and native potato starch granules (**b**) (height difference 18.8 nm, scan size 1 × 1 μm)

8. The obtained image is first converted into a grayscale image, and then important texture characteristic parameters such as energy, contrast, uniformity, entropy, and fractal dimension are extracted from the two-dimensional morphology map of starch particles using the grayscale co-occurrence matrix and differential box-counting algorithm in the Image J software.

4 Notes

1. A polarizing microscope is required to be equipped with a transparent illumination light source (halogen 100 W or above), a polarizing mirror (polarizing mirror) on the illumination side, and a polarizing mirror (polarizing device) on the

observation side, which can be orthogonal. The loading platform can rotate 360° and move. The eyepiece equipped should have a cross marking for measurement.

2. Confocal laser scanning microscopy dye selection: Fluorescent dye should be selected according to the instrument, each channel can only select one dye, and the wavelength is within the wavelength range of the laser. In addition, the excitation light in the "Currently available Lasers" option is fixed, so if there is no expected excitation light, the excitation light close to or similar should be selected, and the wavelength to be received should be avoided for detection.
3. The amount of sample powder should not be too much, and the thickness should be uniform. It prevents the particles from overlapping; otherwise, it is easy to cause poor electrical conductivity of the bottom part of the gold-spraying sample, which may lead to charged artifacts of the particles after gold spraying.
4. The specific gold film thickness and sputtering time need to be determined according to the sample and the sputtering time. Settings of different sputtering coating instruments and the corresponding sputtering gold film thickness will be different. Vacuum evaporation and sputtering coatings are two common technologies. In vacuum evaporation, the sample is rotated inside a vacuum bell jar, while the coating material is continuously evaporated from different angles. Nitrogen is controlled to enter, providing a low concentration of nitrogen molecules in the bell jar, which the evaporated atoms collide with and thus deflect, coating the sample from all directions. In sputtering coatings, the coating material (gold) is the cathode and the specimen is placed on the anode. Sputtering takes place in an argon atmosphere. In 2–3 keV, argon ionization, positive ion bombarding cathode (coating), the cathode erosion, ejects particles from all directions in the samples. The amount of metal needed to be sputtered is related to the porosity of the starch, and the more porous the sample, the greater the amount of gold needed to form a uniform conductive layer on the sample surface.
5. The specific magnification also needs to be determined according to the actual size of the sample. In most cases, resolution rises as the acceleration voltage rises, but most food-related work is performed at 5–20 kV.
6. There are many kinds of supporting films, such as commonly used carbon supporting film (10–20 nm), micro-gate film (15–30 nm), porous film (5–20 nm), and ultrathin carbon

Table 3
Carrier nets used in the study of starch by transmission electron microscopy

Carrier net	Thickness (nm)	Sample thickness (nm)	Material	Crystal structure	References
Carbon-coated grids	10–20	–	Waxy maize starch nanoparticles	A	[49]
Formvar/ carbon-coated 200 mesh copper grid	10–20	–	Waxy maize starch	A	[55]
Carbon-coated microscopy grid	10–20	50–100	Wheat starch nanoparticles	B	[56]
Formvar-supported gold grids	~10	~70	Potato starch films	B	[57]
Collodion-coated copper grids	–	90–100	Waxy hull-less barley starches, normal hull-less barley starches, and high-amylose hull-less barley starches	–	[48]
Carbon-coated grid	10–20	–	Pea starch	C	[58]
Carbon-coated copper grids	10–20	–	Maize amylose-stearic acid complex	V	[59]

film (3–5 nm), that can be selected according to demand and slice thickness. Some carrier nets were used in the TEM study of starch, as shown in Table 3.

7. In the operation panel, the "Intensity" button can adjust the spot size. The "trackball" button is used to adjust the spot position. "Multifunction" can be used for magnification. The buttons "+" and "−" on the far right of the panel are used to adjust the *Z* value, and the "Wobbler" button indicates that the height adjustment is complete.

8. Confocal laser scanning microscopy: Once the preview starts (Live is turned on) and the laser starts shining at the sample, after the mode is turned on, setting parameters requires turning off the preview mode. In order to reduce damage to the sample, it should be operated quickly and the preview time should be minimized.

9. Precautions should be taken before using confocal laser scanning microscopy: (1) The sample was labeled by a fluorescent probe; (2) fixed or living tissue; (3) fixed or live adherent cells should be cultured on confocal special Petri dishes or cover slides; (4) suspend the cells, and seal the film with a cover glass after throwing or dropping the film; (5) the thickness of the slide should be between 0.8 and 1.2 mm, the cover slide should be smooth, and the thickness should be about 0.17 mm; (6) the specimen should not be too thick, such as too thick excitation bright part consumed in the lower part of the specimen, and the upper part directly observed by the objective lens could not be fully stimulated; (7) try to remove unless specific fluorescent signals; and (8) multipurpose glycerin for sealing tablets: PBS mixture (ratio 9:1).
10. Suggest cooperating with the Nanoscope controller for operation.
11. Before starting the controller, adjust the laser, strike the laser at the front end of the cantilever, and adjust the detector position.
12. In the atomic force microscopy field of vision, the probe must be found first, otherwise the needle may be struck.
13. Software with the same functionality can be used.

5 Conclusion

Starch plays an important role in food processing, materials science, biomedicine, textile industry, and cosmetics. Different sources and different modification methods will cause differences in the morphology, physicochemical properties, and molecular structure of starch. The morphological structure detection of starch particles is often the basis of multi-scale structure detection, and the application range of starch can be initially inferred through the most intuitive morphological characterization. This chapter introduces several important starch particle morphology characterization methods (Table 4). Readers can refer to some of the above methods or combine several detection methods according to the actual detection needs, to achieve the purpose of describing the morphological characteristics of the detection object.

Table 4
Five important methods for characterizing starch particle morphology

Method	Experimental environment	Application	Instrument principle
SEM	Vacuum	Starch surface morphology	According to the interaction between electron and matter, use secondary electron signals microscopic imaging to observe the sample surface morphology.
TEM	Vacuum	Ultrastructure of the starch granules	Using the electron beam with a very short wavelength as the lamp source, electromagnetic lens focus imaging is used to show the surface or internal features of the object.
PLM	Atmosphere	Starch crystal structure	A polarizer and analyzer are inserted in the polarizer, with single wavelength light as the light source, to detect some ordered structures in the organism and their refractive optical properties.
CLSM	Atmosphere	Structure and properties of starch particles, and properties of modified starch and starch materials	Using laser, electronic camera, and other high-tech means, according to the principle of conjugate focusing imaging.
AFM	Atmosphere	Starch indicates nanostructural characteristics	By detecting the microscopic force between the sample surface and the force-sensitive element, the surface structure and surface roughness at nanoscale resolution.

References

1. Zhao M, Bao Y, Liu P (2018) Progress in research on fine microstructure of starch granules. Food Sci 39(11):284–294. https://doi.org/10.7506/spkx1002-6630-201811044
2. Sujka M, Jamroz J (2009) α-Amylolysis of native potato and corn starches – SEM, AFM, nitrogen and iodine sorption investigations. LWT Food Sci Technol 42(7):1219–1224. https://doi.org/10.1016/j.lwt.2009.01.016
3. Manca M, Woortman AJJ, Loos K, Loi MA (2015) Imaging inclusion complex formation in starch granules using confocal laser scanning microscopy. Starch – Stärke 67(1–2):132–138. https://doi.org/10.1002/star.201400118
4. Cabálková J, Přibyl J, Skládal P, Kulich P, Chmelík J (2008) Size, shape and surface morphology of starch granules from Norway spruce needles revealed by transmission electron microscopy and atomic force microscopy: effects of elevated CO(2) concentration. Tree Physiol 28(10):1593–1599. https://doi.org/10.1093/treephys/28.10.1593
5. Zhao B, Sun S, Lin H, Chen L, Qin S, Wu W et al (2019) Physicochemical properties and digestion of the lotus seed starch-green tea polyphenol complex under ultrasound-microwave synergistic interaction. Ultrason Sonochem 52:50–61. https://doi.org/10.1016/j.ultsonch.2018.11.001
6. Wang Q, Li J, Shi Y, Cong T, Liu H, Gao Y (2023) Preparation of lauric acid esterified starch by ethanol solvothermal process and its Pickering emulsion. Int J Biol Macromol 248:

125941. https://doi.org/10.1016/j.ijbiomac.2023.125941

7. Perea-Flores MJ, Mendoza-Madrigal AG, Chanona-Pérez JJ, Alamilla-Beltrán L, Gutierrez-López GF (2011) Microscopy techniques and image analysis for the quantitative evaluation of food microstructure. In: Food processing handbook. Wiley-VCH, Weinheim, pp 667–692
8. Pomeranz Y (1976) Scanning electron microscopy in food science and technology*. In: Chichester CO, Mrak EM, Stewart GF (eds) Advances in food research. Academic Press, New York, pp 205–307
9. Arzate-Vázquez I, Chanona-Pérez JJ, Calderón-Domínguez G, Terres-Rojas E, Garibay-Febles V, Martínez-Rivas A et al (2012) Microstructural characterization of chitosan and alginate films by microscopy techniques and texture image analysis. Carbohydr Polym 87(1):289–299. https://doi.org/10.1016/j.carbpol.2011.07.044
10. Oldenbourg R (2013) Polarized light microscopy: principles and practice. Cold Spring Harb Protoc 2013(11). https://doi.org/10.1101/pdb.top078600
11. Canette A, Briandet R (2014) MICROSCOPY | Confocal laser scanning microscopy. In: Batt CA, Tortorello ML (eds) Encyclopedia of food microbiology, 2nd edn. Academic Press, Oxford, pp 676–683
12. van de Velde F, van Riel J, Tromp RH (2002) Visualisation of starch granule morphologies using confocal scanning laser microscopy (CSLM). J Sci Food Agric 82(13): 1528–1536. https://doi.org/10.1002/jsfa.1165
13. Zhu F (2017) Atomic force microscopy of starch systems. Crit Rev Food Sci Nutr 57(14):3127–3144. https://doi.org/10.1080/10408398.2015.1094650
14. Ebnesajjad S (2014) Chapter 4 – surface and material characterization techniques. In: Ebnesajjad S (ed) Surface treatment of materials for adhesive bonding, 2nd edn. William Andrew Publishing, Oxford, pp 39–75
15. Youssef G (2022) Chapter 11 – characterization of polymers. In: Youssef G (ed) Applied mechanics of polymers. Elsevier, Amsterdam, pp 273–299
16. Chakraborty I, Pallen S, Shetty Y, Roy N, Mazumder N (2020) Advanced microscopy techniques for revealing molecular structure of starch granules. Biophys Rev 12(1): 105–122. https://doi.org/10.1007/s12551-020-00614-7
17. Joy DC (2019) Scanning electron microscopy: theory, history and development of the field emission scanning electron microscope. In: Biological field emission scanning electron microscopy. John Wiley & Sons Ltd, Hoboken, pp 1–6
18. Mohammed A, Abdullah A (2019) Scanning Electron Microscopy (SEM): a review. In: Proceedings of International conference on hydraulics, pneumatics, sealing elements, tools, precision mechanics, specific electronic equipment & mechatronics, Romania
19. Ul-Hamid A (2018) A beginners' guide to scanning electron microscopy. Springer, Cham
20. Reimer L (1989) Transmission electron microscopy, Springer series in optical sciences, 2nd edn. Springer, Berlin, Heidelberg
21. Sciau P (2016) Chapter two – transmission electron microscopy: emerging investigations for cultural heritage materials. In: Hawkes PW (ed) Advances in imaging and electron physics. Elsevier, Amsterdam, pp 43–67
22. Li JZ, Wang Y, Wang ZH, Mei D, Zou W, Chang AM et al (2010) Phase-selective staining of metal salt for scanning electron microscopy imaging of block copolymer film. Ultramicroscopy 110(10):1338–1342. https://doi.org/10.1016/j.ultramic.2010.06.006
23. Bozzola JJ, Russell LD (1999) Electron microscopy: principles and techniques for biologists, 2nd edn. Jones and Bartlett, Boston
24. Changoor A, Tran-Khanh N, Méthot S, Garon M, Hurtig MB, Shive MS et al (2011) A polarized light microscopy method for accurate and reliable grading of collagen organization in cartilage repair. Osteoarthr Cartil 19(1): 126–135. https://doi.org/10.1016/j.joca.2010.10.010
25. Xiao H, Wang S, Xu W, Yin Y, Xu D, Zhang L et al (2020) The study on starch granules by using darkfield and polarized light microscopy. J Food Compos Anal 92:103576. https://doi.org/10.1016/j.jfca.2020.103576
26. Kihm KD (2011) Confocal laser scanning microscopy (CLSM). In: Kihm KD (ed) Near-field characterization of micro/nano-scaled fluid flows. Springer Berlin Heidelberg, Berlin, Heidelberg, pp 55–79
27. Allgeier S, Bartschat A, Bohn S, Peschel S, Reichert K-M, Sperlich K et al (2018) 3D confocal laser-scanning microscopy for large-area imaging of the corneal subbasal nerve plexus. Sci Rep 8(1):7468. https://doi.org/10.1038/s41598-018-25915-6
28. Thaarup IC, Gummesson C, Bjarnsholt T (2021) Measuring enzymatic degradation of

degradable starch microspheres using confocal laser scanning microscopy. Acta Biomater 131: 464–471. https://doi.org/10.1016/j.actbio.2021.06.042

29. Chanzy H, Putaux J-L, Dupeyre D, Davies R, Burghammer M, Montanari S et al (2006) Morphological and structural aspects of the giant starch granules from Phajus grandifolius. J Struct Biol 154(1):100–110. https://doi.org/10.1016/j.jsb.2005.11.007
30. Funami T, Nakauma M, Noda S, Ishihara S, Asai I, Inouchi N et al (2008) Effects of some anionic polysaccharides on the gelatinization and retrogradation behaviors of wheat starch: soybean-soluble polysaccharide and gum arabic. Food Hydrocoll 22(8):1528–1540. https://doi.org/10.1016/j.foodhyd.2007.10.008
31. Zhao B, Wang B, Zheng B, Chen L, Guo Z (2019) Effects and mechanism of high-pressure homogenization on the characterization and digestion behavior of lotus seed starch–green tea polyphenol complexes. J Funct Foods 57:173–181. https://doi.org/10.1016/j.jff.2019.04.016
32. Huber KC, Bemiller JN (2000) Channels of maize and sorghum starch granules. Carbohydr Polym 41(3):269–276. https://doi.org/10.1016/S0144-8617(99)00145-9
33. Bae J-E, Hong JS, Baik M-Y, Choi H-D, Choi H-W, Kim H-S (2020) Impact of starch granule-associated surface and channel proteins on physicochemical properties of corn and rice starches. Carbohydr Polym 250: 116908. https://doi.org/10.1016/j.carbpol.2020.116908
34. Borén M, Glaring MA, Ghebremedhin H, Olsson H, Blennow A, Jansson C (2008) Molecular and physicochemical characterization of the high-amylose barley mutant Amo1. J Cereal Sci 47(1):79–89. https://doi.org/10.1016/j.jcs.2007.01.009
35. Gray JA, BeMiller JN (2004) Development and utilization of reflectance confocal laser scanning microscopy to locate reaction sites in modified starch granules. Cereal Chem 81(2): 278–286. https://doi.org/10.1094/CCHEM.2004.81.2.278
36. Peroni-Okita FHG, Gunning AP, Kirby A, Simão RA, Soares CA, Cordenunsi BR (2015) Visualization of internal structure of banana starch granule through AFM. Carbohydr Polym 128:32–40. https://doi.org/10.1016/j.carbpol.2015.04.019
37. Sarkar A (2022) Biosensing, characterization of biosensors, and improved drug delivery approaches using atomic force microscopy: a review. Front Nanotechnol 3. https://doi.org/10.3389/fnano.2021.798928
38. Li Y, Yang J, Pan Z, Tong W (2020) Nanoscale pore structure and mechanical property analysis of coal: an insight combining AFM and SEM images. Fuel 260:116352
39. Devin E, Marc-Andre L, Thomas P (2021) Modulation of a protein-folding landscape revealed by AFM-based force spectroscopy notwithstanding instrumental limitations. Proc Natl Acad Sci U S A 118(12): e2015728118. https://doi.org/10.1073/pnas.2015728118
40. Kolbeck PJ, Dass M, Martynenko IV, van Dijk-Moes RJA, Brouwer KJH, van Blaaderen A et al (2023) DNA origami fiducial for accurate 3D atomic force microscopy imaging. Nano Lett 23(4):1236–1243. https://doi.org/10.1021/acs.nanolett.2c04299
41. Puckert C, Tomaskovic-Crook E, Gambhir S, Wallace GG, Crook JM, Higgins MJ (2020) Molecular interactions and forces of adhesion between single human neural stem cells and gelatin methacrylate hydrogels of varying stiffness. Acta Biomater 106:156–169. https://doi.org/10.1016/j.actbio.2020.02.023
42. Obořilová R, Šimečková H, Pastucha M, Klimovič Š, Víšová I, Přibyl J et al (2021) Atomic force microscopy and surface plasmon resonance for real-time single-cell monitoring of bacteriophage-mediated lysis of bacteria. Nanoscale 13(31):13538–13549. https://doi.org/10.1039/d1nr02921e
43. Arredondo-Tamayo B, Cárdenas-Pérez S, Méndez-Méndez JV, Arzate-Vázquez I, Torres-Ventura HH, Chanona-Pérez JJ (2023) Chapter 9 – advances in food material nanomechanics by means of atomic force microscopy. In: Zhong J, Yang H, Gaiani C (eds) Fundamentals and application of atomic force microscopy for food research. Academic Press, London, pp 263–306
44. Lou Z, Zhang Y, Li Y, Xu L (2023) Study on microscopic physical and chemical properties of biomass materials by AFM. J Mater Res Technol 24:10005–10026. https://doi.org/10.1016/j.jmrt.2023.05.176
45. Waduge RN, Xu S, Bertoft E, Seetharaman K (2013) Exploring the surface morphology of developing wheat starch granules by using Atomic Force Microscopy. Starch – Stärke 65(5–6):398–409. https://doi.org/10.1002/star.201200172
46. Barrera GN, Calderón-Domínguez G, Chanona-Pérez J, Gutiérrez-López GF, León AE, Ribotta PD (2013) Evaluation of the mechanical damage on wheat starch granules by SEM,

ESEM, AFM and texture image analysis. Carbohydr Polym 98(2):1449–1457. https://doi.org/10.1016/j.carbpol.2013.07.056

47. Zhuang Y, Wang Y, Yang H (2023) Characterizing digestibility of potato starch with cations by SEM, X-ray, LF-NMR. FTIR Food Chem 424:136396. https://doi.org/10.1016/j.foodchem.2023.136396
48. Li JH, Vasanthan T, Hoover R, Rossnagel BG (2004) Starch from hull-less barley: V. In-vitro susceptibility of waxy, normal, and high-amylose starches towards hydrolysis by alpha-amylases and amyloglucosidase. Food Chem 84(4):621–632. https://doi.org/10.1016/S0308-8146(03)00287-5
49. Ji N, Li X, Qiu C, Li G, Sun Q, Xiong L (2015) Effects of heat moisture treatment on the physicochemical properties of starch nanoparticles. Carbohydr Polym 117:605–609. https://doi.org/10.1016/j.carbpol.2014.10.005
50. Zhao X, Wang Y, Li D, Wang L (2021) Insight into the biphasic transition of heat-moisture treated waxy maize starch through controlled gelatinization. Food Chem 341(Pt 1):128214. https://doi.org/10.1016/j.foodchem.2020.128214
51. Du J, Yang Z, Xu X, Wang X, Du X (2019) Effects of tea polyphenols on the structural and physicochemical properties of high-hydrostatic-pressure-gelatinized rice starch. Food Hydrocoll 91:256–262. https://doi.org/10.1016/j.foodhyd.2019.01.035
52. Ma Y, Zhang W, Pan Y, Ali B, Xu D, Xu X (2021) Physicochemical, crystalline characterization and digestibility of wheat starch under superheated steam treatment. Food Hydrocoll:106720. https://doi.org/10.1016/j.foodhyd.2021.106720
53. Zhu J, Li L, Chen L, Li X (2012) Study on supramolecular structural changes of ultrasonic treated potato starch granules. Food Hydrocoll 29(1):116–122. https://doi.org/10.1016/j.foodhyd.2012.02.004
54. Jiang X, Wang J, Li L, Bodjrenou DM, Lu X, Zheng B (2023) Effect of chlorogenic acid on lotus seed starch gelatinization behavior and complexation mode during microwave treatment. Food Hydrocoll 144:108925. https://doi.org/10.1016/j.foodhyd.2023.108925
55. Lin Q, Wu D, Singh H, Ye A (2021) Improving solubility and stability of β-carotene by microencapsulation in soluble complexes formed with whey protein and OSA-modified starch. Food Chem 352:129267. https://doi.org/10.1016/j.foodchem.2021.129267
56. Huang J, Wang Z, Fan L, Ma S (2022) A review of wheat starch analyses: methods, techniques, structure and function. Int J Biol Macromol 203:130–142. https://doi.org/10.1016/j.ijbiomac.2022.01.149
57. Rindlav-Westling Å, Stading M, Gatenholm P (2002) Crystallinity and morphology in films of starch, amylose and amylopectin blends. Biomacromolecules 3(1):84–91. https://doi.org/10.1021/bm010114i
58. Wang S, Yu J, Yu J (2008) The semi-crystalline growth rings of C-type pea starch granule revealed by SEM and HR-TEM during acid hydrolysis. Carbohydr Polym 74(3):731–739. https://doi.org/10.1016/j.carbpol.2008.03.001
59. Cuthbert WO, Ray SS, Emmambux NM (2017) Isolation and characterisation of nanoparticles from tef and maize starch modified with stearic acid. Carbohydr Polym 168:86–93. https://doi.org/10.1016/j.carbpol.2017.03.067

Chapter 3

Characterization Methods of Amylose and Amylopectin

Aastha Dewan, Shobhit, Vandana Chaudhary, and Priyanka Kajla

Abstract

Amylose and amylopectin are the major constituents of starch that play a significant role in defining the properties of starch and hence its application in the food industry. Exploration of amylose and amylopectin through various analytical techniques is essential to understand its characteristics and properties. Various quantitative analytical techniques are employed for the quantification of amylose and amylopectin such as the iodine binding method, spectroscopic, and potentiometric titration methods. Modern instrumental techniques are also employed for the characterization of amylose and amylopectin such as differential scanning calorimeter and thermal gravitational analysis for thermal properties, scanning electron microscopy for morphological properties, viscoanalyzer for rheological characteristics, and Fourier transform infrared spectroscopy for secondary structure evaluation. Chromatographic techniques, namely, size exclusion chromatography, can be employed for molecular weight determination while anion exchange chromatography for chain length distribution of amylose and amylopectin. This chapter has summarized all the possible analytical and instrumental methods to characterize amylose and amylopectin. Sample preparation, requirements, detailed protocol, observation, and calculations are discussed against each analytical technique.

Key words AM, AP, Degree of branching, Iodine binding methods

1 Introduction

Starch is the second most common biopolymer present on earth that majorly functions as the main energy storage component of plants [1]. It constitutes a vital supply of energy for the nourishment of humans and other animals, thus making it an essential functional constituent in both the food and nonfood industries. Various components contained in starch granules can be grouped into two categories: the major constituents include amylose (AM) and amylopectin (AP), and other constituents, such as proteins, lipids, and minerals [2]. The molecular structure and ratio of the two α-glucans in starch granules impact the properties of starch such as solubility, gelatinization temperature, fluidity, gel-forming ability, and retrogradation. These factors are therefore significant

Sneh Punia Bangar (ed.), *Standardized Procedures and Protocols for Starch*, Methods and Protocols in Food Science, https://doi.org/10.1007/978-1-0716-3866-8_3,

predictors of the consistency, stability, and quality of finished products produced from starch [3].

The proportions of AM to AP and their structure change significantly depending on the botanical source. Waxy starches contain less than 10% AM, while high-AM starches contain more than 40% AM. On the contrary, regular starches consist of AP ranges between 70% and 80% and AM around 20–30% [4]. With the goal to develop starch that possesses particular pasting characteristics and other technologically significant attributes, a number of strategies designed to optimize the granule AM or AP concentration have been explored. Such breeding initiatives may rely on nongenetic alteration, such as conventional breeding and the selection of varieties that are practicable from an agronomic standpoint, or genetic modification, or rather, modulation of the expression of genes involved in the starch-biosynthetic pathway [3]. To assess the AM and AP concentrations of starch and to ascertain molecular characterization, many analytical techniques have been employed [5–11].

Various kinds of starch and its fractions find wide application in the food industry where it gets heat treatment in almost every application. Numerous molecular and structural modifications might take place when starch granules are cooked in the presence of water. The granule-specific gelatinization behavior of starch is significantly impacted by several inherent traits, including granule size, granule composition (AM: AP), granule organization (branching and chain length distribution), and molecular weight of AM and AP. These variations in starch structure have a connection to this granule-specific gelatinization process behavior [12]. The protocol and degree of the thermal gelatinization of starch are impacted by an array of external factors, which involve the heating temperature, moisture content, time, extent of heating, the presence of additives in processed food products, protocol, starch pretreatment, and storage environment [13]. According to one theory, AP induces granule swelling, while the presence of lipid complexes with AM may hinder swelling gelatinization temperature of starch [14].

In order to understand the characteristics of these vital constituents of starch, namely, AM and AP at an individual level, various protocols have been designed and modified. For instance, the International Organization for Standardization (ISO) and the American Association of Cereal Chemists International (AACCI) issued two slightly distinct protocols because of the fast advancement of a standardized approach to assess AM in rice initiatives for development in the 1970s and early 1980s. The conventional curve methodology has been modified in response to each new development of the iodine-binding technique. The approach most employed to measure AM is a colorimetric test, in which iodine combines with AM to result in a blue-purple color that is spectrophotometrically assessed at a single wavelength (620 nm) [15]. AP also interacts with iodine and generates a color, thus making it challenging to accurately assess the color of AM–iodine complexes [16]. In comparison to the Juliani conventional protocol, a dual-wavelength spectrophotometer generates more precise results for both AM–iodine and AP–iodine in a blended version [17]. Nakamura et al. [16] demonstrated that the dual-wavelength spectrophotometer could be used for evaluating AM and AP isolated forms in different varieties of rice. It was established that there are a minimum of five distinct iodine protocols in use, each with its own unique standing time, standard curve formulation, and assessing wavelength. This was made possible by carrying out an assessment of the methodologies employed for quantifying AM in various rice quality assessment laboratories and evaluating the repeatability among laboratories [16]. Therefore, it was a necessity to devise an approach for measuring AM and to concurrently standardize a methodology for computing the true quantity of the component. Furthermore, very few researchers have focused on characterizing these fractions to get the insight at molecular level. Hence, efforts have been made to compile the protocols to illustrate the properties and characteristics of AM and AP. Thus, the mentioned protocols will help researchers to carry out these and get a better understanding of AM and AP at a molecular scale in view to utilizing its concept in an efficient manner at research as well as commercial level.

1.1 AM and AP: Characterization and Functional Role

1.1.1 Overview of AM

AM is a linear polysaccharide formed from α-D-glucose units joined by α-1,4-glycosidic linkages (Fig. 1) and constitutes around 20–30 of the total value of starch. The straight, helical shape of AM's molecular structure, which results from intra- and intermolecular hydrogen bonding amid neighboring glucose units, is what distinguishes it from other sugars. During the protocol of starch gelatinization, when the helical framework unravels, absorbing water to create a three-dimensional matrix, AM assists in building a firm gel network [4]. AM functional qualities, such as its gelatinization

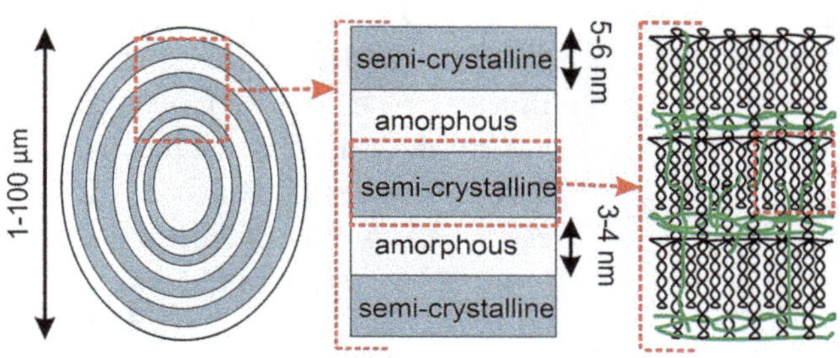

Fig. 1 Representation of starch structure: black lines denote the double helices and branched segments of amylopectin, while green lines indicate the single helices of amylose [18]

temperature, retrogradation behavior, and enzymatic digestibility, are significantly impacted by its physicochemical properties, including its molecular weight, degree of polymerization (DP), and branching arrangement [14]. AM is used in a variety of food systems as a thickening agent, texturizer, and encapsulating material owing to its peculiar structural and functional characteristics [19].

1.1.2 Characterization of AM

The development of a helical inclusion complex with iodine, which possesses a deep blue color, indicates the concentration of AM. However, the AP's longer chains (DP > 60) may additionally establish links with iodine, resulting in a miscalculation of the AM content. on the other hand, renders the iodine combination difficult by establishing connections with phospholipids and free fatty acids, thereby allowing the AM level to be underestimated. The AM content discovered using this method was therefore known as AM equivalents or apparent AM [20]. Thermal characteristics through differential scanning calorimetry (DSC) and thermogravimetric analysis (TGA), while pasting and rheology features employing rapid viscoanaylzer and rheometer, respectively, have been exploited for the characterization of starch and its fractions. AM is often thermally characterized by employing DSC and TGA to understand its thermal stability, retrogradation behavior, and gelatinization characteristics [21]. Structural characterization means evaluating its secondary structure, inter- and intramolecular interactions, and functional groups can be done through X-ray diffraction (XRD), nuclear magnetic resonance (NMR) spectroscopy, and Fourier-transform infrared spectroscopy (FTIR) [22]. A helpful technique for quality control and breeding programs, near-infrared spectroscopy (NIRS) paired with chemometric analysis, enables quick and nondestructive assessment of the AM content in starch samples [23].

1.1.3 Overview of AP

Most of the starch is made up of AP, a highly branched polysaccharide made of α-D-glucose units joined by α-1,4-glycosidic bonds and α-1,6-glycosidic branches (Fig. 1). This polymer normally makes up 70–80% of the total amount of starch [19]. The structure of AP is extensively branched, with both short and long branches coming from the α-1,6-glycosidic connections. Because of its branching structure, AP has special physicochemical characteristics, such as the capacity to form highly hydrated granules and a higher resistance to gelatinization than AM [4]. AP molecular weight, frequency of branching, and distribution of branch lengths are all structural factors that have a big impact on how it functions. These characteristics include the manner in which starch matrices gelatinize, digestibility, retrogradation, and the development of intricate networks within them [24].

1.1.4 Characterization of AP

Analytical approaches are used to characterize AP to ascertain its molecular weight, degree of branching, and chain length distribution. Examples of these techniques include gel permeation chromatography (GPC), size exclusion chromatography (SEC), and enzymatic hydrolysis followed by high-performance anion exchange chromatography (HPAEC). In order to analyze the branching pattern, crystalline structure, and chain organization of AP, structural analysis methods include X-ray diffraction (XRD), nuclear magnetic resonance spectroscopy (NMR), and enzymatic degradation accompanied by chromatographic separation. The branching density, chain length distribution, and molecular architecture of AP have been studied using techniques such nuclear magnetic resonance spectroscopy (NMR), X-ray diffraction (XRD), and enzyme digestion experiments [25]. The understanding of branching patterns and the position of α-1, 6-glycosidic linkages inside the AP molecule are made possible by the enzymatic digestion and hydrolysis of AP using α-amylase and isoamylase. Enzymatic techniques are frequently used to examine the structure and branching patterns of AP, giving information about the degree of clustering and distribution of α-1,6-glycosidic branches. Examples of such methods are isoamylase digestion and phosphorylase digestion [26]. By combining chemometric modeling with techniques such as Fourier-transform infrared spectroscopy (FTIR) and near-infrared spectroscopy (NIRS), it is feasible to do a quick and nondestructive examination of AP features, including branching degree and gelatinization characteristics [25]. These descriptions reveal a substantial understanding of the functions of AM and AP in starch as well as their uses in diverse fields.

2 Materials and Methods

2.1 Sample Preparation

For the characterization of AM and AP, starch must be isolated which needs to be further fractionated into purified AM and AP to analyze it further. Starch isolation from cereals such as wheat, barley, and rice has been discussed with further fractionation into AM and AP.

2.1.1 Starch Isolation from Wheat [27]

Sample: Wheat kernels

Equipments: Centrifuge

2.1.2 Starch Isolation from Barley Kernels [28]

Sample: Barley

Reagents: 0.05% NaOH solution—Add 0.125 g of sodium hydroxide in 250 mL of distilled water, protease (Type XIV, from *Aspergillus*), toluene, and methanol.

Equipment: Vitamix blenders

2.1.3 Starch Extraction from Potato Flour [29]

Sample: Potato

Reagents: Potassium metabisulfite

2.1.4 Fractionation of AM and AP [28]

Sample: Barley starch

Reagents:

Glycerol (70%)—Mix 70 mL of glycerol with 30 mL of distilled water.

Phosphate buffer—Prepare 800 mL of distilled water in a suitable container. Add 20.21 g of sodium phosphate dibasic heptahydrate to the solution. Again, add 3.39 g of sodium phosphate monobasic monohydrate to the solution. Add distilled water until the volume is 1 L.

1-Butanol: Mix one part of 1-butanol with three parts of distilled water.

Ethanol.

Equipments:

Centrifuge and lyophilizer

2.1.5 Separation and Purification of AM and AP [30]

AM and AP were fractionated from rice starch and purified following Schoch's method [31]. The scientific concept involved in this separation states that AM forms a complex with butanol and gets precipitated. On the contrary, AP does not form a complex with n-butanol and thus remains in supernatant.

Sample: Rice starch

Reagents:

Anhydrous ethanol.

0.5 mol/L sodium hydroxide solution—Weigh 20 g of sodium hydroxide and dissolve it in 500 mL of distilled water and make up the volume to 1 L with distilled water.

2 mol/L hydrochloric (HCl) acid—Add 167 mL of concentrated HCl in 500 mL distilled water and make up the volume to 1 L with distilled water.

n-Butanol.

Isoamyl-alcohol.

Equipments:

Centrifuge

Refrigerator

Water bath

2.2 Determination of AM and AP Structure

2.2.1 Spectrophotometric Method [32]

Sample: Corn starch

Reagents:

0.5 N potassium hydroxide (KOH) solution—Dissolve 2.8 g KOH in a mixture of equal volumes of ethanol, double the distilled water, and make up to 100 mL.

0.1 N HCl—Take 8.3 mL of concentrated hydrochloric acid into a 1000-mL volumetric flask. Dilute to the mark with water, and mix well.

Iodine reagent—Dissolve 20 g of potassium iodide and 2 g resublimed iodine in 100 mL of distilled water. After that, dilute 10 mL of this solution to 100 mL which makes it an iodine reagent.

2.2.2 Potentiometric Titration Method

The AM content is determined by the formation of a helical inclusion complex with iodine, which shows a deep blue color. However, the longer chains in AP (DP > 60) can also combine with iodine to produce complexes that lead to an inaccurate estimation of the AM concentration. On the contrary, AM hinders iodine combination by forming complexes with phospholipids and free fatty acids, which results in an underestimate of the AM level. As a result, the AM content found using this method is referred to as AM equivalents or apparent AM [20].

Reagents:

Potassium iodate solution for titration (0.0005 N).

Hydrochloric acid (1 N).

Potassium hydroxide (0.5 N).

Potassium iodide (0.4 N).

Ascorbic acid is used as the primary standard for checking the titer value of potassium iodate solution.

Equipment:
Auto-titrator equipped with DM140 platinum electrode, burette, and LabX light titration software (version 2.1) for titrating dispersed starch solutions at room temperature (21 °C) (*see* **Note 2**)

2.3 Determination of AM Content

The AM content was also determined by gel permeation chromatography (GPC) [33].

Samples: Wheat, rye, barley (normal AM, waxy, high AM), waxy maize, potato (normal-AM (Desired)), normal-AM (Prevalent), high AP, and pea

Sample Preparation:
Prepare the samples (1.5 mg/mL) essentially as described by Torneport et al. [34] (*see* **Note 4**). For debranching, use 5 μL of isoamylase from *Pseudomonas amyloderumosa* (EC 3.2.1.68, 71,000 U/mg protein) and inactivate the enzyme using a boiling water bath for 5 min.

Reagents:
0.25 M KOH—Dissolve 14.025 g of KOH in 600 mL of distilled water and make up the volume to 1 L with distilled water.

Equipment:
Gel permeation chromatography with column Sepharose CL 6B column (1.6 × 70 cm)

2.4 Molecular Weight Determination of AM and AP

The molecular weight (M_w) of the AM and AP is characterized using SEC [35].

Sample: Rice starch samples

Requirements:
Dimethyl sulfoxide (DMSO), lithium bromide, hydrophilic Teflon membrane filter, Styragel HMW 6E column, and refractive index (RI) detector

2.5 Chain Length Distribution Analysis of AM and AP [36]

Sample: Rice starch samples

Reagents:

- DMSO (90% solution)—Take 90 mL of DMSO make up the volume to 100 mL distilled water and mix well.
- Acetate buffer (0.1 M; pH 3.5)—Take 7.72 g of sodium acetate to 800 mL of distilled water. Furthermore, add 352.5 mg of

acetic acid to the solution. Adjust the solution to the final desired pH using either HCl or NaOH. Make up the volume to 1 L using distilled water. Sodium azide and isoamylase (150–300 units).

Equipments:
High-performance anion-exchange chromatography (Dionex, Sunnyvale, California) and Water Bath (NSW, India)

2.6 Rheological Analysis of AM and AP [37]

Sample: Corn starches with different AM/AP ratios (80/20, 50/50, 23/77, and 0/100)

Equipment:
Haake Rheocord Polylab

Sample preparation:
Premix the starches with water to achieve designated moisture contents. After adding the required amount of water, use a Haake parallel, co-rotating twin-screw extruder with a rod-shaped die (nozzle L/D ¼ 3, diameter ¼ 3 mm) to produce gelatinized starch pellets, employing a screw speed of 90 rpm. In order to prevent foaming, the die temperature should be kept at 105 °C during compounding and the highest temperature should be maintained at 160 °C.

2.7 Thermal Characterization of AM-AP Using Differential Scanning Calorimeter (DSC) [38]

Equipment: Differential Scanning Calorimeter (DSC)

Sample preparation: Weigh various samples of starch in stainless-steel pans and then add water with the help of a micro syringe. Seal the pans hermetically and store them overnight for moisture equilibration at 20 °C.

2.8 Thermo-gravimetry Analysis (TGA) [11]

Sample: Rice starch

Equipment:
Thermogravimetric analyzer

2.9 Morphological Characterization Using Scanning Electron Microscopy (SEM) [39]

Sample: Rice starch

Equipments: Scanning electron microscope (SEM)

2.10 Functional Characterization Using FTIR [40]

Sample: Rice starch

Equipment:
Fourier-transform infrared spectroscopy

2.11 Pasting Properties of AM and AP [41]

Sample: Waxy and high AM maize starch samples

Equipment:
Rapid Viscoanalyzer supported by Thermocline for Windows software (Ver. 2.2; Newport Scientific Private Ltd., Warriewood, NSW, Australia)

3 Methods

The protocol adopted for every step starting from starch isolation which further was fractionated into AM and AP followed by the characterization protocols is illustrated as follows.

3.1 Protocol for Starch Isolation from Wheat

1. Soak wheat grains in distilled water with a ratio of 1:5 for overnight.
2. Grind the soaked grains into a fine slurry.
3. Wash the slurry using distilled water on 200-mesh sieves and collect the wash suspension.
4. Centrifuge the collected washed suspension in 50 mL falcon tubes at 4000 rpm for 10 min.
5. Carefully scrape off the protein layer to obtain a maximum pure sample. Washing of the starch sample should be done in order to remove maximum impurity from the starch samples.
6. Oven-dry the collected starch at 42 °C and sieve through 100-mesh size.

3.2 Protocol for Starch Isolation from Barley Kernels

1. Blend the whole barley grains in a propeller that has been slightly cracked while soaking for 24 h in a 0.05% NaOH solution (pH 12).
2. Throw away the soaking solution and give the kernels a good rinse in water.
3. Using a Vitamix blender set to its highest speed, mix the sample for 5 min.
4. The sample was treated with protease (Type XIV, from Aspergillus, 0.35 units/mg solid), at a pH of 7.5.

5. Run the blender to form a uniform slurry for 1 min at maximum speed, and then incubate it for 2 h in a shaker water bath at 37 °C at 50 rpm.
6. Strain the mixture via a nylon screen (30 nm) adding extra distilled water after incubation so that the fiber fraction can be cleaned.
7. Add more water to the fiber fraction and blend until all of the starch comes out.
8. The particles retrieved by centrifugation were purified four times using the toluene shaking method and an extra protease XIV treatment (5 mg/g of starch in 30 mL of incubation mixture).
9. With methanol and water, clean the white layer.
10. Dry for 48 h in a convection oven at 40 °C to obtain dried and purified starch.

3.3 Protocol of Starch Extraction from Potato

1. Wash, brush, and peel the potatoes.
2. Cut the potatoes into small pieces (4 cm^2) immediately after peeling.
3. Submerge in water that has 35 g/100 L of potassium metabisulfite.
4. Remove the parts with defects.
5. Utilize a lab-scale juicer to obtain juice from potato chunks.
6. To prevent browning, add a small amount of potassium metabisulfite (5 g/L) to the juice.
7. Use muslin over a sieve to filter the juice.
8. Use distilled water to clean the remaining substance from the muslin cloth.
9. Strain the filtrate into a glass beaker, and then throw away any leftover material on the muslin fabric.
10. Overnight, leave the beakers containing the filtrate untouched. Starch settles down in the form of a thick layer.
11. Remove the liquid supernatant. Remix the distilled water with the starch layer. Let the starch settle.
12. Four to five repetitions were performed in this manner till the supernatant became translucent.
13. Dry the starch cake obtained from the above method in a hot-air cabinet dryer at a temperature of 40 °C.

3.4 Protocol Fractionation of AM and AP from Barley Starch

1. About 5% of hydrated starch should be mixed with a 70% glycerol solution for enhanced extraction.
2. While bubbling nitrogen dioxide and slowly stirring the resultant suspension, heat it to 89 °C at a rate of 0.8 °C/min.
3. Hold the temperature steady for 1 h. The suspension should then be cooled followed by filtration and washing with ethanol for removal of solvent residues.
4. For further removal of residues, cool the solution using an ice bath (0–2 °C) followed by centrifugation for 1 h.
5. Use phosphate buffer to set the pH of pretreated starch aqueous solution (2%) to 6.0–6.3, followed by stirring at 200 rpm, and subject to heating at 98 °C.
6. The AM fraction is present in the resultant supernatant.
7. However, AP is made up of solid fractions that are produced during the separation process. These fractions are collected and lyophilized.
8. Combine supernatants with 1-butanol (1:3 v/v), and keep them at room temperature for 8 h.
9. The above solution is then centrifuged for 10 min.
10. Lyophilize the solid portion obtained, which is identified as purified AM.

3.5 Protocol for Separation and Purification of AM and AP from Rice Starch

I. Crude Extraction

1. Dissolve 10 g of rice starch in 10 mL of anhydrous ethanol +300 mL of 0.5 mol/L sodium hydroxide solution.
2. Heat the starch solution for 30 min until it becomes homogenous.
3. Furthermore, allow to cool the hot starch solution at room temperature.
4. Centrifuge the cooled solution at low speed for 20 min to remove precipitates.
5. Neutralize the supernatant using 2 mol/L HCl.
6. Prepare a 200-mL blend of n-butanol and isoamyl-alcohol (3: 1, v/v), and mix in a neutralized solution. Heat the solution in boiling water for 20–30 min till it becomes clear and transparent.
7. Cool down the hot solution to room temperature, followed by further cooling at 4 °C for 24 h.
8. Centrifuge the above solution at 3000 rpm for 20 min to collect crude AM in the form of precipitate and "crude AP" in the form of supernatant.

II. Purification of AM

1. Mix the crude AM obtained in Subheading 3.4.3 (point-I h) in 120 mL of n-butanol solution where the volume of n-butanol to water is kept 1:9.
2. Heat this solution in boiling water to obtain a clear transparent solution (*see* **Note 1**).
3. Furthermore, allow to cool down the hot solution at room temperature, and then place it in the refrigerator at 4 °C for 24 h.
4. To collect AM in the form of precipitates, centrifuge the cool solution at 3000 × *g* for 20 min.
5. This cycle of collecting precipitates needs to be repeated five times, and all the precipitates are combined.
6. Soak and wash the collected precipitates with anhydrous ethanol five times, and dry at 40 °C to yield purified AM.

III. Purification of AP

1. Add 20 mL of n-butanol into the crude AP obtained from the protocol of crude extraction and heat and stir this solution in boiling water to obtain a clear transparent solution.
2. Cool down the above solution at 4 °C for 24 h.
3. Centrifuge the cool solution at 3000 rpm for 20 min to collect the supernatant.
4. This experimental cycle of heating, cooling, and centrifuging must be repeated five times to collect all supernatants and combine them.
5. Concentrate the collected supernatant by adding two volumes of anhydrous ethanol to precipitate out AP.
6. For further purification of AP, wash the precipitates with anhydrous ethanol about five times and dry it at 40 °C.

The spectrophotometric approach measures the interaction between AM and light at specific wavelengths and can be used to estimate the quantity of AM in food-derived starch. The AM quantity from extracted starch is assessed as per Williams et al. (1970).

3.6 Protocol for the Determination of AM and AP Structure Using Spectrophotometer Method

1. Vortex the starch sample in 10 mL of 0.5 N potassium hydroxide for 5 min. Make up to 100 mL with distilled water.
2. Take 10 mL of aliquot.
3. Add 5 mL of hydrochloric acid (HCl) (0.1 N) and 0.5 mL of iodine reagent and make up to 50 mL.
4. Measure the absorbance at 625 nm.
5. Determine the AM content (%) by plotting a standard curve using blends of AM and AP.

6. Calculate the AP content by the difference between total starch (*see* **Note 1**) and AM content.

Interpretation

AM content in corn starch samples varied from 24.23% to 19.69%, while AP content ranged from 64.86% to 61.94%. A higher AM/AP ratio was observed for the samples with the highest starch content.

3.7 Protocol for the Determination of AM and AP Structure Using Potentiometric Titration Method

The apparent and absolute AM content is estimated by the method of Schoch [31] and Kasemsuwan [42].

1. Weigh the barley starch sample (20–50 mg, or 5–10 mg of pure AM depending on the starch and AM content of barley variety) into a 50-mL beaker.
2. Include 10 mL of 0.5 N potassium hydroxide (KOH) and agitate the mixture magnetically at room temperature for 20 min, or until starch is completely dispersed.
3. Transfer the dispersed solution of starch, using 50 mL of distilled water to quantitatively transmit the solution, to a 100-mL titrator beaker.
4. Prepare the beaker on the auto-titrator for titration with 0.0005 N potassium iodate by adding 10 mL of 1 N HCl and 5 mL of 0.4 N potassium iodide solutions (*see* **Note 3**). Calculate iodine affinity (I_{AM}) of 20% for AM [43].
5. Determine the iodine affinity of AP (I_{AP}) using pure AP and following the method of Schoch [31].
6. Calculate the absolute AM contents using the formula $(I_{starch} - I_{AP})/(I_{AM} - I_{AP}) \times 100\%$ [44].

Interpretation

Iodine potentiometric titration was used for assessing the apparent and absolute AM contents of barley starches, and the findings revealed the concentration of absolute AM in glacier, high-AM hull-less glacier, high-AM glacier, and W.B. Merlin barley starches as 29.5%, 43.4%, 44.7%, and 9.1%, respectively. AP contents from W.B. Merlin haven't shown any sort of iodine affinity.

3.8 Protocol for Determination of AM Content Using GPC

1. Inject the samples on a Sepharose CL 6B column (1.6 × 70 cm) using 0.25 M KOH as eluent at a flow rate of 13 mL/h (*see* **Notes 5** and **6**).
2. Collect 2 mL fractions and detect the elution profile by the phenol-sulfuric acid method of Dubois et al. (1956) [45].
3. The AM, consisting of the long-chain α-1,4-glucans after debranching of the starch, comprised the first fraction eluted, including the void peak and the intermediate material between the AM and the AP unit chains.

Interpretation

According to GPC, the total AM contents for potato starch, high-AM, barley Waxy, normal-AM, and normal-AM (Prevalent), normal AM (Desired), and High-AP were 23, 7.1, 29.3, 39, 1, 4, and 20, respectively. Compared to those determined by iodine staining, the AM contents determined by gel permeation chromatography (GPC) for the potato starches were likely more accurate. When the empty peak and the intermediate material before the AP unit chains in AM fraction were considered for gel permeation chromatography analysis of normal and high-AM barley, the cumulative curves generated were nearly comparable. This outcome demonstrated that the amount of AM in the intermediate material was equivalent in both samples. This conclusion is consistent with earlier studies published by Tester et al. [46].

3.9 Protocol for Molecular Weight Determination of AM and AP

1. Dissolve the dried AM and AP in dimethyl sulfoxide (DMSO) along with 0.5% (w/w) lithium bromide at a concentration of 2 mg/mL (*see* **Note 7**).
2. Filter through a 0.45-μm hydrophilic Teflon membrane filter and analyze by SEC.
3. Inject the prepared sample of AM and AP into the SEC system equipped with a Styragel HMW 6E column and refractive index detector (RI).

The molecular weight-retention time (T_R) equation developed from this calibration curve

$$\text{Retention time } (T_R) = -2.0874 \times \lg\ (\text{Mw}) + 32.634\ \left(R^2 = 0.9998\right)$$

Interpretation

The size exclusion chromatography of debranched rice starch samples is displayed in Fig. 2, where the graph is plotted between weight distribution represented by w (log$X$), which is the weight of chains as a function of their degree of polymerization (DP) denoted by X. The highest percentage and proportion of peak were observed for AP, while, on the contrary, the lowest was displayed with AM. Chains with DP less than 100 are considered AP branches and those longer than DP 100 are notified as AM [47].

3.10 Protocol for Chain Length Distribution Analysis

1. Precipitate AP using 1 mL of 90% DMSO solution along with 4 mL of methyl alcohol.
2. Dissolve the precipitate in hot distilled water while continuously stirring in a boiling water bath for 1 h.

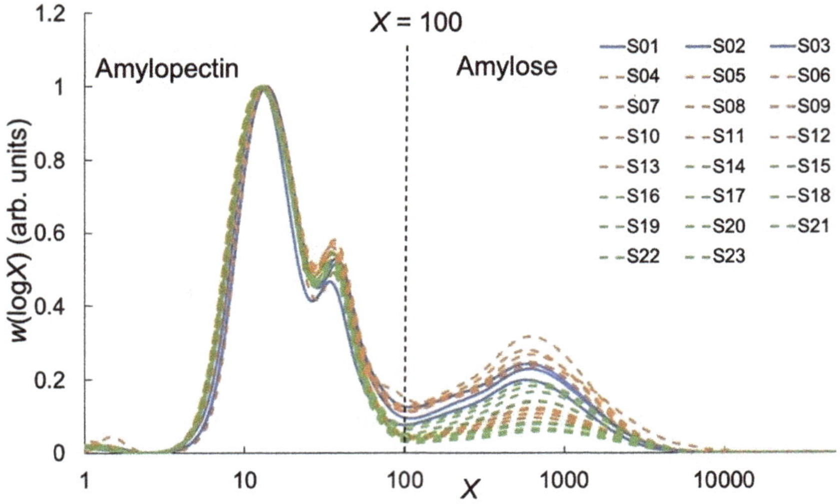

Fig. 2 The size exclusion chromatography spectra of the AM and AP from rice starch (23 varieties) [47]

3. After bringing the solution to room temperature (25 °C), add 1 mL of iso-amylase and the acetate buffer. Utilizing 120 strokes per minute, incubate the solution for 48 h in a shaker water bath at 40 °C.
4. Use sodium hydroxide solution to gradually get the pH to 6 and heat the mixture for 15 min in a boiling water bath to inactivate the enzyme. Add sodium azide to the sample solution (0.02%) to prevent microorganisms from growing.
5. Use high performance anion exchange chromatography outfitted with an amylo-glucosidase reactor and a pulsed amperometric detector to analyze the branch chain length distributions of AP (Dionex, Sunnyvale, CA).

Interpretation

Figure 3 illustrates the branch chain length distributions of debranched AP extracted from rice starch samples (four Thai commercial varieties) that were examined employing HPAEC-PAD. The instrument resolves the AP branch chains with degree of polymerization (DP) not more than or equal to 56. The calculated peak area ratios were divided into A chains where DP ranged from 6 to 12, B1 chains with DP ranging from 13 to 24, B2 chains with DP varying from 25 to 36, and B3+ chains with DP not more than 37. The highest portion of A chains was observed with RD29 with 33.72%, while the lowest was found in RD57 with 27.42%. RD57 exhibited the highest B1 chains, and KDML105 revealed the lowest percentage of B1 chains. The highest B3+ chains were observed with KDML105 with 7.46%. However, major chain length with DP ranging from 13 to 24 was found to be prominent among all four rice varieties with variation from 50.85% to 57.31% [48].

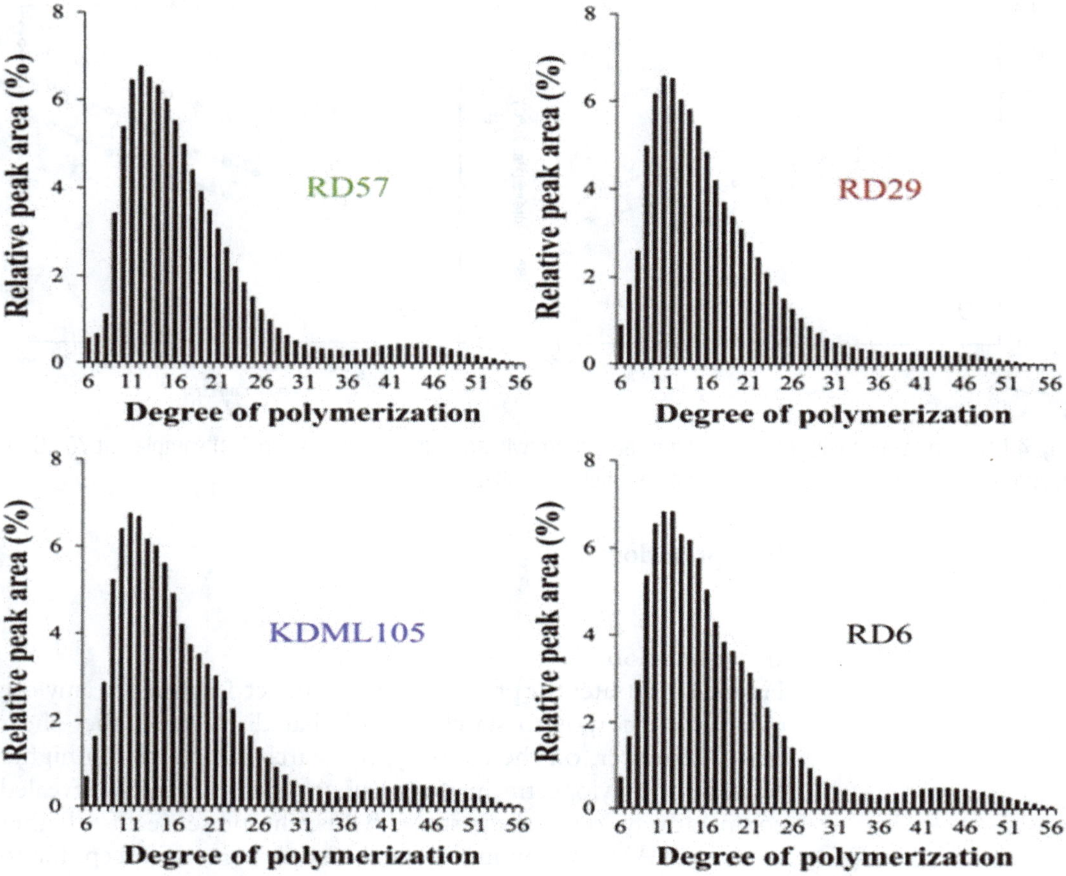

Fig. 3 The chain length distribution of AP branches of RD57, RD29, KDML105, and RD6 four Thai commercial rice starches [48]

3.11 Protocol for Rheological Analysis of AM and AP

1. A slit capillary die with dimensions of 20 × 1 mm will be used to gauge the apparent viscosity and shear stress of the samples at various shear rates at particular temperatures.
2. The screw speed of Rheocord is to be adjusted within a range of 30–180/min, and typically six data points are to be recorded for each sample using the cutting and manual entry measurement mode.

   For calculating the apparent shear rate:

$$\gamma = \frac{6Q}{WH^2}$$

   where γ is the shear rate, Q is the volumetric flow rate (in cm^3/s), W is the slit width, and H is the slit height.

   For shear stress:

$$T = \frac{H \cdot \Delta P}{2L}$$

   where τ is the shear stress, ΔP is the pressure drop over the capillary, and L is the slit length.

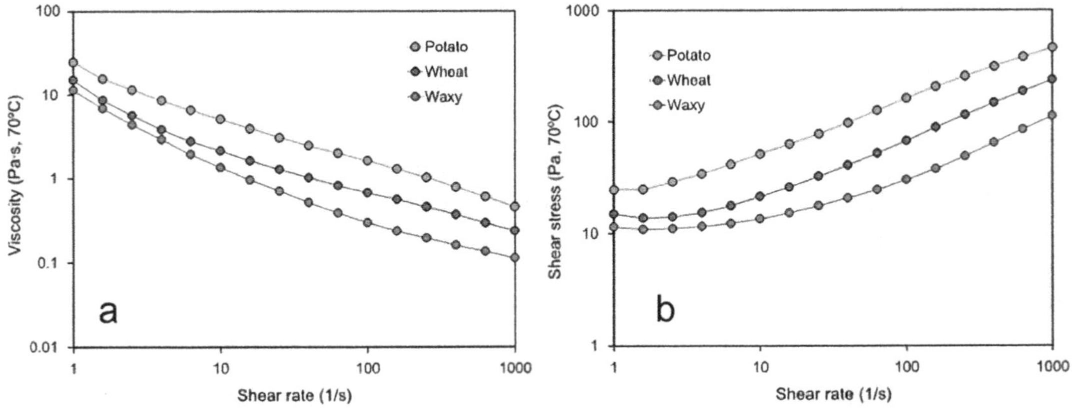

Fig. 4 Rheological behavior of pure gelatinized starch (wheat, potato, and waxy maize) samples at 70 °C (**a**) apparent viscosity and (**b**) shear stress as a function of shear rate [49]

Interpretation

Interpretation

Figure 4 illustrates the presence of the enhanced amount of amylose content in the potato starch sample that displayed higher shear stress. However, on the contrary, the starch sample with a higher content of amylopectin in the waxy corn starch sample revealed comparatively lower shear stress. Thus, the presence of a higher amount of AM fraction in the starch sample is more susceptible to shear stress.

3.12 Protocol for DSC Analysis of AM and AP

1. Equilibrated samples are to be heated to 120–159 °C at a scanning rate of 10–160 °C/min.
2. An empty pan is to be used as an inert reference.

I. **For Detecting the Incremental Change in Heat Capacity Cp Associated with Glass Transition (Tg)**

  1. Place the samples (20 mg) into a vial already having 2 μL of distilled water.
  2. Allow them to remain for 72 h until they are fully moistened.
  3. Next, enclose about 2 mg of hydrated samples in aluminum pans and subject to heating at a rate of 15 °C per minute, starting from 25 °C and reaching 270 °C.
  4. Commence scanning of samples to eliminate the thermal memory of the AM and AP, obtained during the starch separation protocol.
  5. Once cooled, perform a second scan with the same heating rate, this time spanning from 50 to 270 °C.

6. All scans should be conducted under a flow of 50 mL per minute of nitrogen gas (N_2).

II. **For Evaluating the Melting Endotherm of the AM and AP Fractions**

1. Begin by placing 2 mg of each sample into aluminum pans and then add 10 μL of distilled water to each pan.
2. Once the pans are sealed, allow them to rest for 1 h to ensure the complete hydration of starch.
3. Continue heating the hydrated samples in the calorimeter by increasing the temperature by 10 °C per minute, from 30 to 100 °C.

Interpretation

The observed DSC parameters include the onset temperature (T_o), peak temperature (T_p), conclusion temperature (T_c), and gelatinization enthalpy (ΔH gel). These endotherm values were acquired using the TA-60® software, with all scans conducted under a nitrogen (N_2) flow of 50 mL/min.

The loss of molecular order is reflected by the enthalpy of gelatinization, and the gelatinization temperature is considered an indicator of the perfection of crystalline structures. The presence of AM reduces the melting temperature of crystalline regions and the energy required for gelatinization initiation, as AP plays a significant role in starch granule crystallinity. In the absence of AM-rich amorphous regions, more energy is needed to initiate melting. This relationship suggests that starch with a higher AM content has fewer crystalline regions and more amorphous regions, resulting in a lower gelatinization temperature and enthalpy of the endothermic process.

Interpretation

The DSC thermogram shown in Fig. 5 represents the thermal behavior of corn starch varieties with varying AM and AP content. It is clearly observable from the thermogram that the sample with 0% AM exhibited higher melting temperature, while the decline in the samples with an increasing proportion of amylose content in the varieties. The enthalpy of melting is a vital feature that assesses the amount of energy necessary to break the molecular double-helix structure. The decline in crystallinity, which causes starch granules less resistant to gelatinization, clarifies the reason for the lowering of melting enthalpy with the rise in amylose content. As illustrated, the melting enthalpies of the starches with the extended-branch chain length of amylopectin were higher, indicating that more energy was needed to melt the crystallites with the long branch length [50].

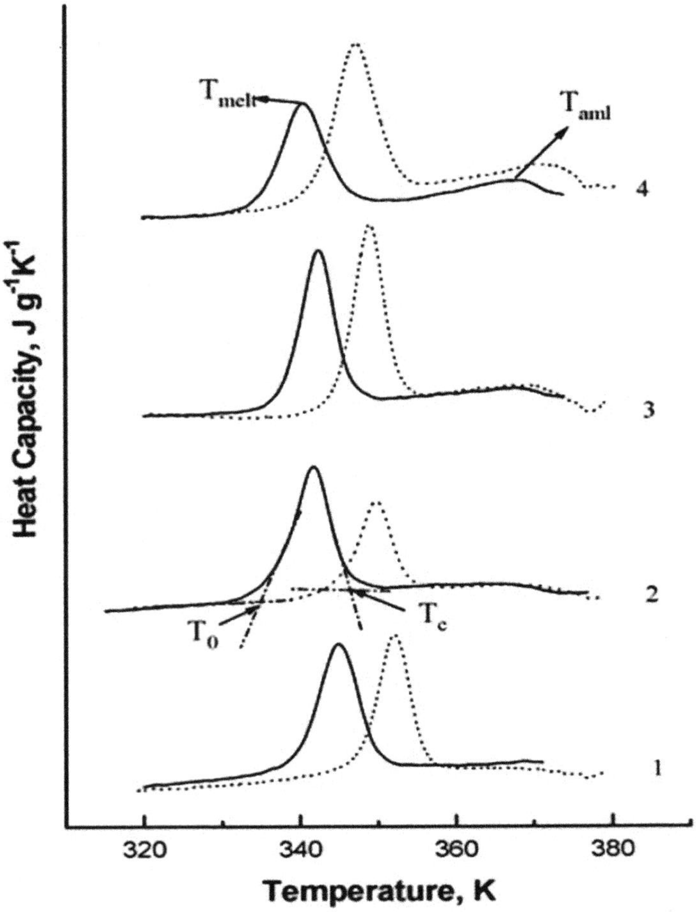

Fig. 5 DSC thermogram of starch samples aqueous solution (0.3% w/w) (straight line) and 1.6 M KCl starch solution (dotted line) of different corn genotypes (1) with 0% AM content, (2) with 15% AM content, (3) with 38% AM content, and (4) 25.5% AM content. T_m represents the melting temperature of AP, T_o and T_c represent the onset and conclusion temperatures of AP, and Tami is the melting temperature of the AM–lipid complex [50]

3.13 Protocol TGA Analysis of AM and AP

1. Use a platinum crucible in the specified experimental determinants: sample mass of approximately 5 mg, N_2 flow rate of 40 mL min^{-1}, heating rate of 15 °C min^{-1}, and temperature range of 25–600 °C.
2. Utilize the Pyris1 TGA® software to determine the moisture content of the samples.
3. Observe and record the mass lost during the initial thermal event of each TG curve, which occurs between 25 and 200 °C.
4. Employ the same software to calculate various measurements associated with degradation events, such as T_{onset} DTG, T_{50}%, and T_{peak} DTG.

5. T_{onset} DTG represents the temperature at which degradation onset occurs, while T_{50}% signifies the temperature at which 50% of the mass is lost. Calculate the slopes using the T_{onset} DTG as the left limit and T_{50}% as the right limit.

Interpretation

In the TGA curve, there are two primary stages involved in weight loss. The initial step involves evaporation and dehydration, which commences as soon as the temperature increases and concludes at around 110 °C. The amount of weight lost in this stage is influenced by the moisture level present in the starch samples. As anticipated, samples with higher moisture content exhibit a greater degree of weight loss until approximately 110 °C. The second phase of weight loss commences around 320 °C and involves thermal degradation. The initial amount of water present in the sample does not impact the decomposition temperature since all water evaporates from the sample before it reaches that temperature. Dehydration and decomposition are generally regarded as distinct processes associated with the degradation mechanisms.

Figure 6 illustrates the thermal behavior of AP. A higher AP content is associated with an increased decomposition temperature as the sample containing 70% AP requires a higher temperature range from 450 to 600 °C. The former peaks indicate the evaporation of water from AM/AP (Fig. 6a); however, the latter peaks reveal the degradation and decomposition of AM/AP (Fig. 6b).

3.14 Protocol for Morphological Characterization Using Scanning Electron Microscopy (SEM)

1. Acquire micrographs of desiccated, isolated AM and AP fractions using a Tescan-Vega Lmu field emission scanning electron microscope (Oxford Instruments X-Act) (*see* **Note 8**).
2. Capture a set of images of gold-coated samples at 15 kV and assess structures at a magnification of 1200×.

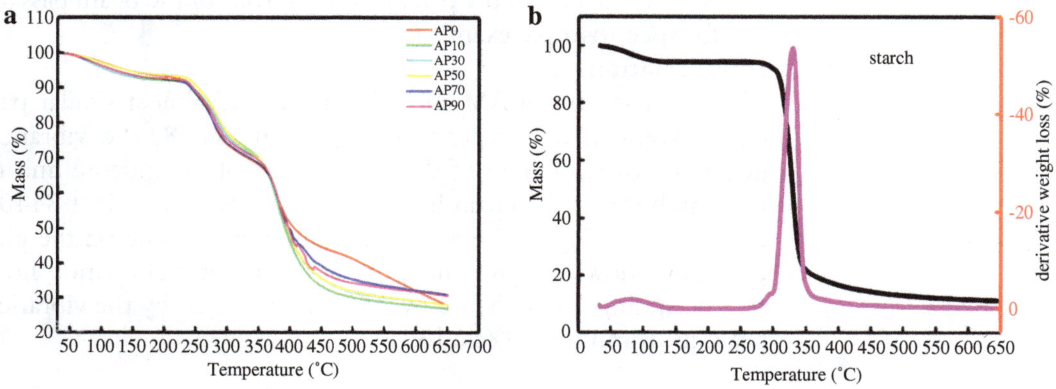

Fig. 6 (**a**) TGA curve of AM/AP and starch gel samples and (**b**) TG degradation curve [39]

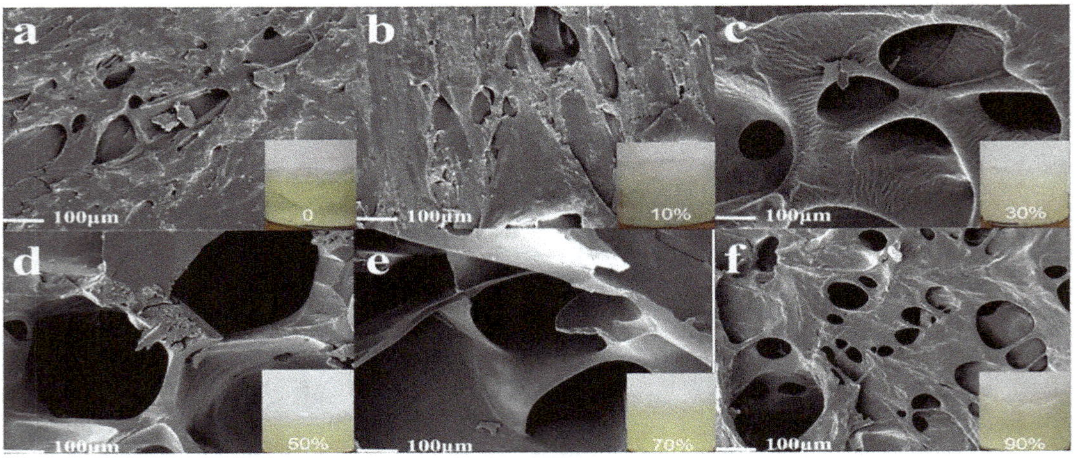

Fig. 7 Micrographs of starch gel samples with (**a**) 0% amylose (AM), (**b**) 10% AM, (**c**) 30% AM, (**d**) 50% AM, (**e**) 70% AM, and (**f**) 90% AM [39]

Interpretation

Figure 7 reveals the influence of increasing content of amylopectin in the morphology of starch gel analyzed through scanning electron microscopy. An increase in the number of holes in the starch gel can be observed in the micrographs with the increasing percentage of AP in the gel samples. The possible reason indicated by Luo et al. [39] is that diluting of the hole wall is taking place with the rising content in AM proportion; thus, higher grafting ratio increases the relationship between AM and AP chains. This resulted in fewer formation of molecular chains leading to more porosity of the gel network.

3.15 Protocol for Functional Characterization Using FTIR

1. Blend dried samples with dried spectroscopic grade KBr (1%, w/w) (*see* **Note 9**).
2. Compress the portion of the blended sample with KBr in a 15-mm dye at 1×10^5 kPa to obtain a transparent disk.
3. Mount the disk in the plate placed in front of the beam passage for spectroscopic examination.

Interpretation

The FTIR spectra of AM and AP represented almost similar patterns except at a few peak points. From Fig. 8, the vibration generated by stretching of the O–H group of the glucose unit of the starch molecular chain has a substantial peak for AP at 3440/cm and stretching of the group containing methylene on the glucose unit shows a significant peak at 2930/cm. The amorphous area bounding the moisture by AM is represented by the vibration peak, measuring 1643/cm [39].

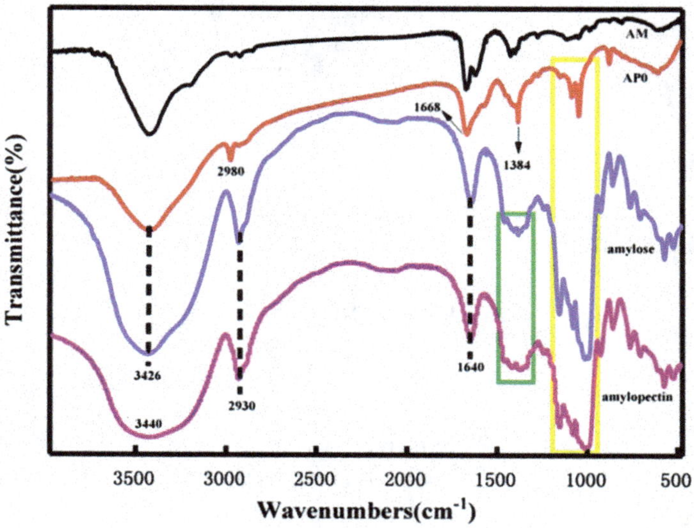

Fig. 8 FTIR spectra of acrylamide (AM), starch gel (APO), AM, and AP [39]

3.16 Protocol for Pasting Properties Evaluation of AM and AP

1. Standard profile measurement was employed in accordance with ICC standard protocol no 162. The viscosity was measured in centipoise (cp).
2. Accurately weighed, 3 g of the sample was dissolved in 25 mL of distilled water and allowed to stir in an RVA canister at 960 rpm for 10 s, followed at 160 rpm for the remaining time of the protocol.
3. The standard temperature profile method involved the stages mentioned below:
   - At 50 °C, which is the initial temperature, hold the sample for 1 min.
   - Heat sample canister to 95 °C for 3.42 min.
   - Hold for 2.30 min at 95 °C.
   - Cool within 3.48 min at 50 °C.
   - Hold for 2 min at 50 °C.
4. The total time taken for the test to complete is 13 min and 4 s of the interval was set between viscosity and temperature reading.
5. Values to be measured from the viscosity profile: peak viscosity (cp), Trough (cp), final viscosity (cp), pasting temperature (°C), peak time (s), and breakdown (cp).

Interpretation

AP absorbs water rapidly and easily; therefore, there is an observance of low pasting temperature with the increased content of the AP. However, on the contrary, AM has a higher gelatinization temperature than 100 °C and thus may increase the pasting temperature in case added to starch, which was not impacted by the addition of AP. Therefore, it has been observed that the pasting characteristics are primarily affected by the content of AP [41].

4 Conclusion

Characterization of AM and AP, crucial components of starch, provides valuable insights into the structural and functional properties. Employing various conventional and advanced analytical techniques, researchers have determined molecular weight, chain length distribution, branching patterns, and physicochemical properties of both starch fractions. These characterizations enhance the understanding of the role of AM and AP in starch gelatinization, retrogradation, enzymatic digestibility, and hence, overall functionality. Additionally, the characterization of AM and AP has contributed to the development of a range of starch-based food products with tailored properties, such as improved texture, stability, gelation, and nutritional attributes. Furthermore, research in this area is required to develop advanced standardized methodologies and protocols with rapid and reproducible results. This will strengthen the in-depth knowledge to explore versatile applications of AM and AP in different industries, including food, pharmaceuticals, and biotechnology.

5 Notes

1. For starch estimation, the method of Malick and Singh [51] can be employed.
2. Titrant addition is based on the kinetics of the titration curve (dynamic addition), and potential measurement can be equilibrium controlled. Recognition of the equivalence point (EQP) can be determined by adjusting the threshold to suitable values.
3. The titration is terminated either when one equivalence point is reached, or the maximum volume of the burette (10.0 mL) is dispensed.
4. Dissolve the native and oxidized/reduced potato starch samples in aqueous urea/dimethyl sulfoxide and precipitate with ethanol.
5. Dextran standards are used for column calibration.
6. Two buffer elution systems, namely, sodium phosphate buffer, 0.2 M, pH 7, and sodium acetate buffer, 0.2 M, pH 5, can also be tested.

7. DMSO/LiBr is used as a solvent combination both to dissolve starch completely and minimize degradation.
8. It is important to ensure that the sample should be dried for SEM analysis as this hampers the SEM vacuum and may cause contamination in the instrument chamber.
9. It is important to ensure the samples are completely dried; otherwise, water forms bonds with AM and AP, broadens polar group absorptions such as C–O–H and C–O–C, and creates a broad high absorption background and rough peaks on overall FTIR spectra.

References

1. Eliasson AC (2004) Starch in food. CRC Press LLC, England
2. Copeland L, Blazek J, Salman H, Tang MC (2009) Form and functionality of starch. Food Hydrocoll 23:1527–1534
3. Blazek J, Copeland L (2008) Pasting and swelling properties of wheat flour and starch in relation to AM content. Carbohydr Polym 71:380–387
4. Tester RF, Karkalas J, Qi X (2004) Starch composition, fine structure and architecture. J Cereal Sci 39:151–165
5. Herrero-Martínez JM, Schoenmakers PJ, Kok WT (2004) Determination of the AMeAP ratio of starches by iodine-affinity capillary electrophoresis. J Chromatogr A 1053:227–234
6. Mestres C, Matencio F, Pons B, Yajid M, Fliedel G (1996) A rapid method for the determination of AM content by using differential-scanning calorimetry. Starch – Stärke 48:2–6
7. Moorthy SN, Andersson L, Eliasson A-C, Santacruz S, Ruales J (2006) Determination of AM content in different starches using modulated differential scanning calorimetry. Starch – Stärke 58:209–214
8. Wang J, Li Y, Tian Y, Xu X, Ji X, Cao X et al (2010) A novel triple-wavelength colorimetric method for measuring AM and AP contents. Starch – Stärke 62:508–516
9. Wang JP, Yu B, Xu X, Yang N, Jin Z, Kim JM (2011) Orthogonal-function spectrophotometry for the measurement of AM and AP contents. Food Chem 127:102–108
10. Grant LA, Ostenson AM, Rayas-Duarte P (2002) Determination of AM and AP of wheat starch using high-performance size-exclusion chromatography (HPSEC). Cereal Chem J 79:771–773
11. Stawski D (2008) New determination method of AM content in potato starch. Food Chem 110:777–781
12. Lindeboom N, Chang PR, Tyler RT (2004) Analytical, biochemical and physicochemical aspects of starch granule size, with emphasis on small granule starches: a review. Starch – Stärke 56(3–4):89–99
13. Beck M, Jekle M, Becker T (2011) Starch re-crystallization kinetics as a function of various cations. Starch – Stärke 63:792–800
14. Jane J, Chen YY, Lee LF, McPherson AE, Wong KS, Radosavljevic M et al (1999) Effects of AP branch chain length and AM content on the gelatinization and pasting properties of starch 1. Cereal Chem J 76:629–637
15. Juliano BO (1971) A simplified assay for milled-rice AM. Cereal Sci Today 12:334–360
16. Nakamura S, Satoh H, Ohtsubo KI (2015) Development of formulae for estimating AM content, AP chain length distribution, and resistant starch content based on the iodine absorption curve of rice starch. Biosci Biotechnol Biochem 79(3):443–455
17. Yamashita S, Baba N, Moriyama H (1993) Application of dual-wavelength spectrophotometry to determination of AM and AP contents. Nippon Shokuhin Kogyo Gakkaishi 40(5):365–369
18. Dome K, Podgorbunskikh E, Bychkov A, Lomovsky O (2020) Changes in the crystallinity degree of starch having different types of crystal structure after mechanical pretreatment. Polymers 12(3):641
19. Singh J, Dartois A, Kaur L (2010) Starch digestibility in food matrix: a review. Trends Food Sci Technol 21(4):168–180
20. Chen MH, Bergman CJ (2007) Method for determining the AM content, molecular weights, and weight- and molar-based distributions of degree of polymerization of AM and

fine-structure of AP. Carbohydr Polym 69(3): 562–578

21. Regina A, Blazek J, Gilbert E, Flanagan BM, Gidley MJ, Cavanagh C, Ral J-P, Larroque O, Bird AR, Li Z (2012) Differential effects of genetically distinct mechanisms of elevating AM on barley starch characteristics. Carbohydr Polym 89(3):979–991
22. Liu B, Zhong F, Yokoyama W, Huang D, Zhu S, Li Y (2020) Interactions in starch co-gelatinized with phenolic compound systems: effect of complexity of phenolic compounds and AM content of starch. Carbohydr Polym 247:116667
23. Pandiselvam R, Thirupathi V, Vennila P (2016) Fourier transform–near infrared spectroscopy for rapid and nondestructive measurement of AM content of paddy. Sci J Agric Eng 2:93–100
24. Bangar SP, Ashogbon AO, Dhull SB, Thirumdas R, Kumar M, Hasan M, Chaudhary V, Pathem S (2021) Proso-millet starch: properties, functionality, and applications. Int J Biol Macromol 190:960–968
25. Pandiselvam R, Thirupathi V, Vennila P (2016) Fourier transform–near infrared spectroscopy for rapid and nondestructive measurement of AM content of paddy. UDK, Berlin
26. Jobling S (2004) Improving starch for food and industrial applications. Curr Opin Plant Biol 7(2):210–218
27. Hu H, Qiu M, Qiu Z, Li S, Lan L, Liu X (2023) Variation in wheat quality and starch structure under granary conditions during long-term storage. Food Secur 12(9):1886
28. Song Y (2000) Characterization of biopolymers: starch and soy protein. Iowa State University. Retrospective Theses and Dissertations. 12361. https://lib.dr.iastate.edu/rtd/12361
29. Sandhu KS, Kaur M (2010) Studies on noodle quality of potato and rice starches and their blends in relation to their physicochemical, pasting and gel textural properties. LWT-Food Sci Technol 43(8):1289–1293
30. Guo B, Wang Y, Pang M, Wu J, Hu X, Huang Z et al (2020) Annealing treatment of AM and AP extracted from rice starch. Int J Biol Macromol 164:3496–3500
31. Shobhit A, Sindhu R, Khatkar BS (2020) Assessment of physicochemical properties of popular brands of corn starch for custard preparation. Int J Chem Stud 8(4):3631–3637
32. Schoch TJ (1945) The fractionation of starch. In: Advances in carbohydrate chemistry, vol 1. Academic Press, Amsterdam, pp 247–277
33. Fredriksson H, Silverio J, Andersson R, Eliasson AC, Åman PJCP (1998) The influence of amylose and amylopectin characteristics on gelatinization and retrogradation properties of different starches. Carbohydr Polym 35(3–4): 119–134
34. Torneport LJ, Salomonsson BAC, Theander O (1990) Chemical characterization of bromine oxidized potato starch. Starch – Stärke 42(11): 413–417
35. Bai Y, Wu P, Wang K, Li C, Li E, Gilbert RG (2017) Effects of pectin on molecular structural changes in starch during digestion. Food Hydrocoll 69:10–18
36. Wong K, Jane J-L (1997) Quantitative analysis of debranched amylopectin by HPAEC-PAD with a postcolumn enzyme reactor. J Liq Chromatogr Relat Technol 20(2):297–310
37. Xie F, Yu L, Su B, Liu P, Wang J, Liu H, Chen L (2009) Rheological properties of starches with different AM/AP ratios. J Cereal Sci 49(3):371–377
38. Kou T, Gao Q (2019) A study on the thermal stability of AM-AP and AP-AP in cross-linked starches through iodine binding capacity. Food Hydrocoll 88:86–91
39. Luo H, Dong F, Wang Q, Li Y, Xiong Y (2021) Construction of porous starch-based hydrogel via regulating the ratio of AP/AM for enhanced water-retention. Molecules 26(13): 3999
40. Rashid I, Omari MHA, Leharne SA, Chowdhry BZ, Badwan A (2012) Starch gelatinization using sodium silicate: FTIR, DSC, XRPD, and NMR studies. Starch – Stärke 64(9):713–728
41. Juhász R, Salgó A (2008) Pasting behavior of AM, AP and their mixtures as determined by RVA curves and first derivatives. Starch – Stärke 60(2):70–78
42. Kasemsuwan T, Jane J, Schnable P, Stinard P, Robertson D (1995) Characterization of the dominant mutant amylose-extender (Ae1-5180) maize starch. Cereal Chem 72: 457–464
43. Takeda Y, Hizukuri S, Juliano BO (1987) Structures of rice amylopectins with low and high affinities for iodine. Carbohydr Res 168(1):79–88
44. Takeda C, Takeda Y, Hizukuri S (1983) Physicochemical properties of lily starch. Cereal Chem 60(3):212–216
45. DuBois M, Gilles KA, Hamilton JK, Rebers PA, Smith F (1956) Colorimetric method for determination of sugars and related substances. Anal Chem 28(3):350–356

46. Tester R, South J, Morrison W, Ellis R (1991) The effects of ambient temperature during the grain-filling period on the composition and properties of starch from four barley genotypes. J Cereal Sci 13(2):113–127
47. Zhao Y, Henry RJ, Gilbert RG (2022) Testing the linearity assumption for starch structure-property relationships in rices. Front Nutr 9: 916751
48. Sangwongchai W, Tananuwong K, Krusong K, Natee S, Thitisaksakul M (2023) Starch chemical composition and molecular structure in relation to physicochemical characteristics and resistant starch content of four Thai commercial rice cultivars differing in pasting properties. Polymers 15(3):574
49. Chipón J, Ramírez K, Morales J, Díaz-Calderón P (2022) Rheological and thermal study about the gelatinization of different starches (potato, wheat and waxy) in blend with cellulose nanocrystals. Polymers 14(8): 1560
50. Khatefov EB, Goldstein VG, Krivandin AV, Wasserman LA (2023) Main characteristics of processed grain starch products and physicochemical features of the starches from maize (Zea mays L.) with different genotypes. Polymers 15(8):1976
51. Malik CP, Singh M (1980) Plant enzymology and histo-enzymology. Kalyani Publications, New Delhi

Chapter 4

Molecular Structure Characterization of Starch

Priyanka Kajla, Parveen Kumari, Shobhit, and Ankur Luthra

Abstract

The physicochemical, functional, and digestibility properties of starch are greatly influenced by the characterization of the molecular structure. There is a correlation between little variations in the amylopectin structure and substantial variations in the functional characteristics of starch. The systematic design of starch-based products is heavily dependent on the development of appropriate methods for characterizing the structural features of starch components. Chromatographic methods have been investigated to characterize the structure of starch chains accurately. In order to accurately determine the starch structure, sample preparation and the solubilization process are crucial. Both starch components can be quantitatively eluted using the chromatographic system and the process for structural characterization of starches, which is effective for solubilizing starches. Following the solubilization of the starch samples, size exclusion chromatography or asymmetrical field flow fractionation is used in chromatographic columns to separate the starch components. Both of these separation technologies are frequently connected to pulsed amperometric, refractive index, and multi-angle light scattering detectors, which provide details on the structure and concentration of the starch constituents.

Key words Electron microscopy, Size exclusion chromatography, Multi-angle scattering, Refractive index, PAD, Fractionation

1 Introduction

The molecular composition of amylose (AM) and amylopectin (AP) greatly influences the functional characteristics of the starch. Since amylopectin has a more complex structure than amylose, even small variations in its structural amylopectin produce changes in the functional characteristics of starch. In general, AM is part of starch with the lowest average molar mass (MM), whereas the polysaccharide fraction with the highest MM, AP, is made up of linear branch chains. The starch polysaccharides are polydisperse; hence, the MM is typically expressed as mean average values, such as weight average (M_w), or provided as molar mass distribution (MMD). Even though the MM is a property that depends on the starch source, the growing conditions, and the extraction procedure, the

Sneh Punia Bangar (ed.), *Standardized Procedures and Protocols for Starch*, Methods and Protocols in Food Science,
https://doi.org/10.1007/978-1-0716-3866-8_4,

determined values are highly dependent on the conditions of experimentation, including sample preparation and the type of analytical technique employed [1]. The requirement to completely disperse the sample (disperse as independent molecules) without its degradation presents the biggest challenge in estimating the molecular mass of starch. Enzymatic debranching, however, likely offers a great substitute for labor-intensive starch fractionation techniques based on AM complexing, which require a lot of organic solvents for AM and AP recovery [1]. By forming hydrogen bonds with a protonic solvent like water or a dipolar aprotic solvent, namely, dimethyl sulfoxide (DMSO), or by hydrating the ionized hydroxyl groups of starch in the presence of a base like NaOH, starch can be dispersed [2]. However, it should be emphasized that the amylopectin molecules are extremely vulnerable to shear destruction even under low-shear conditions, such as mild agitation. As a result, a key challenge for the systematic design of products based on starch is the development of appropriate methods for the structural characterization of starch components [3].

Chromatography methods have been investigated to characterize the structure of starch chains accurately. In order to accurately determine the starch structure, sample preparation and the solubilization process are crucial. The method and solvent both affect how effectively the material can be dissolved. By analyzing solubilized polymers using various chromatographic techniques, namely, size exclusion chromatography multi-angle light scattering (SEC-MALS), namely, high-performance size exclusion chromatography (HPSEC)-MALS-refractive index (RI) and asymmetrical field flow fractionation (AF4)-MALS-RI systems, structural characteristics, such as weight average MM and radius of gyration, can be computed. However, the separating process of the starch fractions, amylose, and amylopectin differentiates among chromatographic techniques [4]. The structural characterization, including gyration radius, molar mass, molar mass dispersity, apparent molecular density, and fractal dimension, might significantly alter depending on the solvents used and the operating temperature. The aforementioned variables offer important insight into the size, shape, and order of arrangement of starch fractions. The starch components are separated using chromatographic columns after the starch samples have been solubilized using size exclusion or asymmetrical field flow fractionation (AF4). A multi-angle light scattering (MALS) detector and a refractive index detector (RID) are typically linked to both separation methods, and these instruments offer information on the composition and concentration of the starch elements [5]. Multi-angle laser light scattering (MALLS) can be used to measure starch's weight-average molecular weight and average radius of gyration without size separation; with the right sample handling, shear scission would not occur. However, without size separation, neither dynamic nor static light scattering can produce exact size distributions. The characterization of high molecular weight macromolecules has recently drawn much more attention to the field-flow fractionation (FFF) group of related techniques. Due to its capability to separate macromolecules without the use of a stationary phase or minimum shear scission, this approach can reveal the full-size distribution of natural starch that has not been degraded. Because it is challenging to get a sufficient signal-to-noise ratio for starch in an optimal eluent, it is not possible to detect the size distribution of fully dispersed and intact starch using FFF techniques. It is essential to comprehend that SEC separates by molecule size rather than molecular weight [6, 7].

The poor stability of starch chains in neutral aqueous solutions makes it difficult to do a structural investigation of starch in aqueous media, which leads to inaccurate molar mass calculations due to chain aggregation or inadequate dispersion. The SEC is typically used since it is inclusive and more accurately descriptive [5]. To characterize molecular starches, size exclusion chromatography (SEC) in combination with SEC-MALS and differential refractive index (DRI) detection is an effective technique. Although both low- and high-branching molecule configurations are eluted simultaneously, the separation theory based on the hydrodynamic volume of the starch polymers in solution accounts for this. SEC is typically utilized to determine the molecular size of AM and AP or their debranched chains when describing the hydrodynamic radius of starches derived from various plant sources. AM chains are generally reported to have hydrodynamic radius $R_h < 100$ nm, while AP chains have $R_h > 100$ nm [8]. The SEC method is used to separate starch fractions based on the distributions of corresponding hydrodynamic volumes (V_h) or hydrodynamic radii (R_h, $V_h = 4/3R_h^3$), with the larger molecules eluting more quickly than the smaller molecules [9]. Regardless of the structure and composition of any polymers, SEC analysis solely considers their size or V_h [10, 11]. It is also challenging to discern between the partially hydrolyzed AM and the degraded AP fractions in the SEC chromatograms, especially while examining acid-thinned starch. Isoamylase or pullulanase effectively separates AM and AP branch chains during the enzymatic debranching of a native or acid-thinned starch sample by cleaving the majority of α-1,6-glycosidic bonds. The α-1,3-glycosidic linkages, the existence of an intermediate fraction, and the inaccuracy of the substance identification are drawbacks in the SEC-chromatogram of a sample of fully debranched starch [12]. This chapter discusses standardized protocols of chromatographic techniques, sample preparation methods for different starch sources, and data interpretation of the results obtained.

1.1 Overview of Size Exclusion Chromatography (SEC) and Asymmetrical Flow Field-Flow Fractionation (AF4) for Molecular Characterization of Starch

Size exclusion chromatography separation is based on molecular size differences influenced by shape and weight variation. Presently, it has wide application in assessing the molar mass distribution of homopolymers having a linear arrangement, estimation of molecular properties of branched polymers, size pattern of macromolecules isolated from solution, and their possible association and compounds. SEC is utilized in the preparative and analytical study of different starches [13, 14]. Size exclusion chromatography (SEC) components include one separation column with a stationary phase of porous polymer packing material. Mobile phase containing larger hydrodynamic volume polymers will elute earlier as they cannot penetrate pores; hence, smaller polymers will take more time to elute as these have contact with extra volume inside the pores. Furthermore, detectors give the required information on the distribution of polymer molecules. Analytes must be non-reactive to the stationary phase to give proper size segregation of polymers (Fig. 1). SEC is also utilized to characterize the second level of starch structure, with average molecular weight, number, multiple detections, and z-average radius of gyration distribution. SEC successfully explains the characteristics of large chained amylose, although it is highly affected by some issues like band broadening, which means the target molecule of a volume. The current SEC system has a drawback of unavoidable shear scission of amylopectin; because of this, accurate size characterization of amylopectin molecules is not possible [11, 15]. MALLS (multi-angle laser light scattering) measures the average radius of gyration and weight-average molecular weight without size separation. As in MALLS, the sample is treated carefully, and there is no shear scission. However, neither static nor dynamic light scattering reliable size distribution is impossible without size separation [7].

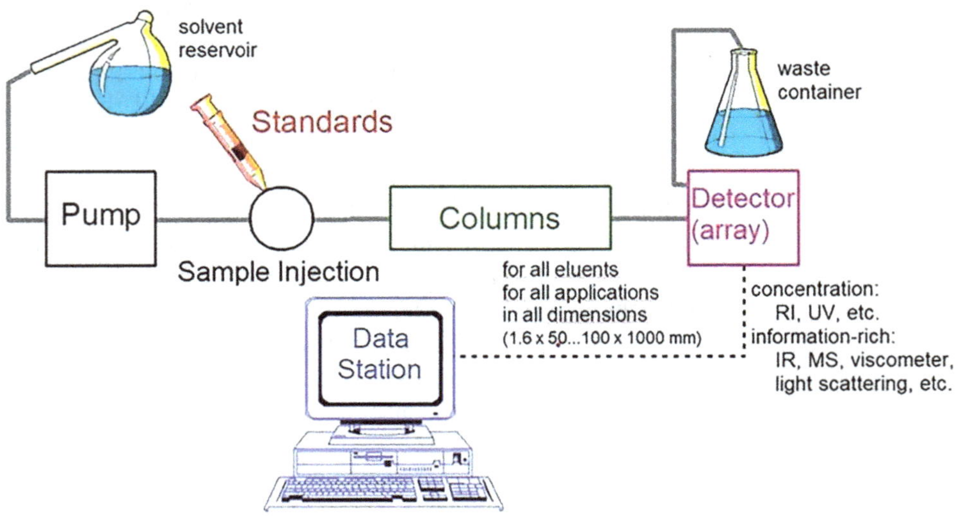

Fig. 1 Schematic representation of size exclusion chromatography laboratory setup [16]

Asymmetrical flow field-flow fractionation (AF4), a sub-type of the larger family of FFF techniques [17], is an excellent technique to characterize starch materials due to minimal shear forces [18–20]. It offers the benefit of an open channel separation without shear degradation or column adsorption effects that arise with size exclusion chromatography (SEC). In AF4, resolution can be regulated by changing the cross-flow. The resolution can be improved using an asymmetric channel of enhanced design and flow control. The higher resolution may allow better molecular characterization and differentiation of starch samples with smaller variations. The distribution of starch glucans which is obtained with SEC and AF4 varies remarkably as the separation of molecules in SEC is based on excluded volume, while in AF4, it is due to differences in translational diffusion coefficients (DT) [18, 19].

2 Materials

2.1 Buffers and Solutions

Tricine Buffer (250 mM, pH 7.5)

1. Add 44.80 g of tricine to 1 L of deionized water.
2. Prepare 1 M NaOH by dissolving 4 g of NaOH in 100 mL of deionized water.
3. Adjust the pH of the buffer prepared to 7.5 by adding ~20 mL of NaOH sol prepared above (*see* **Note 1a**).

Acetate Buffer (0.1 M, pH ~3.5)

1. Acetic acid (0.1 M)—Add 1.15 mL of 17.4 M acetic acid with deionized water and make the volume to 200 mL.
2. Sodium acetate (0.1 M)—Dissolve 1.64 g of anhydrous sodium acetate (or 2.72 g of sodium acetate trihydrate) in deionized water and dilute to 200 mL.
3. Adjust the pH of the acetic acid solution (0.1 M) with sodium acetate solution (0.1 M) to pH 3.5 (18.9:1) (*see* **Note 1b**).

Sodium Bisulfite Solution (0.45% w/w)

1. Add 0.45 g of sodium bisulfite to 100 mL of deionized water. Store at room temperature (RT) (*see* **Note 2a**).

Dimethyl Sulfoxide (DMSO)/LiBr

1. Add 13.75 g of LiBr to 2.5 L of pure DMSO with dissolved LiBr (0.5% w/w). Store at room temperature (*see* **Note 2b**).

2.2 Enzymes and Standards

Store isoamylase at 4 °C, proteinase K, lichenase, and β-xylanase.

Pullulan standards with molecular weight M_w (5.9×10^3, 1.2×10^4, 2.2×10^4, 2.0×10^5, and 5.0×10^5 g mol^{-1}) are used for standard calibration (*see* **Note 3**).

2.3 Equipment

Membrane filter, cryogrinder, thermomixer, centrifuge, shaking water bath, vacuum dryer, Teflon cup, and microwave oven.

1. SEC-PAD (Size Exclusion Chromatography-Pulsed Amperometric Detection).
   - (a) Gradient pump GP50.
   - (b) Electrochemical detector ED50.
   - (c) Thin layer amperometric cell fitted with the gold electrode and silver/silver chloride reference electrode.
   - (d) PAD (pulsed amperometric detection).
   - (e) TSK gel size-exclusion columns G6000PWxl and G4000PWxl.
   - (f) Chromeleon software (calibration curve and integration of data).
   - (g) PeakFit software (automatically locates hidden peaks).
2. SEC-MALS-DRI (Size Exclusion Chromatography-Multi-angle Laser Light Scattering-Differential Refractive Index).
   - (a) SEC-1100 series (Agilent) with autosampler injector (1000 μL piston).
   - (b) Isocratic pump and degasser.
   - (c) RID (refractive index detector, PN3140 at 45 °C).
   - (d) Columns arranged in series: GRAM PreColumn, 30 and 3000 analytical columns; GRAM PreColumn, 1000 and 10,000 columns.
   - (e) Software: Astra (Wyatt).
3. HPSEC-MALS-RI (High-Performance Size Exclusion Chromatography-Multi-angle Light Scattering-Refractive Index).
   - (a) Programmable HPLC Iso Pump.
   - (b) An autosampler and degasser.
   - (c) Ultra hydrogel 6 × 40 mm guard column (Waters, Milford, MA).
   - (d) Two columns in series: OHpak SB-807 HQ and OHpak SB-804 HQ.
   - (e) MALS Wyatt DAWN® HELEOS II and Optilab T-rEx.
   - (f) Reflective index (RI) (Waters 410).
   - (g) ASTRA software (v. 5.3.1.5).

4. AF4-MALS-RI (Asymmetric Flows Field-Flow Fractionation-Multi-angle Light Scattering-Refractive Index).
   (a) A vacuum degasser.
   (b) A 0.22-mm inline filter.
   (c) A Flow Box P 2.1.
   (d) A Savebox V5.
   (e) An AF4 channel (comprising an impermeable upper channel wall (320 × 100 × 145 mm^3) and a permeable lower accumulation channel wall (36.09 cm^2).
   (f) Polysulfone membrane (M_w—cutoff of 10 kDa).
   (g) A 190-μm spacer.
   (h) A 100-μL injector sample loop.
   (i) Detection units (multi-angle laser light scattering Wyatt Dawn®EOS) (λ 690 nm) with toluene-based calibration constant of 9.04×10^6.

3 Methods

3.1 Sample Preparation Methods

The preparation of the sample in a molecularly dispersed (unaggregated) form and without loss and degradation is a prerequisite for characterizing molecular starch. Isolation of starch from its native form is difficult compared with modified or processed starch due to high crystallinity and the presence of non-starch material, namely, proteins, lipids, cell wall material, and non-starch polysaccharides. In addition, for instance, starch isolation from a natural source, cereal grain needs to be grounded, which results in molecular degradation [21, 22]. However, because extraction, degradation, and processing likely result in smaller molecules, it is simple to dissolve previously extracted or degraded starch without further degradation. In addition, it partially or eliminates non-starch elements.

3.1.1 Milling of Grains [6]

1. Fill the cryo-grinder with enough liquid nitrogen (*see* **Note 4**).
2. Place samples of barley grain in the plastic vials with grinding beads (*see* **Note 5**).
3. Cryo-grinder vials should be locked.
4. Cover the cryo-grinder with its cover. Freeze the barley grain samples in liquid nitrogen for 2 min.
5. Grind the samples at a speed of 10 per second for 5 min, open the lid, and remove the samples.
6. Pour the barley flour into another tube.
7. Keep samples at room temperature before using them.

3.1.2 Starch Isolation from Barley Flour [23, 24]

Steep coarsely ground barley kernel in 0.02 N HCl (100 mL) solution and swirl gently overnight at 4 °C (*see* **Note 6**).

1. Neutralize with 0.2 N NaOH and centrifuge at 4000 *g*.
2. Scrap the precipitate into a mortar with 30 mL of 0.1 M Tris–HCl buffer containing 0.5% $NaHSO_3$ (pH 7.0).
3. Treat with enzymes proteinase K, lichenase, and β-xylanase.
4. Digest the samples overnight in a shaking water bath (shaking speed = 1) at 25 °C.
5. After incubation, sieve the contents on a 75-μm sieve and centrifuge (28,000 *g*, 10 min) the slurries.
6. Re-suspend the precipitate with 4 mL of water, layered over 100 mL of 80% cesium chloride solution, and then centrifuge at 28,000 *g* for 30 min.
7. After centrifuging, wash the starch pellet with water and filter on a 0.45-μm membrane.
8. Filter 10 mL of acetone through the starch to remove the remaining water.
9. Lastly, allow the starch to air-dry overnight.

3.1.3 Starch Extraction from Normal, High Amylose, and Waxy Corn Flour [6]

1. Transfer 7–9 mg of flour to a 2-mL microcentrifuge (Eppendorf) tube (*see* **Note 7**).
2. Stir using a vortex mixer at 37 °C for 30 min while adding 0.5 mL of protease in tricine buffer (2.5 units/mL) to the flour.
3. For 10 min, centrifuge at 4000 *g*; discard the supernatant (*see* **Note 8**).
4. Using a vortex mixer, combine the precipitate with 0.5 mL of sodium bisulfite solution (0.4%) and incubate for 30 min at 37 °C.
5. Spin the sample at 4000 *g* for 10 min and then discard the supernatant.
6. Shake the 1.5 mL DMSO/LiBr solution gently by hand to create a homogeneous mixture after suspending the precipitate in the 0.5% w/w solution (*see* **Note 9**).
7. Heat in the thermo mixer at 80 °C, 350 rpm for 24 h, regularly inverting by hand to ensure that the mixture is homogeneous.
8. Combine thoroughly, centrifuge at 4000 *g* for 10 min, and then collect the supernatant in a 15-mL centrifuge tube (*see* **Note 10**).
9. Remove any remaining proteins and non-starch soluble substances by centrifuging at 4000 *g* for 10 min and discarding the supernatant (*see* **Note 11**).

10. Add an additional 10 mL of 100% ethanol and gently mix. Centrifuge at 4000 × *g* for 10 min and discard the supernatant (*see* **Note 12**).

Invert the tube on a paper towel to remove the remaining bit of ethanol.

(For molecular analysis of fully branched starch, dissolve the samples in a solution of 2 mg/mL of DMSO/LiBr in a thermomixer at 80 °C with 2 h of shaking. This follows the preceding process for starch extraction. The samples should be centrifuged at 4000 *g* for 10 min, and the supernatant should be transferred to SEC vials to be stored until SEC analysis (*see* **Note 13**).

3.1.4 Starch Extraction from Millets [25]

1. Grind millet seed stored in liquid nitrogen to prepare flour.
2. Take 100 g millet flour and stir with sodium borate buffer (2 L) for 5 min for protein extraction.
3. Centrifuge at 900 *g* for 5 min. Repeat the extraction process with residue.
4. Wash the residue with distilled water thrice.
5. Suspend washed residue in distilled water overnight and stir to liberate bound protein from the target, i.e., starch.
6. Pass starch slurry through muslin cloth and 70-μm nylon mesh.
7. Centrifuge the filtered starch slurry.
8. Scrap the brown layer from the top of the starch layer using a spatula.
9. Again, suspend water in starch and centrifuge at 1600 *g* for 10 min.
10. Repeat the procedure until brown particles are separated from the starch fraction.
11. Express starch yield as dry wt, per wt. of millet flour.

3.1.5 Dimethyl Sulfoxide Treatment and Starch Solubilization [26, 27]

Dissolve 1 g of the sample first in 20 mL of 95% dimethyl DMSO using a magnetic stirrer for 3 days at room temperature. Precipitate the samples with 60 mL of ethanol and store overnight at 4 °C. Filter the precipitate using a glass filter (G4) and wash successively with 10 mL of acetone and diethyl ether each. Air-dry the precipitate under a hood for a few hours to eliminate the solvents, and finally dry in a vacuum dryer at 45 °C for 18–20 h.

- Begin the solubilization procedure by adding 20 mL of water to 10 mg of dried and dispersed starch weighed in a Teflon cup and filter through a 0·1-μm membrane filter.
- Maintain the sample for 15 min under N_2 flow. Incorporate the Teflon cup with the cap in the bomb; the bomb had a pressure cap of its own. In a 900-W microwave, position this assembly in

the middle. Heat samples for 35 s (maximum temperature: 143 °C), 50 s (177 °C), 70 s (201 °C), or 90 s (211 °C) at 900 W, and record the maximum temperature (in brackets) inside a home-made Teflon cup with an optic fiber. To test the temperature of the solution during the microwave treatment, make a Teflon cup. Following cooling for 30 min in an ice bath, centrifuge the sample at 31,200 *g* for 10 min at 20 °C. Inject the supernatant solution right away into an HPSEC column after filtering it through a 5-μm cellulose acetate membrane.

- To prepare Pullulan solutions, disperse around 0.2 g of the solid material in 100 mL of water. The resulting mixture should have a concentration of about 2 mg/mL and be utilized for the initial operational testing of AF4 systems.

3.1.6 Debranching [28]

Debranched starch, a linear polysaccharide (α-1,4-linked D-glucose units), is analogous to amylose [29]. It is obtained upon hydrolysis of starch with debranching enzymes, namely, isoamylase or pullulanase. The debranching processes lead to the breakdown of α-1,6 glycosidic bonds at the branch of amylopectin and amylose, thus enabling better separation of amylose and amylopectin branch chains [30, 31]. Furthermore, enzymatic debranching of amylopectin and dextrins, as well as molecular characterization using either size exclusion chromatography-multi angle laser light scattering (SEC-MALS) or high-performance size exclusion chromatography-multi angle light scattering (HPSEC-MALS) enables detection of molecular weight and amount of amylose fraction. In addition, the mathematical peak separation of the SEC and HPSEC chromatograms was appropriate for the molecular characterization of amylose fraction. This enzymatic debranching process is an excellent substitute for time-consuming starch fractionation methods requiring more organic solvents for amylose and amylopectin recovery.

Barley Starch

1. Incubate the starch solution (3 mL) with isoamylase (500 units) in 1 mL of acetate buffer (0.1 M, pH 3.5) for 24 h at 40 °C.
2. After incubation with the enzyme, neutralize the digested starch solution with 1 N NaOH and heat in a boiling water bath for 5 min to stop the enzyme activity.
3. Filter through a 0.45-μm membrane, and finally inject in the HPSEC-MALS-RI system.
4. Consider the debranched zero amylose waxy starch profiles to confirm complete debranching using isoamylase.

5. After 24 h of incubation with the enzyme, check the elution profile for the absence of any high molecular weight material, which indicates full hydrolysis of amylopectin (zero amylose starch).

Millet Starches [25]

1. Take a starch sample of approximately 4 mg and dissolve in 100 μL of 90% DMSO and heat in lukewarm water for 5 min. Keep at room temperature and stir slowly overnight.
2. Add 750 μL warm water to the sample and then add sodium acetate buffer (0.01 M, pH 5.5).
3. Then add 1 μL of pullulanase M1 (925 U/mL) and 1 μL of isoamylase (465 U/mL) to the solution and incubate it with slow stirring at room temperature.
4. To inhibit enzymes at the last incubation stage, keep the sample in a boiling water bath for 5 min.
5. After that, add 5 M NaOH (50 μL) and finally dilute the sample to about 1.5 mL with deionized water.
6. A volume of 1 mL of sample is employed on a Sepharose CL 6B column, and elution can be done at 1 mL/min with NaOH (0.5 M).
7. Analysis of fractions is done with phenol–sulfuric acid reagent, and amylopectin/amylose content can be estimated by dividing the SEC chromatogram at the lowest point between two peaks.
8. To study the detailed chain arrangement (short and long) of amylose, further fractions can be performed to study amylose fractions by analyzing elution at the void of gel and the elution point between amylose and amylopectin chains.

3.1.7 Modified Starch

Native starches are fundamentally unsuited for the majority of uses; therefore, these must be altered physically, chemically, and enzymatically to maximize their positive aspects and minimize their disadvantages. These modifications degrade the amorphous portion of starch granules and solubilize both crystalline and amorphous regions, thereby modifying the chain length.

Enzyme Modification Process [32]

The majority of enzymes employed to modify starch are hydrolyses, which produce carbohydrates with low molecular weights [33]. However, through transglycosylation processes, transferases such as amylosucrase (EC 2.4.1.4) and amylomaltase (AMase, EC 2.4.1.25) can significantly alter the molecular structure. For instance, by modifying the reaction conditions, the transglycosylation reaction carried out by amylomaltase may change the length of the side chains of amylopectin, thereby altering the molecular weight and resulting in starches exhibiting various structural characteristics [34, 35]. The starch modified using an enzyme like AMase led to the elimination of amylose while expanding amylopectin's chain length distribution [36]. Therefore, enzymes are mainly utilized to carry out structural modifications, including the debranching of amylopectin and amylose molecules in order to release linear fragments, reducing or increasing the length of amylopectin chains, the introduction of distinctive linkage patterns, and development of cyclic structures [37].

Procedure

1. Prepare a 40% (w/w) suspension by dissolving 80 g of dry native potato starch in 320 mL of water. Then immerse the above-prepared suspension for 15 min in a water bath kept at 40 °C with continuous stirring at the rate of 250/min.
2. Use a suitable amount of the α-amylase enzyme solution to begin the hydrolysis. Stop hydrolysis after 15 min through fast water removal and further washing. Use a suction filter fitted with filter paper (125 mm/100 units) for the purpose of dewatering.
3. Wash the residue promptly using three times distilled water (250 mL) and then wash the residue twice with (96% v/v) ethanol to deactivate the enzyme completely.
4. Dry the modified starch for 24 h at room temperature (20 ± 2 °C). After that, grind dry starch samples using a 200-mm mesh and keep them at 20 ± 2 °C in sealed plastic containers.
5. Add varying quantities of the α-amylase solution, namely, 0.04, 0.08, 0.16, and 0.33 mL to the starch suspension to systematically prepare enzyme-modified potato starch by grading the specific enzyme concentration.

Alkaline Modification Process [38]
Alkali predominantly damages the amorphous fraction of starch granules, increasing granule swelling and solubility, which causes significant degradation of amylose and amylopectin. However, for amylopectin, the activity is mainly restricted to the short range (double helices and aggregated helices), leaving the crystallites partially intact [39]. It is also found that adding Na_2CO_3 and NaOH inhibits the retrogradation of old starch gels [40].

Procedure

1. Weigh about 0.250 g (dry mass) of potato starch into a glass bottle with a screw cap and then fill it with water until it reaches a quantity of 5.0 g. Place the suspension in an oil bath and stir it

in a magnetic stirrer at 25 °C maintained at a speed of 400 per minute.

2. Immediately add 5 mL of 2 M NaOH or 2 M KOH dropwise for the purpose of gelatinization and then cap and stir the dispersion at 400/min at 25 °C for half an hour (or 02 and 24 h).
3. Add 10 mL of 1 M HCl dropwise once gelatinization is finished and then mix the paste at 400 rpm for an additional 5 min.
4. Dilute 2 g of aliquot with preheated (80 °C) 1:5 w/v DMSO.
5. Before the dilution in hot DMSO, apply severe shear treatment to the previously neutralized paste (0.625 or 1.25% w/v), combine two independently made identical starch solutions (40 mL), and process it in an Ultra-Turrax T25 at 11,000 rpm (24,000 rpm) for 2 and 4 min at 20 ± 2 °C.
6. Place the post-treated samples in an oil bath and agitate them @ 400 per minute by maintaining the temp at 25 °C for 24 h.
7. Before molecular analysis, filter the samples with a concentration of 2.5 ppm (or 1.25 ppm) through 5.0-mm PTFE filters at 20 ± 2 °C.

Acid Modification Process [41]

The process of acid modification is also known as acid hydrolysis [42]. Hydrolysis occurs randomly, shortening the polymeric chain and rupturing the α-1, 4, and α-1,6 linkages. It occurs in two distinct stages: an initial stage, during which the hydrolysis selectively attacks the amorphous areas of granules at a fast reaction rate, and a later stage, during which hydrolysis occurs in the crystalline region at a slower speed [43]. The rate of hydrolysis and modification of starch depends on the ratio of amylose to amylopectin and the shape and size of the granule [44]. Following hydrolysis, starch's molecular weight gets reduced with a subsequent increase in crystallinity [45]. First of all, the acid hydrolysis damages the outer portion, i.e., the surface of the starch granule, followed by the attack on loosely organized amorphous regions enclosed inside the starch granular structure mostly made up of centralized amorphous regions and outer amorphous growth rings. The hydrolysis of the amorphous portion in a selective manner causes an increase in the percentage of crystalline areas and double helical structure [46].

Procedure

1. Suspend around 80 g (dry matter) of native starches in dilute mineral acid (hydrochloric acid, sulfuric acid) to form a 40% (w/w) suspension.

2. Using 2 N acid (aqueous phase), make the acid concentration 0.36 and 0.72 N, respectively.
3. Stir the solution thoroughly at 200 per minute and maintain the temp at 40 °C in a water bath for 4 and 24 h, respectively.
4. Add 2 N NaOH dropwise to stop the hydrolysis until the required pH reaches between 5.4 and 5.7.
5. Using cellulose-based filter paper in a suction filter, repeatedly wash the neutralized slurry using deionized water (at least thrice the sample volume), followed by dewatering.
6. Grind the starch product using a 200-mm mesh and store it in closed containers at a temperature of 20 ± 2 °C after drying the moist starch in a climate-controlled cabinet for 24 h at 40 °C and 35% relative humidity.

3.2 Molecular Characterization of Starch Using SEC-PAD [47]

Principle SEC, combined with PAD, is successfully utilized to separate molecules such as starch and other carbohydrates. By applying a positive potential at the surface of the gold electrode at high pH, carbohydrates are electrocatalytically oxidized because the current created is proportional to the content of carbohydrates, thus helpful in the identification and measurement of carbohydrates. When an electrode is subjected to a single potential, oxidation products slowly contaminate the electrode surface. The analyte signal is lost as a result of this electrode surface poisoning. The electrode surface is cleaned by a succession of potentials applied for predetermined amounts of time after the detecting potential to prevent signal loss. A waveform is a set of potentials applied repeatedly for a predetermined period. Pulsed amperometry is based on the repeated application of a waveform. A waveform's potentials are identified as E1, E2, E3, etc., where E1 is the detecting potential. The leftover potentials repair and clean the electrode in preparation for further detection. Potentials recorded at time intervals t1, t2, and t3. Tdel and Tdet are subsets of the initial time interval (t1). To ensure that only current from analyte oxidation is recorded during the detection time, Tdet, changing current (generated during potential changes) is given time to decay during the delay period Tdel [48, 49]. To oxidize the working electrode's surface and remove any adsorbed oxidation products from the previous detection cycle, pulsed amperometry raises the electrode to an oxidizing potential (voltage). The working electrode is then changed to a reducing potential, reducing the gold oxide back to gold at the surface. The working electrode's potential is then stepped to a level suitable for oxidizing carbohydrate aldehyde or hydroxyl groups (usually around 0.05 V relative to the potential of a silver/silver chloride reference electrode). Current in the cell is sampled for a defined amount of time after a brief delay (to allow the charging current in the cell to dissipate), and the charge is estimated as a variable (time-dependent). This triple-pulse sequence's total cycle length of approximately 1 s provides sufficient temporal resolution for conventional HPLC analysis times [48].

Significance Pulsed amperometric is a recent technique that makes possible detection of carbohydrates in acidic or alkaline solutions with reasonable reliability and specificity [50]. Pulsed amperometry is a potent carbohydrate detection technique that doesn't require sample derivatization. PAD is a strong detection method with a wide linear range and extremely low detection limits.

Procedure

- PAD generally operates at a repeating sequence of three potentials that are employed for specific time durations.
- This detector is highly sensitive to carbohydrate detection for size exclusion at pH 12 or > 12 because this high pH destroys the columns used.
- Mobile phase (buffered) with added NaOH is applied to the post column to increase pH.
- The columns are arranged in series maintained at (G6000PWxl and G4000PWxl) at 40 °C.
- Acetate buffer (10 mM) eluent is employed at a flow rate of 0.35 mL/min. This buffer should be stored with argon in a pressurized bottle.
- Use 300 mM sodium hydroxide as a post-column reagent maintained at a flow rate of 0.65 mL/min.
- Adjust the sample volume to 25 μL for injection volume.
- Use the software Chromeleon to integrate chromatographic data and prepare calibration curves.
- Another software PeakFit is used for the deconvolution of chromatograms that automatically detect hidden peaks using a filtering algorithm/ a Fourier deconvolution.

Data Interpretation
Molecular weight identification of eluted molecules resulting from starch hydrolysis was performed after the deconvolution of peaks as per Pullulan standards (Fig. 2). Pullulanase was used to enzymatically hydrolyze the starch, producing hydrolysates with low molecular weight starch polymers that primarily matched the P-10, P-5, P-10, P-5, and P 5 Pullulan standards [47].

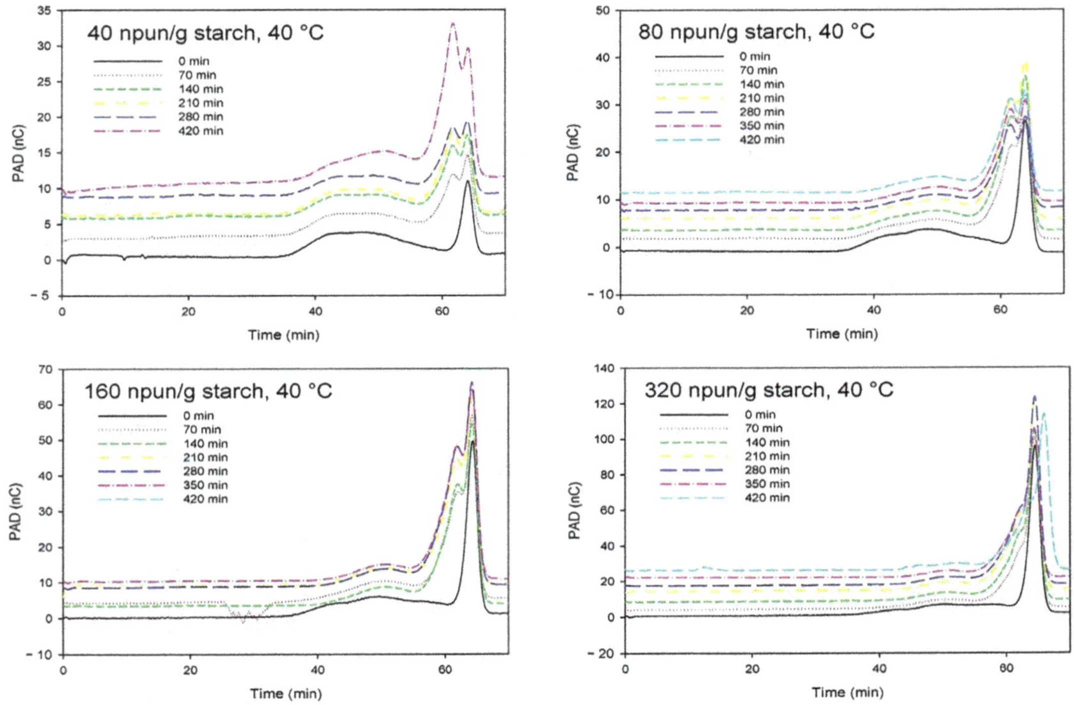

Fig. 2 Effects of Pullulanase concentration (40, 80, 160, and 320 npun/g) on the kinetic profile of enzymatic hydrolysis of starch. (Adapted from Ref. [47]).

3.3 Molecular Characterization of Modified Starch Using SEC-MALS/MALLS-DRI [15]

Principle: The most popular method for determining molecular mass is SEC coupled with a UV (ultraviolet) or RI (reflective index) detector. SEC isolates molecules according to their size or hydrodynamic volume, using filtration via columns packed with beads with numerous pores with a specified diameter range. The movement of small molecules via the column is slowed down when they permeate into these column pores. On the contrary, big molecules that do not fit through the pores are eluted via the column's void space. As a result, bigger molecules are eluted faster than smaller ones during SEC separation. Typically, an RI detector is used to identify molecules that have been eluted from the column. The sample's M_n, M_w, and peak molecular mass (M_p) are then determined using calibration curves. SEC in combination with multi-angle laser-light scattering (SEC-MALLS detector is now employed widely for the molecular mass determination of various polysaccharides particularly starches [51, 52], cellulose, and hemicellulose. With the use of this combined analytical technique, it is possible to measure the absolute M_n, M_w, and molar mass dispersity ($Đ_M$) without using a calibration curve developed by molecular mass makers. A laser monitor in a MALS detector measures the intensity of the incident laser light (I_0), and a forward monitor measures the intensity of the transmitted light (I_t). Rayleigh scattering takes place when the incident ray strikes a sample in a measuring cell, and the dispersed intensity is picked up by photodiodes at various angles. Regardless of the light's wavelength, these photodiodes can detect it. M_w is determined as follows based on the magnitude of the Rayleigh light scattering:

$$R(\theta) = I_\theta r^2 / I_0 V$$

$$K^* c / R(\theta) = 1 / M_w P(\theta) + 2\mathrm{A}_2\mathrm{c}$$

$$K* = 4\pi^2 n_0^2 (dn/dc)^2 / [\lambda_0^4 N_A]$$

$$K^* c / R(\theta) = 1 / M_w P(\theta)$$

where $R(\theta)$ is the excess Rayleigh ratio; I_θ is the scattered intensity at angle θ; I_0 is the incident intensity; r is the distance between the photodiode and the center of the sample cell; V is the illuminated volume of the scattering medium; c is a sample concentration (g/mL); A_2 is the second virial coefficient (a measure of the solvent–solute interaction); K^* is an optical parameter; n_0 is the solvent refractive index; dn/dc is the refractive index increment of specimen solution; N_A is Avogadro's number, and λ_0 is the wavelength of the incident laser light. $P(\theta)$ is a function depending on the scattered light's angular, which represents the gyration radius ($<Rg_2>$) of the specimen molecule [53].

Procedure

- Prepare standard of Pullulan with varied molecular weight by dissolving in DMSO/LiBr solution at a concentration (2 g/L) at room temperature. (Mark–Houwink features for the same eluent at 80 °C are $K = 2.424 \times 10–4$ dL/g and $\alpha = 0.68$.)
- To perform SEC separation of starch sample use SEC-1100 series (Agilent) with autosampler injector (1000 μL piston), isocratic pump, degasser, RID (refractive index detector, PN3140 at 45 °C), without temperature control.
- Conduct separation using two different systems enclosed in SEC-1100 series column oven at 80 °C: GRAM PreColumn, 30 and 3000 analytical columns; GRAM PreColumn, 1000 and 10,000 columns.
- Capillary length from oven to detector should be minimum to avoid heat loss and band broadening.
- Use mobile phase of DMSO/LiBr (0.5% w/w) solution filtered using a 0.45-μm Teflon membrane (hydrophilic).
- Elute starch sample with flow rate, i.e., 0.5 mL/min.
- Pullulan standard is used for calibration.

- Use suitable software like Astra (Wyatt) or WinGPC (PSS) to analyze SEC chromatograms.

The separation of molecules in SEC is based on their hydrodynamic volume (V_h) and their radius, while weight distribution is presented by w ($\log V_h$) and of debranched sample is by wde ($\log V_h$). SEC is also used to calculate the degree of polymerization. A calibration curve is prepared using the relation of elution time to V_h and finally to R_h ($V_h = 4/3\ \pi Rh$). The detector signal is used to calculate size distribution w ($\log R_h$) [54].

Data Interpretation

A typical example of molecular characterization of modified potato starch using SEC-MALLS-DRI.

SEC analysis of corn and potato starch dispersed in urea/NaOH aqueous media [5].

Experimental Conditions: 30 mg of each starch sample was allowed to fully disperse in 4.5 mL of 2.3 M urea and 1 M NaOH overnight at 0 °C followed by neutralization with 0.5 M H_3PO_4 and dilution with the mobile phase utilized for SEC separation. All samples had a final concentration of 1.2 mg/mL. PTFE filters with a thickness of 5 μm were used to clarify the samples, and a volume of 50 μL was injected.

A conventional dual cell refractometer (RI) detector was employed. Two aqueous SEC columns and a guard column of the SB-G type were used for the aqueous SEC analysis. Eluent was employed at a flow rate of 0.4 mL/min and included a total of 50 mM phosphate buffer (pH 8.0) and 100 mM urea. The oven operating at 50 °C consisted of both the detectors and the column. Pullulan (11.3 kDa) was used as the calibration reference. The solvent had a refractive index of 1.334. The OmiSEC 4.7 software was used to do the calculations.

Results: Retention volumes between 12.7 and 23.3 mL were used to elute the starch molecules. For the corn starch sample, a trimodal peak distribution (RI signal) was seen. Amylopectin was represented by the first peak (15–18.7 mL), while amylose was represented by the second and third peaks. Waxy corn starch RI chromatogram showed a single peak (15–22.5 mL) with a distinct tailing towards larger retention volumes (Fig. 3). The peaks of amylopectin and amylose, which were detected in the 15–22 mL range, did not appear to be clearly suppressed in the potato starch profile. The pattern of the waxy potato starch resembled that of the waxy corn starch. Right-side peak tailing was seen along with a weak signal that indicated a significant amount of hydrodynamic volume that was recorded in the retention volume range of 12.7–15.4 mL. Since the samples were just neutralized and diluted after dispersion rather than precipitated, strong non-starch signals

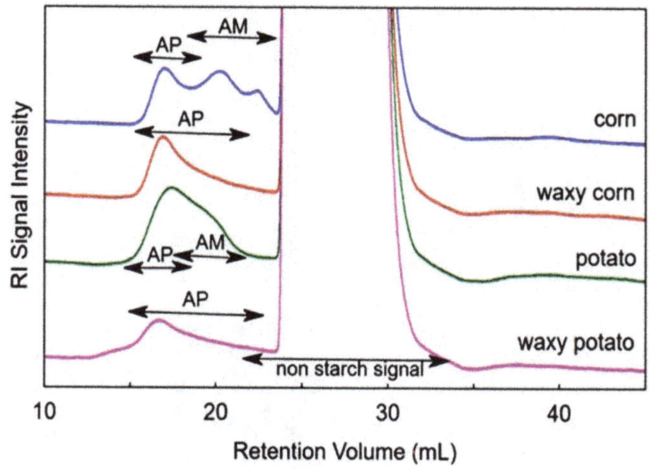

Fig. 3 SEC chromatograms of starches, namely, potato, waxy potato, corn, and waxy corn using RI detector. (Adapted from Ref. [5])

were also seen. It was discovered that there were too many urea and phosphorus salts. Both the precipitation in vials and the SEC profiles of the samples diluted in the eluent (pH 8.0) did not occur for 3 days or change during 24 h.

3.4 Molecular Characterization of Starch Using HPSEC-MALS-RI System [4]

Principle: The working principle of the HPSEC-MALS-RI system is quite similar to the SEC-MALS RI system already described in detail. HPSEC-MALS-RI uses high-pressure and smaller particles to achieve faster and more efficient separation. In addition, it offers more accurate information about the molecular weight and structure of starch components in comparison to SEC-MALS-RI.

Samples: Different corn varieties.

- The HPSEC system consists of OHpak SB-807 HQ and OHpak SB-804 HQ columns serially connected to the injector and maintained at 60 or 70 °C.
- Filter the mobile phase (0.15 M $NaNO_3$), which has a flow rate of 0.4 mL/min, through 0.2 μm and then 0.1 μm of cellulose acetate membranes.
- Carry out all the separations under similar experimental conditions with a volume injection of 50 μL at 0.5 mL per min (*see* **Note 14**).
- By using the scattering data from eight angles, compute the M_w and R_G with the Astra software (v. 5.3.1.5).
- The Astra software provides the M_w (MM) and R_G (gyration radius) in a fractogram or Debye plot.
- Calculate starch recovery by integrating RI signal by a refractive index (dn/dc) of 0.146 g/mL.

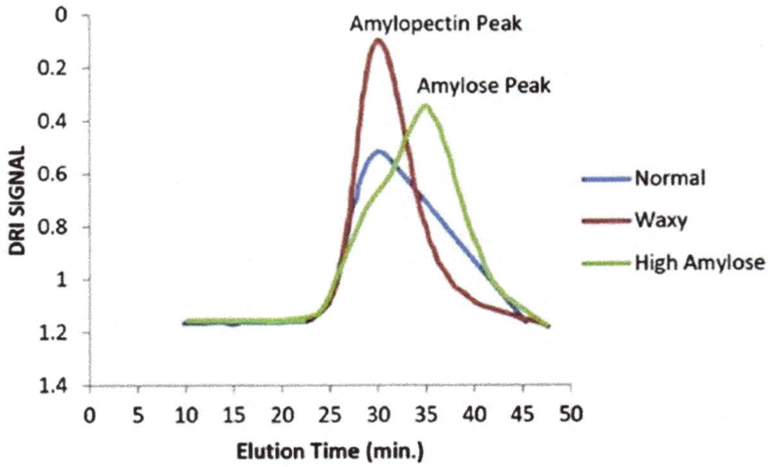

Fig. 4 HPSEC elution profiles for corn starches having a variable amylose content

Data Interpretation

The raw signals from one or more detectors must undergo data processing to be transformed into one or more distribution functions, such as a number or weight. The signals depend on a particular arrangement of separation and detection systems. These functions are highlighted as V_h or R_h distributions that specifically depend on the samples, not the separation/detection system. An arbitrary normalization constant is comparable to the *w* (log *Vh*) and *w* (log *Rh*). Since researchers are more familiar with the size (diameter) than the volume of the molecular entity, it is preferred to report results in terms of *Rh* rather than *Vh*.

Elution profiles for corn starches with different amylose content are presented in Fig. 4. At 30 and 37 min elution times, respectively, the first peak corresponds to amylopectin chains and the second to amylose chains. The chromatogram shows excellent resolution of both components, and the region indicates the amylose/amylopectin ratio beneath the peak. The normal corn starch displays big and small peaks of amylopectin and amylose with 70% and 30% area, respectively, while waxy starch exhibited only an amylopectin peak. The high amylose corn starch had a similar pattern with 55% amylose. Furthermore, it was observed that the molar mass of amylopectin varied from 1.0 to 2.5×10^8 g/mol, and the values decreased with increased amylose content. This pattern is related to the percentage of branch connections and starch biosynthesis. In actuality, waxy maize starch showed roughly 16.4% branch connections, while high amylose maize starch (53% and 70%) showed almost 21.5%. The molar mass disparity of nearly 1.0 indicated the uniformity of both the polymers, namely, amylopectin and amylose.

Normal and waxy maize starches had molar masses (M_w) and gyration radii (R_G) that were, respectively, 1.4 and 2.5108 g/mol and 178 and 212 nm. High amylose maize starch was found to have the lowest molar mass and gyration radius values. The relationship between RG-Mw on a double logarithmic plot is linear for higher molar mass values, indicating that the data may be explained by the power law behavior RG = *kMv*. It can be expressed as under [55]:

$$M = K\ R_G\ 1/\nu = K\ R_G\ df.$$

The exponent ν, calculated from the slopes of straight lines, relates to the global structure of the sample. The normal and waxy maize starches displayed similar ν values, increasing with the highest amylose content. While higher values for high amylose maize starch revealed a more linear shape, the v values of these starches suggest a high branching structure. The amylopectin chains are heavily branched and packed densely at a value around 0.33, and a value between 0.5 and 0.6 is indicative of a linear random coil.

3.5 Molecular Characterization of Starch Using the AF4-MALS-RI System [56]

Principle: The AF4 separation is performed in an open channel where two plates are separated by a spacer, determining the thickness and breadth of the channel. The upper plate is impermeable, and the bottom comprises a porous frit. An ultrafiltration membrane covering the bottom plate is the accumulation wall for preventing sample penetration through the bottom of the channel. Asymmetric flow in AF4 varies from symmetric flow in which channel flow is continuously lost through a single porous wall of the channel, resulting in a decrease in longitudinal flow velocity along the channel, compensated in part by the trapezoidal shape of the channel spacer [57]. The inflow solvent is introduced at the channel inlet at a volumetric flow rate (V_{in}), which is further divided into channel or outlet flow rate (V_c) and cross-flow rate (V_x). The cross-flow is the excess flow created by limiting the channel outflow to detectors leading to $V_c < V_{in}$. The V_x is zero at the upper wall and increases non-linearly toward the accumulation wall [58]. Furthermore, the cross-flow departs from the channel through a porous accumulation wall, thus, creating a flow field perpendicular to the longitudinal flow and forcing analyte molecules downward. Under laminar flow conditions, V_c has a parabolic profile under laminar flow conditions where velocity is highest at the center of the channel and lowest at the walls [17]. Diffusional forces balance the downward force induced by V_x, and the differences in diffusion coefficients of analyte molecules form the basis for selecting the technique. Lower masses with higher diffusion coefficients establish a steady state closer to the center of forward flow, thus eluting first, whereas higher masses pass through more slowly. The retention time (t_r) in field-flow fractionation is computed as under:

$$\frac{t_r}{t_0} = \frac{w^2 V_x}{6DV_0}$$

where w denotes the channel thickness, V_x and V_0 represent the volumetric cross-flow and channel volume, respectively, and D stands for the diffusion coefficient of the sample.

Procedure:

1. In AF4, sort the molecules and particles in accordance with the variations in the translational diffusion coefficient (DT) (Eq. 1), which can be translated to the hydrodynamic radius (Rh) assuming a spherical geometry (Eq. 2) [18, 59]:

$$D_T = \frac{\left(t^0 F_{cr} w^2\right)}{6\ V_0} \frac{1}{t_T} \tag{1}$$

$$R_h = \frac{k_B T\ V^0}{\pi \eta\, t^0 F_{cr} w^2} t_r \tag{2}$$

where t_0: void time, F_{cr}: cross flow, w: channel thickness, V_0: void volume, t_r: detected retention time, k_B: Boltzmann's constant, t_T: absolute temperature, and h: viscosity coefficient of the solvent.

2. Utilizing an AF4-MALS/RI system, analyze the starch solutions by injecting 100 mL of each sample solution with degassed 0.035 M KSCN that contains 0.05% NaN_3 and filtering the eluent through a Millipore 0.2-mm cellulose nitrate filter.

3. The best focusing time was found to be 7 min at 2.5 mL/min for the channel flow rate (Fch) and 2.4 mL/min for the cross-flow rate (Fcr) (as measured by direct observation of the amylopectin-azure solution).

4. Take note of the scattering intensities at 690 nm and at 18 scattering angles (u), which range from 26° to 143°.

5. Using Berry's equation (Eq. 3), extrapolate the excess Rayleigh factors to scattering angle 0 and concentration 0 to produce apparent molar mass (Mw) and radii of gyration (R_G) values [18, 60]:

$$\frac{\sqrt{K\ c}}{R_\theta} = \frac{\sqrt{1}}{M_w} + \frac{16\,\pi^2}{3\lambda^2} R_G{}^2\ \sin^2(\theta/2) \tag{3}$$

where M_w: weight-average molar mass

R_θ: excess Rayleigh factor

K: optical and instrumental constant

λ: laser wavelength

M_w is determined from the intercept of the $c \rightarrow 0$ and $u \rightarrow 0$ double extrapolation, while R_G is determined from the slope of the $u \rightarrow 0$ extrapolation.

Data Interpretation

Chromatograms obtained exhibited a small peak that eluted around 8.5 min which could be ascribed to a drop in pressure at the beginning of elution [61]. On the other hand, the second peak, eluted at about 12.5 min, corresponds to amylose as these chains get separated before the component with a high molecular weight. The amylose peak in maize was seen in high-amylose starch in abundance, followed by normal starch and low-waxy starch. However, the AM/AP ratio of normal corn, barley, and potato starches resembles the signal of the RI detector, which was somewhat different for amylose. The M_w of amylose in these starches varied from 2.7 to 6.4×10^5 g/mol, and R_G varied from 154 to 194 nm. The R_G values of amylose are associated with a polydispersity (1.9–4.1), representing the heterogeneity of chain lengths. The huge peak generally refers to the larger fraction, which represents amylopectin. The M_w of amylopectin was reported to be 1.2 and 6.7×10^8 g/mol with R_G values from 211 to 331 nm and polydispersity from 1.0 to 2.1. In maize, the high-amylose starch was found to have an abundant amylose peak, which was then followed by regular starch and was missing in waxy starch.

3.6 Conclusion

Techniques described in this chapter can be employed to characterize starch samples from various botanical sources at the molecular level. Size exclusion chromatography and different detectors can be employed to study the detailed characterization of starch for an in-depth understanding of its structure and to explore its applications in different food systems. SEC has some drawbacks, such as its restricted applicability to studying smaller, less branched amylose molecules that impose high shear on amylopectin. Nevertheless, these drawbacks can be overcome using additional size-separation methods that impose very low shear, like field-flow fractionation, for the amylopectin fraction. SEC is further hindered from adequately analyzing starch structure because of the poor solubility of starch in aqueous solutions. This leads to the widespread use of DMSO as a solvent, but this reduces the sensitivity of RI detectors. This bolsters the investigation into the potential application of innovative, aqueous solvents for the SEC analysis of starches intended for food applications. It can be said that chromatographic analytical techniques are beneficial for characterizing starch based on molecular structure identification and correlation with its functional applications.

4 Notes

1. (a) Tricine buffer needs to be stored at 4 °C.

   (b) Acetate buffer needs to be stored at 4 °C. It will expire in 6 months.

2. (a) Store at room temperature.

   (b) Store at room temperature.

3. Dissolve the Pullulan standards in distilled water having 0.02% NaN_3 at 25 °C and filter with 0.2-mm syringe filter nylon.

4. As the cryo-grinder utilizes a significant amount of liquid nitrogen, use it in a well-ventilated environment.

5. The amount of time needed for the grinding process will depend on the kind and quantity of grain.

6. When pure starch, steps 1 and 3–6 can be skipped.

7. If samples are less than 30, starch extraction and de-branching process are performed in a single day. If samples are greater than 30, it is preferable to separate them into multiple sets and do the starch extraction and debranching as necessary.

8. Steps 4 and 5 are recommended only for samples with high protein content with high disulfide bonds.

9. Starch samples are typically frozen-dried overnight, and then the dried samples are dissolved in DMSO/LiBr (0.5% w/w) solution over an additional 24 h.

10. In this stage, lignin and non-starch polysaccharides are primarily removed.

11. In this stage, non-starch soluble substances such as lipids and any leftover proteins are eliminated.

12. This procedure is to get rid of any leftover DMSO/LiBr.

13. This approach is equally applicable to measure fully branched starch molecular structure, with the exception that the debranching step is obviously skipped and the columns are typically different.

14. Last but not least, the SEC separation time for a single starch sample is about 1 h.

References

1. Ulbrich M, Salazar ML, Floeter E (2017) Separation and molecular characterization of the amylose-and amylopectin-fraction from native and partially hydrolyzed potato starch. Starch-Stärke 69(7–8):1600228
2. Hu J, Cheng F, Lin Y, Zhao K, Zhu P (2016) Dissolution of starch in urea/NaOH aqueous solutions, vol 133. J. Appl. Polym, Sci, p 133
3. Gilbert RG, Gldley MJ, Hill S, Kilz P, Rolland-Sabaté A, Stevenson DG, Cave RA (2010) Characterizing the size and molecular weight distribution of starch: Why it is important and why it is hard. Cereal Foods World 55:139–143
4. Bello-Perez LA, Agama-Acevedo E, Lopez-Silva M, Alvarez-Ramirez J (2019) Molecular

characterization of corn starches by HPSEC-MALS-RI: A comparison with AF4-MALS-RI system. Food Hydrocoll 96:373–376
5. Szwengiel A, Kubiak P (2020) Molecular dispersion of starch as a crucial parameter during size-exclusion chromatography. Food Secur 9(9):1204
6. Li C, Li H, Gilbert RG (2019) Characterizing starch molecular structure of rice. Rice Grain Quality: Methods and Protocols, pp 169–185
7. Gilbert RG (2011) Size-separation characterization of starch and glycogen for biosynthesis structure-property relationships. Anal Bioanal Chem 399:1425–1438
8. Syahariza ZA, Sar S, Hasjim J, Tizzotti MJ, Gilbert RG (2013) The importance of amylose and amylopectin fine structures for starch digestibility in cooked rice grains. Food Chem 136(2):742–749
9. Hernández JM, Gaborieau M, Castignolles P, Gidley MJ, Myers AM, Gilbert RG (2008) Mechanistic investigation of a starch-branching enzyme using hydrodynamic volume SEC analysis. Biomacromolecules 9(3):954–965
10. Xu H, Zhou J, Liu X, Yu J, Copeland L, Wang S (2021) Methods for characterizing the structure of starch in relation to its applications: a comprehensive review. Crit Rev Food Sci Nutr:1–18
11. Vilaplana F, Gilbert RG (2010) Characterization of branched polysaccharides using multiple-detection size separation techniques. J Sep Sci 33(22):3537–3554
12. Ulbrich M, Wiesner I, Flöter E (2015) Molecular characterization of acid-thinned wheat, potato and pea starches and correlation to gel properties. Starch-Stärke 67(5–6):424–437
13. Berek D (2010) Size exclusion chromatography: A blessing and a curse of science and technology of synthetic polymers. J Sep Sci 33: 315–335
14. Hagel L (2011) Gel filtration: size exclusion chromatography. In: Jansson J-C (ed) Protein purification. Principles, high resolution methods, and applications, 3rd edn. Wiley, Hoboken, NJ, pp 51–91
15. Cave RA, Seabrook SA, Gidley MJ, Gilbert RG (2009) Characterization of starch by size exclusion chromatography: the limitations imposed by shear scission. Biomacromolecules 10: 2245–2253
16. Held D, Kilz P (2021) Size-exclusion chromatography as a useful tool for the assessment of polymer quality and the determination of macromolecular properties. Chem Teach Int 3(2): 77–103
17. Giddings JC (1993) Field-flow fractionation: analysis of macromolecular, colloidal, and particulate materials. Science 260(5113): 1456–1465
18. Juna S, Williams PA, Davies S (2011) Determination of molecular mass distribution of amylopectin using asymmetrical flow field-flow fractionation. Carbohydr Polym 83(3): 1384–1396
19. Juna S, Huber A (2012) Characterisation of normal corn starch using asymmetrical flow field-flow fractionation. Starch-Stärke 64(1): 18–26
20. Rolland-Sabaté A, Colonna P, Mendez-Montealvo MG, Planchot V (2007) Branching features of amylopectins and glycogen determined by asymmetrical flow field flow fractionation coupled with multiangle laser light scattering. Biomacromolecules 8(8): 2520–2532
21. Syahariza ZA, Li E, Hasjim J (2010) Extraction and dissolution of starch from rice and sorghum grains for accurate structural analysis. Carbohydr Polym 82(14–20):27
22. Hasjim J, Jane JL (2009) Production of resistant starch by extrusion cooking of acidmodified normal-maize starch. J Food Sci 74:C556–C562
23. South JB, Morrison WR (1990) Isolation and analysis of starch from single kernels of wheat and barley. J Cereal Sci 12(1):43–51
24. You S, Izydorczyk MS (2002) Molecular characteristics of barley starches with variable amylose content. Carbohydr Polym 49(1):33–42
25. Annor GA, Marcone M, Bertoft E, Seetharaman K (2013) In vitro starch digestibility and expected glycemic index of Kodo millet (*Paspalum scrobiculatum*) as affected by starch–protein–lipid interactions. Cereal Chem 90(3):211–217
26. Planchot V, Colonna P, Saulnier L (1996) Dosage des glucides et des amyloses. In: Godon B, Loisel W (eds) Guide Pratique d'Analyses dans les Industries des Ce´reales. Publication Lavoisier, Paris, p 11
27. Bello-Perez LA, Roger P, Baud B, Colonna P (1998) Macromolecular features of starches determined by aqueous high-performance size exclusion chromatography. J Cereal Sci 27(3): 267–278
28. Ramesh M, Mitchell JR, Jumel K, Harding SE (1999) Amylose content of rice starch. *Starch-Stärke* 51(8–9):311–313
29. Chung HJ, Liu Q, Lee L, Wei D (2011) Relationship between the structure, physicochemical properties and in vitro digestibility of rice starches with different amylose contents. Food Hydrocoll 25(5):968–975
30. Ulbrich M, Beresnewa-Seekamp T, Walther W, Flöter E (2016a) Acid-thinned corn starch—

impact of modification parameters on molecular characteristics and functional properties. Starch-Stärke 68(5–6):399–409

31. Ulbrich M, Lampl V, Flöter E (2016b) Impact of modification temperature on the properties of acid-thinned potato starch. Starch-Stärke 68(9–10):885–899
32. Asiri SA, Ulbrich M, Flöter E (2019) Partial hydrolysis of granular potato starch using α-amylase–effect on physicochemical, molecular, and functional properties. Starch-Stärke 71(5–6):1800253
33. Bangar SP, Ashogbon AO, Singh A, Chaudhary V, Whiteside WS (2022) Enzymatic modification of starch: A green approach for starch applications. Carbohydr Polym 287: 119265
34. Gangoiti J, Corwin SF, Lamothe LM, Vafiadi C, Hamaker BR, Dijkhuizen L (2020) Synthesis of novel α-glucans with potential health benefits through controlled glucose release in the human gastrointestinal tract. Crit Rev Food Sci Nutr 60(1):123–146
35. Korompokis K, Verbeke K, Delcour JA (2021) Structural factors governing starch digestion and glycemic responses and how they can be modified by enzymatic approaches: a review and a guide. Compr Rev Food Sci Food Saf 20(6):5965–5991
36. Van Der Maarel MJ, Capron I, Euverink GJW, Bos HT, Kaper T, Binnema DJ, Steeneken PA (2005) A novel thermoreversible gelling product made by enzymatic modification of starch. Starch-Stärke 57(10):465–472
37. Miao M, BeMiller JN (2023) Enzymatic approaches for structuring starch to improve functionality. Annu Rev Food Sci Technol 14: 271–295
38. Ulbrich M, Schwurack B, Flöter E (2017a) Alkaline dissolution of native potato starch—impact of the preparation conditions on the solution properties determined by means of SEC-MALS. Starch-Stärke 69(5–6):1600256
39. Karim AA, Nadiha MZ, Chen FK, Phuah YP, Chui YM, Fazilah A (2008) Pasting and retrogradation properties of alkali-treated sago (Metroxylon sagu) starch. Food Hydrocoll 22(6):1044–1053
40. Lai LN, Karim AA, Norziah MH, Seow CC (2004) Effects of Na_2CO_3 and NaOH on pasting properties of selected native cereal starches. J Food Sci 69(4):FCT249-FCT256
41. Ulbrich M, Natan C, Flöter E (2014) Acid modification of wheat, potato, and pea starch applying gentle conditions—impacts on starch properties. Starch-Stärke 66(9–10):903–913
42. Amaya-Llano SL, Martínez-Alegría AL, Zazueta-Morales JJ, Martínez-Bustos F (2008) Acid thinned jicama and maize starches as fat substitute in stirred yogurt. LWT-Food Sci Technol 41(7):1274–1281
43. Wang L, Wang YJ (2001) Structures and physicochemical properties of acid-thinned corn, potato and rice starches. Starch 53(11): 570–576
44. Hoover R (2000) Acid-treated starches. *Food reviews international* 16(3):369–392
45. Zuo Y, Gu J, Tan H, Qiao Z, Xie Y, Zhang Y (2014) The characterization of granule structural changes in acid-thinning starches by new methods and its effect on other properties. J Adhes Sci Technol 28(5):479–489
46. Wang S, Copeland L (2015) Effect of acid hydrolysis on starch structure and functionality: a review. Crit Rev Food Sci Nutr 55(8): 1081–1097
47. Chorfa N, Nlandu H, Belkacemi K, Hamoudi S (2022) Physical and enzymatic hydrolysis modifications of potato starch granules. Polymers 14:2027. https://doi.org/10.3390/polym14102027
48. Hardy MR, Rohrer JS (2007) High-pH anion-exchange chromatography (HPAEC) and pulsed amperometric detection (PAD) for carbohydrate analysis. Compr Glycosci 2:303–327
49. Rohrer J (2013) Optimal settings for pulsed amperometric detection of carbohydrates using the dionex ed40 electrochemical detector. Thermo Fisher Sci, Sunnyvale
50. White DR Jr, Hudson P, Adamson JT (2003) Dextrin characterization by high-performance anion-exchange chromatography–pulsed amperometric detection and size-exclusion chromatography–multi-angle light scattering–refractive index detection. J Chromatogr A 997(1–2):79–85
51. Han JA, Lim ST (2004) Structural changes of corn starches by heating and stirring in DMSO measured by SEC-MALLS-RI system. Carbohydr Polym 55(3):265–272
52. Ono Y, Hiraoki R, Fujisawa S, Saito T, Isogai A (2015) SEC–MALLS analysis of wood holocelluloses dissolved in 8% LiCl/1, 3-dimethyl-2-imidazolidinone: challenges and suitable analytical conditions. Cellulose 22:3347–3357
53. Wang L (2020) Analysis of solution structure of isolated lignins and their related compound with size-exclusion chromatography equipped with a multi-angle light scattering detector. Doctoral dissertation, Hokkaido University
54. Wang K, Hasjim J, Wu AC, Henry RJ, Gilbert RG (2014) Variation in amylose fine structure of starches from different botanical sources. J Agric Food Chem 62(19):4443–4453

55. Hanselmann R, Burchard W, Ehrat M, Widmer HM (1996) Structural properties of fractionated starch polymers and their dependence on the dissolution process. Macromolecules 29(9):3277–3282
56. Bello-Pérez LA, Rodriguez-Ambriz SL, Lozano-Grande MA (2017) Molecular characterization of starches by AF4-MALS-RI: an alternative procedure. J Cereal Sci 75:132–134
57. Litzen A, Wahlund KG (1991) Zone broadening and dilution in rectangular and trapezoidal asymmetrical flow field-flow fractionation channels. Anal Chem 63(10):1001–1007
58. Wahlund KG, Giddings JC (1987) Properties of an asymmetrical flow field-flow fractionation channel having one permeable wall. Anal Chem 59(9):1332–1339
59. Schimpf ME, Caldwell K, Giddings JC (eds) (2000) Field-flow fractionation handbook. Wiley
60. Sudut GSCLP, Othman Z, Al-assaf SAPHWAN, Hassan O (2010) Molecular characterization of sago starch using gel permeation chromatography multi-angle laser light scattering. Sains Malays 39(6):969–973
61. Chiaramonte E, Rhazi L, Aussenac T, White DR Jr (2012) Amylose and amylopectin in starch by asymmetric flow field-flow fractionation with multi-angle light scattering and refractive index detection (AF4–MALS–RI). J Cereal Sci 56(2):457–463

Chapter 5

Starch Paste Properties

Ana Flávia Coelho Pacheco, Flaviana Coelho Pacheco, Fabio Ribeiro dos Santos, Jeferson Silva Cunha, and Paulo Henrique Costa Paiva

Abstract

Starch plays an important role in the food industry. However, the pasting properties of starch are still not fully understood, and many times, its native structure limits its applications. In this context, understanding its paste formation process is of great importance to provide improvements in terms of food texture and nutrition. In this chapter, studies on the pasting properties of starch have been compiled. In this way, current information will be provided that discusses the paste properties of starch, as well as future perspectives and challenges to optimize its application in various food products.

Key words Rapid viscoanalyzer (RVA), Starch, Protocols, Structure

1 Introduction

Starch is known as a complex polysaccharide, widely found in nature. In addition to being a very important carbohydrate in human life, it is the most important source of energy for human nutrition [1]. Starch has a wide availability and multifunctionality attributed to its diverse technological functions. It can be used in different applications in most diverse industrial fields such as food, oil, pharmaceuticals, paints, and cosmetics, among others [2]. However, the food industry is the sector that uses starch the most (around 60%), applying it as a functional ingredient in a wide variety of foods, and for a variety of purposes, such as replacing gluten, gelling agent, thickener, stabilizing, vitrifying, and encapsulating, among others [3, 4].

Starch is a granule or particle composed mainly (around 98% on a dry basis) of two macromolecules: amylose and amylopectin (Fig. 1) [5]. Amylose is an essentially linear polymer formed by α-D-glucopyranosyl units linked by glycosidic bonds (1→4) containing few branches connected by α-D-(1→6) bonds,

Sneh Punia Bangar (ed.), *Standardized Procedures and Protocols for Starch*, Methods and Protocols in Food Science, https://doi.org/10.1007/978-1-0716-3866-8_5,

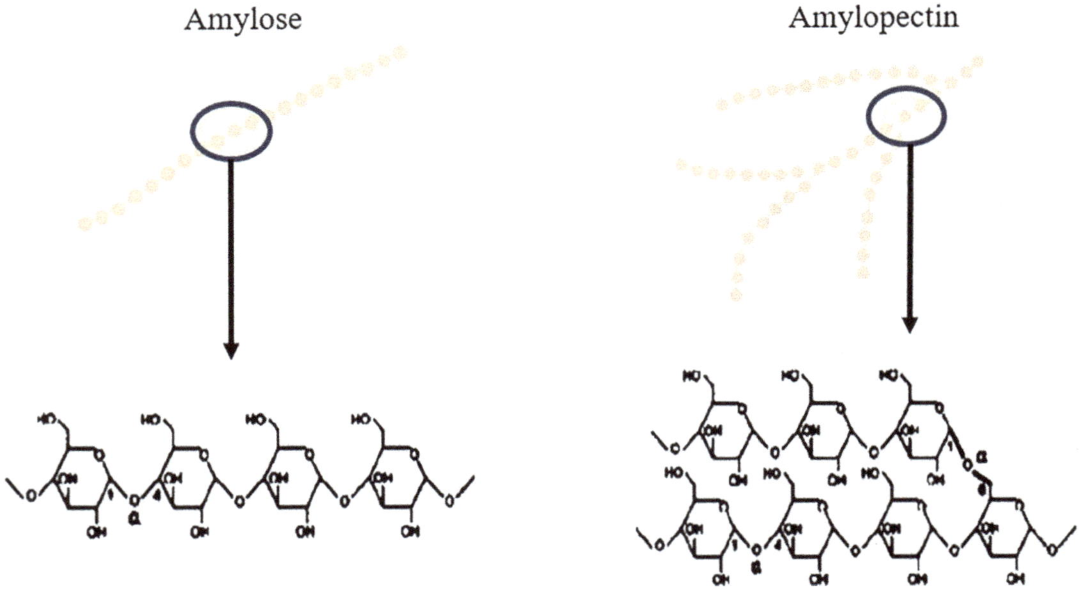

Fig. 1 Amylose and amylopectin polymers

approximately 0.3–0.5% of total connections [6]. On the other hand, amylopectin is a branched polymer (between 4% and 5% of the total bonds) composed of linear chains with α-(1–4) bonds and α-(1–6) bonds [7]. Amylopectin is the main polysaccharide of most starches: between 70% and 75% of native starches and 98% and 99% of so-called waxy starches [1, 7].

The properties, structure, and concentrations of amylose and amylopectin have a great impact on the various properties of starch, including the pasting properties [8]. For example, long amylose chains and large amylopectin sizes have been shown to contribute to the lower peak viscosity of rice flour [9]. Furthermore, granule surface characteristics have significant impacts on starch swelling and amylose/amylopectin leaching during starch gluing, due to the conversion of slow swelling to rapid swelling with the extraction of proteins/lipids from the granule surface [10]. Studies have reported that wheat and rice starches with high amylose content, particularly with the presence of lipids and which have "thicker" granule surface protein layers, tend to intertwine with amylopectin and restrict granule swelling of starch [11, 12]. Thus, cereals with high amylose content, for example, wheat [13] and barley [14] have lower peak viscosity. On the other hand, starches with a high amylopectin content contribute to swelling power and the development of starch viscosity during cooking [15].

Sizing is a process that measures changes in sample viscosity during the heating of a starch suspension [16]. The tendency of a given starch to retrograde can also be studied from its sticking behavior, typically by observing changes in viscosity during programmed heating and cooling of a starch suspension, using a variety of instruments [13]. A starch paste is characterized by its mechanical and rheological properties [17].

2 Materials

The rapid viscosity analyzer (RVA) is a rotational viscometer that continuously records the viscosity of a sample under controlled temperature and shear conditions [18]. The basic principle of RVA testing is to heat a starch suspension while continuously recording the torque applied to the rotating blade in the fluid and through calibration factors, the applied torque can be used to measure viscosity [19]. The ability of RVA to suspend samples in a solvent, maintain them in suspension throughout testing, and apply an appropriate amount of shear to match processing conditions makes it particularly valuable in many process and research applications [19]. It is suitable for a variety of applications that require accurate viscosity information, such as testing starch-based products for quality control. Standardized test profiles are available, including those approved by the AACC International: Approved Methods of Analysis, 11th Ed. Method 61-02.01 [20].

The combination of shearing, heating, and cooling, applied over time, creates a viscosity curve for the material [19]. During a standard starch analysis, starch is heated in an aqueous environment (Fig. 2). Heating the starch in water breaks the hydrogen bonds between the polymer chains, thus weakening the granule [21]. This

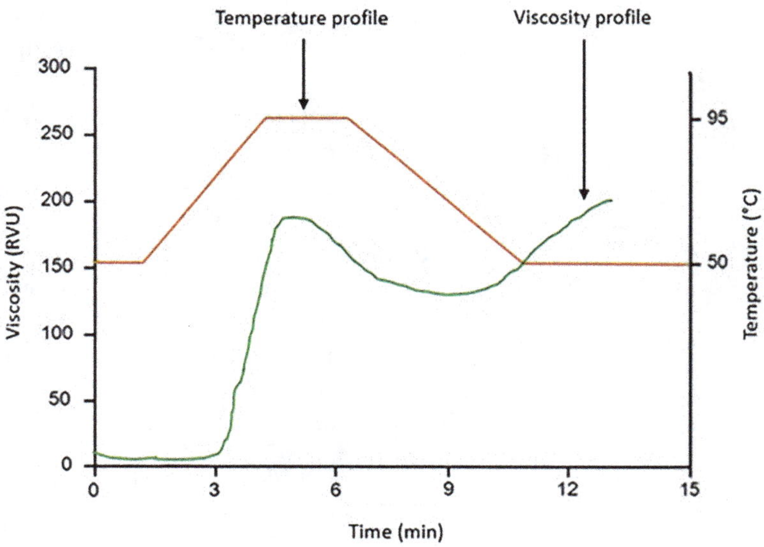

Fig. 2 Typical behavior of starch granules during viscosity analysis in the rapid viscosity analyzer (RVA) equipment

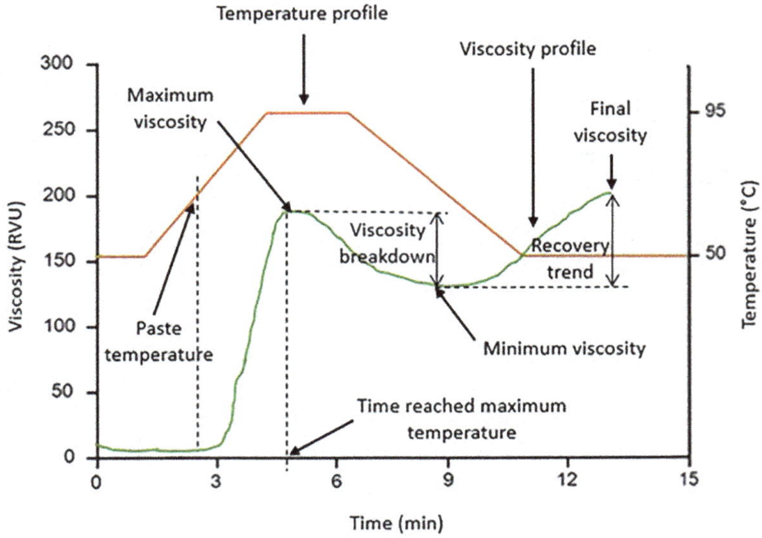

Fig. 3 Response variables obtained in the rapid viscosity analyzer (RVA) equipment and their identification in the viscoamylogram

characterizes swelling and initial expansion, which causes the structure of the starch granule to be weakened [19]. As heating continues, the viscosity reaches a maximum point and then decreases, characterizing the dissolution of the starch granule and the loss of the integrity of the initial structure, which is modified with the decrease in temperature, leading to retrogradation [18].

As can be seen in Fig. 3, during the heating/shearing/cooling steps, the viscosity of the paste is measured in relation to time and the gluing parameters (peak viscosity (PV), peak time (Pt), trough viscosity (TV), final viscosity (FV), breakdown viscosity (BDV), setback viscosity (SBV), and pasting temperature (PT)) are obtained.

The general configuration of the RVA curve can be divided into five successive phases:

(a) *Sample Mixing Stage*

At this stage, the initial viscosity of starch in water at 25 °C is called cold viscosity. When there is a low water absorption capacity of starch granules present in their natural state, the result is low initial viscosity values [21]. The temperature and rotation speed promote friction and homogenize the starch in the water inside the RVA tube, thus forming a gel.

(b) *Heating Stage*

Native starch granules are generally insoluble in water at temperatures below about 50 °C. Heating these granules in the presence of water beyond their critical temperature causes the starch granules to lose their crystalline structure, absorb a large amount of water, and swell, greatly increasing their original size due to the hydrogen bonds between the water molecules and the available hydroxyl groups of starches [19]. This swelling refers to the beginning of gelatinization of the starch granules. Gluing leads to an increase in viscosity and the viscosity continues to increase until the starch granules reach their maximum size [21, 22].

At this stage, the shear forces caused by the swollen starch granules compressing against each other result in an increase in viscosity [21, 23]. The temperature initially recorded before the increase in viscosity during heating is known as the paste temperature (TP). TP is given in °C and is attributed to the beginning of swelling of starch granules and provides an indication of the minimum temperature required to cook a given sample [24, 25].

The heterogeneity of the granules of different starches that swell at different temperatures reflects the slope of the initial increase in viscosity in the sizing curve. Starch modification, such as annealing or cross-linking, will typically reduce this range [26–28].

(c) *Stage in Which the Maximum Temperature Is Maintained*

Paste viscosity is a parameter that measures the resistance of starch granules to collapse resulting from the action of constant temperature (normally 95 °C) and mechanical shear stress in the RVA. This viscosity is known as retention strength, hot paste viscosity, or trough viscosity [19].

As the temperature increases further, the granules continue to swell until a peak in the viscosity curve, known as peak viscosity (PV) [29, 30]. PV is a parameter related to the starch's ability to absorb water and consequently to the degree of swelling of the starch granules during heating. Thus, when there are large quantities of granules with a high swelling capacity, high viscosity peaks occur. This parameter is correlated with different sample quality parameters [24, 31].

PV occurs at the equilibrium point between polymer swelling and leaching [32]. Amylose, which is relatively soluble, is leached into solution, followed, in some cases, at a slower rate, by amylopectin [33, 34]. Due to the mechanical shear applied to the sample, these polymers tend to align. During RVA test analysis, it is possible to quantify these processes (known as gluing) by monitoring the viscosity of a sample as it is heated [19, 35].

The temperature and time corresponding to the viscosity peak are referred to as peak temperature (PT) and peak time (Pt). In detail, PT is the temperature at the point at which the curve begins to form on the graph, due to the beginning of gelatinization of the starch granules, and Pt is the heating time at which maximum paste viscosity occurs [19, 36].

With the breakdown of the starch granules, the viscosity begins to decrease and is known as Trough viscosity (TV), that is, the lowest viscosity value after reaching a constant temperature of 95 °C, the instant at which there is the greatest dissolution of the starch granules starch.

(d) *Cooling Stage*

When the cooling stage begins, the amylopectin and amylose chains change their structure and become more ordered. The reassociation between starch molecules, especially amylose, occurs to a greater or lesser extent; this is called retrogradation [21]. When retrogradation is achieved a gel forms, the viscosity will normally increase and stabilize at a final viscosity and regression can be observed in the curve [19].

In the starch sizing curve, the region between PV and FV is commonly referred to as the setback region. The difference between the two ends of this region is known as breakdown viscosity (BDV). In detail, BDV is the measure of the degree of granule disintegration or paste stability against the application of shear force and temperature [19]. On the other hand, in cases where the final viscosity may not stabilize, as occurs in some starches (e.g., rice), the setback is measured as the difference between FV and TV, known as setback viscosity (SBV) [21]. SBV shows retrogradation of starch molecules, a process in which the amylose and amylopectin chains reassociate into an ordered structure, favoring the recrystallization of starch. It is the difference between the final viscosity and the lowest viscosity value during maintenance at 95 °C – SBV = FV – TV.

(e) *Final Retention Stage*

After the final temperature is reached, viscosity may continue to increase. This is due to temperature differences in the sample; in the middle, it is warmer and the viscosity is lower [18]. When the final temperature is reached throughout the sample, the viscosity will reach a plateau value, known as final viscosity (FV). FV is the viscosity value obtained at the end point of the cooling cycle at a temperature of 50 °C. This parameter is used to define the gluing properties of a given starch sample and indicates the material's ability to form a viscous paste or gel after cooking and cooling [18].

The following materials are required for the analysis of the rapid viscosity analyzer (RVA):

- Rapid viscoaAnalyzer
- Aluminum canisters
- Water
- Analytical balance
- Analytical mill
- Rotating paddle

3 Methods

3.1 Rapid Viscosity Analyzer

(a) *Sample Preparation*

1. Several research articles and books have highlighted proper sample preparation for RVA for determining bonding properties based on the methodology proposed by AACC International: Approved Methods of Analysis, 11th Ed. Method 61-02.01 [20], using a rapid visco analyzer.

(b) *Sample Grinding*

1. Dry grinding can be carried out by mechanical or manual means. On the other hand, wet grinding is done in mechanical mills in the presence of water, for example.
2. Grinding a representative sample quantity (recommended quantity of ~300 g) can be carried out by mechanical or manual means. This is an important step to ensure that the sample being analyzed represents all of the material. A precision splitter or similar device must be used to collect an appropriate subsample.
3. Generally, a mill suitable for passing a 0.5–0.8-mm sieve (*see* **Note 1**) is used to grind samples with large particles, in order to obtain very small particle sizes and improve analysis precision. Clean the grinder thoroughly after each sample is ground. If not using immediately, store the ground sample in a cold room;

(c) *Determination of Sample Moisture*

1. After grinding, shake the ground sample well before weighing a subsample. Weigh the appropriate amount of ground sample into a container to perform moisture content analysis.
2. The moisture content (%) of the sample can be determined by a near-infrared spectrometer or other suitable methods.
3. Generally, the total moisture of samples for RVA testing is corrected to 14%. Use the "Moisture Conversion Table 1" provided below to determine the amount of sample in grams to weigh into a container.

(d) *Sample Weighing*

1. The sample is weighed using an analytical balance in a weighing container (can be aluminum foil) and mixed with an exact volume of distilled water in an RVA sample vessel/can, using a bottle dispenser or other sampling device suitable for volumetric pipetting. The amount of sample to use depends on the nature of the material, with a typical range from 3.00 to 8.00 g db.

Table 1
Weight of sample and added water corrected for moisture content

Moisture content (%)	Weight of sample (14% moisture basis) (g)	Weight of water (g)	Moisture content (%)	Weight of sample (14% moisture basis) (g)	Weight of water (g)
7.0	5.55	25.45	12.0	5.86	5.14
7.1	5.55	25.45	12.1	5.87	5.13
7.2	5.56	25.44	12.2	5.87	5.12
7.3	5.57	25.43	12.3	5.88	5.12
7.4	5.57	25.43	12.4	5.88	5.11
7.5	5.58	25.42	12.5	5.89	5.10
7.6	5.58	25.42	12.6	5.90	5.10
7.7	5.59	25.41	12.7	5.90	5.09
7.8	5.60	25.40	12.8	5.91	5.09
7.9	5.60	25.40	12.9	5.91	5.08
8.0	5.61	25.39	13.0	5.92	5.07
8.1	5.61	25.39	13.1	5.93	5.07
8.2	5.62	25.38	13.2	5.93	5.06
8.3	5.63	25.37	13.3	5.94	5.05
8.4	5.63	25.37	13.4	5.95	5.05
8.5	5.64	25.36	13.5	5.95	5.04
8.6	5.65	25.35	13.6	5.96	5.04
8.7	5.65	25.35	13.7	5.97	5.03
8.8	5.66	25.34	13.8	5.97	5.02
8.9	5.67	25.34	13.9	5.98	5.02
9.0	5.67	25.33	14.0	5.98	5.01
9.1	5.68	25.32	14.1	5.99	5.00
9.2	5.68	25.32	14.2	6.00	5.00
9.3	5.69	25.31	14.3	6.01	4.99
9.4	5.70	25.30	14.4	6.01	4.99
9.5	5.70	25.30	14.5	6.02	4.98
9.6	5.71	25.29	14.6	6.03	4.97
9.7	5.71	25.29	14.7	6.04	4.97
9.8	5.72	25.28	14.8	6.04	4.96
9.9	5.73	25.27	14.9	6.05	4.95

(continued)

Table 1
(continued)

Moisture content (%)	Weight of sample (14% moisture basis) (g)	Weight of water (g)	Moisture content (%)	Weight of sample (14% moisture basis) (g)	Weight of water (g)
10.0	5.73	25.27	15.0	6.06	4.95
10.1	5.74	25.26	15.1	6.06	4.94
10.2	5.75	25.25	15.2	6.07	4.94
10.3	5.75	25.25	15.3	6.08	4.93
10.4	5.76	25.24	15.4	6.08	4.92
10.5	5.77	25.24	15.5	6.09	4.92
10.6	5.77	25.23	15.6	6.10	4.91
10.7	5.78	25.22	15.7	6.11	4.90
10.8	5.78	25.22	15.8	6.11	4.90
10.9	5.79	25.21	15.9	6.12	4.89
11.0	5.80	25.20	16.0	6.13	4.89
11.1	5.80	25.20	16.1	6.14	4.88
11.2	5.81	25.19	16.2	6.14	4.87
11.3	5.81	25.19	16.3	6.15	4.87
11.4	5.82	25.18	16.4	6.16	4.86
11.5	5.83	25.17	16.5	6.16	4.85
11.6	5.83	25.17	16.6	6.17	4.85
11.7	5.84	25.16	16.7	6.18	4.84
11.8	5.85	25.15	16.8	6.18	4.84
11.9	5.85	25.15	16.9	6.19	4.83

2. As mentioned, generally, the total moisture content of samples for RVA testing is corrected to 14%. Thus, the correlation can be determined from Eqs. (1) and (2):

$$S_1 = S_0\, x \frac{(100 - M_0)}{(100 - M_1)} \tag{1}$$

$$W_1 = W_0 + (S_0 - S_1) \tag{2}$$

where:
S_0 = standard sample weight (typically 3.5–4.0 g)
S_1 = corrected sample weight (g)
W_0 = standard water weight (25 g)
W_1 = corrected water weight (g)

M_0 = reference moisture content of the sample (typically 14%)

M_1 = actual moisture content of the sample (%)

(e) *Shaking the Sample*

1. To avoid agglomeration, especially in high concentrations of starch, the sample containing the corrected amount of water and which is inside the RVA vessel/can is shaken vigorously at a specific speed up and down 10 times with the aid of a blade. Rotating paddle to disperse the sample. From these rotational movements, the sample rotates and the sample's resistance to flow creates a torque in the opposite direction to the rotation. This creates a shear rate that is a function of the speed of the rotating blades and the distance of the blades from the walls of the container where the sample is placed.
2. Make sure that there are no lumps left at the bottom of the container where the paddle cannot reach. If there are still lumps remaining, repeat the movements as many times as necessary to make sure that the entire sample is suspended and free of lumps. Use the paddle to scrape the sample off the spatula and place the paddle into the container.

3.2 Protocol

The literature has different methods that have been widely studied to develop a suitable procedure to evaluate the pasting properties of starch. On the other hand, it lacks detailed protocols to evaluate the pasting properties of starch.

Basically, the pasting properties of different types of starches are obtained by simultaneously heating and stirring an aqueous suspension of ground starch in a disposable container placed in an RVA. A precisely controlled heat retention temperature profile is used to cause viscosity changes in the mixture.

In RVA, a specific temperature, time, and blade rotation speed program is selected and the particular data form an RVA profile. The user can choose their own profile, but it is recommended to use standard methods, as it is easier to compare the result with other studies in the literature and use the correct temperature and time for the sample to be analyzed.

3.3 For RVA Analysis

1. Starches are isolated from various sources such as cereals, tubers, legumes, and other starchy products.
2. RVA instrument preparation.
3. Turn on the RVA unit and leave for 30 min to warm up.
4. To ensure measurement accuracy, the calibration offset (zero viscosity) must be checked daily. Follow the instructions in the installation and operation manual.

5. Calibration of the RVA unit should be checked with a standard oil of known viscosity once a month (*see* **Note 2**). (Certified Viscosity Standard can be obtained from Cannon Instrument Company. Note that some RVA units are calibrated at 25 °C. In this case, the calibration check must be carried out with a CAL25 profile and suitable oil.)
6. Heat the oil to between 60 °C and 70 °C and weigh 28.0 ± 0.1 g. Measure the viscosity of the calibration oil using the CAL95 program (profile designed for calibration verification). The viscosity of the calibration oil must be within ±5% of the recommended value given by the standard.

3.4 Analysis of Starch Pasting Properties

Procedure A—Running a Test Sample: 1

1. After turning on the RVA and waiting 30 min for it to warm up, turn on the associated computer.
2. Open the appropriate RVA method in Thermocline Software for Windows (TCW) (*see* **Note 3**) or select the appropriate RVA method using the instrument keyboard if using a standalone RVA.
3. Insert a shovel into the RVA tower.
4. Rotate the paddle paddle until it clicks into place and is secure.
5. Select "Execute" → "Zero."
6. Click on "Start."
7. Allow the value in the box to stabilize and approach 0.
8. Click on "zero" and "stop" and close the window.
9. Perform a quality control test using a test starch package, as described in Procedure B, below.
10. Select "Select configuration to run" in the "Run" menu.
11. Load the "Rice Rapid" test profile using the Thermocline software.
12. Enter a file name and a file path under which to save the data.
13. Click "Execute" and "Yes" when prompted to save the changes.
14. A sample weight calculator should appear.
15. Make sure that "3.5 g" is selected as the sample size, "25.0 g" as the water weight, and "14%" as the moisture base (*see* **Note 4**).
16. Enter the moisture content of the sample to calculate the exact water and sample weights needed. Calculation is used to adjust all samples to a 14% moisture base.
17. Prepare the sample as described in item 6.

Briefly

- Place an aluminum container/can on the scale and press "tare" to tare the scale.
- Weigh the specified amount of deionized/distilled water into the aluminum can/can container.
- Place an aluminum foil to weigh the starch sample on the scale and press "tare" to tare the scale.
- Weigh the specified amount of sample into the container—After use, discard the aluminum foil used to weigh the sample.
- Add the weighed sample to the water and vigorously move the paddle blade across the sample up and down 10 times. First, add the water and then the sample to the container to facilitate dispersion.
- Make sure that any pieces of sample stuck to the inside of the container are pushed into the water.
- Repeat as necessary until the sample is suspended and no lumps remain.

18. S elect the profile to be used to capture rheological information (RVA curves) as per the example shown in Table 2.
19. After selecting a test profile, entering all the necessary sample information, and preparing it, you can load the sample into the RVA.

Table 2
Details of the RVA standard maize profile (temperature and time)

Stage	Standard profile
Initial temperature (°C)	50
Initial holding time (min)	1:00
Heating time (min)	3:42
Max temperature (°C)	95
Hold at max temperature (min)	2:30
Cooling time (min)	3:48
Final temperature (°C)	50
Final holding time (min)	2:00
Total test time (min)	13:00

Fonte: Standard profile [20]

20. To load a sample into the RVA, insert the container containing the sample in water and the paddle into the RVA instrument.
21. Rotate the paddle manually until it fits in the right place and is completely secure to start the analysis.
22. With the paddle fully inserted into the container, slide the upper part of the paddle into the coupling and make sure that the paddle is correctly seated, pressing it against the rear part of the coupling.
23. Gently push the engine tower down until you hear a click and release immediately (*see* **Note 5**).
24. When prompted by the program, press the tower to start the starch paste gluing cycle. The gluing cycle test will begin immediately.
25. The test will end automatically, and the analysis results will be reported.
26. After completing the gluing cycle, remove the container containing the sample (*see* **Note 6**).
27. Wash the shovel and bowl/aluminum can with hot water and a brush and dry completely with suitable paper.
28. The paddle can be used repeatedly, while the aluminum container/can can only be used three times before discarding it.
29. The viscosity of the sample will be represented graphically on the monitor or display if using an independent RVA.
30. The test will end automatically and the analysis results will be reported (*see* **Note 7**).
31. In the "View" menu, select "View analysis results."
32. Copy the values (TP, PV, Pt, TV, BDV, SBV, and PT) and paste them into an Excel spreadsheet (*see* **Note 8**).
33. To start the bond reading of the next sample, go to the "Run" menu and select "Run last test."

Procedure B—Running a Control Sample Using a Standardized Test Starch Packet: 1

1. After turning on the RVA and waiting 30 min for it to warm up, turn on the associated computer and open the RVA control software; TCW.
2. Zero the instrument, as demonstrated in **steps (b–g)** of Procedure A.
3. Select "Select configuration to run" in the "Run" menu.
4. Load the "Standard 1" test profile through the Thermocline software.

5. Place an aluminum container/can on the scale and follow the steps described in item p) of Procedure A.
   - The weight of the deionized/distilled water to be added to the container is 25.00 ± 0.01 g.
   - The weight of the starch sample to be added to the aluminum foil is 3.5 g.
6. Insert the container and paddle into the instrument.
7. Rotate the paddle until it clicks into place and is secure.
8. Upon completion of the cycle, peak viscosity and decomposition values must fall within the ranges specified on the starch package (*see* **Note 9**).
9. Clean the container and paddle as described in Procedure A.
10. To proceed with the test samples, begin at step 10 of Procedure A.

3.5 Viewing Results During a Test

Whenever you begin a test, a new tab is created in the currently open project file (or in a new project file if none is open). Any plots defined in the test configuration file, such as the values of viscosity, temperature, and speed as a function of time, are then displayed and updated on the graph while the test is running.

The test time, temperature, speed, and viscosity will also be shown on the display of your RVA.

3.6 Interpretation of RVA Data

The viscosity curves evaluate the gluing characteristics of the samples during heating due to structural changes in the starch molecules, immediately after retreat during cooling. The properties of starches will affect the shape and size of the gluing curve. For example: native starch exists as microscopic granules of varying sizes that will depend on the characteristics of their botanical source.

The gluing properties determine the functional properties of starch in the raw material and in different industrial applications. These applications may vary according to their behavior, whether cold or hot, depending on the product required by the consumer.

RVA curve parameter results are displayed as viscosity in centipoise (cp) or rapid visco units (RVU).

The basic procedure for interpreting the obtained RVA data is as follows:

- TCW will be used to display and analyze data from one or more tests. Each project file can display an unlimited number of datasets.

- The parameter values obtained by the Software (TP, PV, Pt, TV, BDV, SBV, and PT) can be pasted into an Excel spreadsheet or transferred to other applications, such as a word processing document.
- The data contained in the file will be displayed in the form of a graph with one X-axis and two Y-axes (*see* **Note 10**).
- The X-axis is time (min), while the Y-axes are viscosity (cp) and temperature (°C)/time (min) vs. viscosity (cp) and time (min) vs. temperature (°C).

3.7 Factors Affecting the Starch Sticking Curve

(a) *Amylose and Amylopectin Content*

Amylopectin is the main component of starch responsible for the swelling of the granules and increasing viscosity after heating, while amylose is often intertwined with amylopectin, limiting the swelling of starch granules. Thus, the higher the amylase content, the lower the PV of a starch paste.

(b) *Source of Starch*

Nonwaxy starches have lower PV and higher PT values than waxy cereal starches (He et al., 2023). Furthermore, after cooling, nonwaxy starches present higher SBV than their waxy variety, due to amylose mainly favoring the development of starch gel in the short term, during cooling.

(c) *Test Conditions*

Several test conditions can affect the starch sticking curve, such as the following:

- Sample viscosity and concentration: Samples with high starch concentrations tend to be more impacted by shear effects and competition for moisture, resulting in a deviation from this relationship.
- Heating and cooling rates: May affect the degree of swelling and solubilization differently for different starches.
- Sample preparation: Failure to solubilize the sample in water leads to large and uneven particle sizes, which can significantly alter the results. Furthermore, the time elapsed after grinding can also be another factor that can alter the results.
- Apparent viscosity of starch pastes: Different periods exposed to temperature and shear also affect the results.

(d) *Modification of Starch*

Native starches have limitations in food applications, so their modification becomes an important strategy to adapt to a specific application [37]. However, modifications can have different effects on the bonding curve, such as the following:

- Hydrolysis by acid or amylolytic enzymes: Drastically reduces the viscosity of the paste, as well as oxidation.
- Crosslinking: Delays granule breakdown and, at higher levels, can inhibit granule swelling and reduce viscosity.
- Replacement: Can delay recoil and reduce gluing temperature.
- Annealing: Generally reduces the gelatinization temperature range, causing an increase in PT and a more pronounced increase in viscosity.

(e) *Interactions with Other Components*

The interaction with other components also has significant influences on the starch sticking curve. These components can be endogenous to the sample or added to formulations such as sugars, lipids, emulsifiers, gums, salts, and pH modifiers.

- Lipids and emulsifiers: Amylose, in the presence of polar lipids or emulsifiers, tends to intertwine with amylopectin and restrict the swelling of starch granules during cooking, forming amylose–lipid/emulsifier complexes [38]. The formation of these complexes can sometimes be observed as a small increase in the viscosity curve as the test temperature increases, and again as it drops.
- Gums: Small amounts of gums, such as sodium carboxymethyl cellulose, can markedly increase the viscosity of starch.
- Sugars: Can delay the onset of gelatinization and thus produce gluing curves at higher gluing temperatures.

4 Notes

1. Grinding of samples is extremely important for quality assessment purposes to obtain uniform exposure to particles. Characteristic particle size is preferable because large pore sizes produce several different behaviors, increasing the bias in sample readings. Furthermore, wet grinding methods are preferable to dry grinding methods due to the formation of amorphous regions in the latter grinding method.

2. Certified Viscosity Standard can be obtained from Cannon Instrument Company. Note that some RVA units are calibrated at 25 °C. In this case, the calibration check must be carried out with a CAL25 profile and suitable oil.
3. Thermocline for Windows (TCW) software: programmable instrument control: profile style method configuration with advanced options and automatic functions for flexibility and ease of use. Flexible data analysis: Automatic functions for commonly derived parameters and advanced functions for searching. Real-time graph display: Interactive and overlay functions for greater clarity. Single page report: traceability data, results, and graph, all on a single page. Traceability: meets electronic records/electronic signatures requirements for data security.
4. The sample quantity in grams, water weight, and moisture content are defined according to the protocol established by the study. Normally, the sample weight is around 3.5 g, with water around 25.0 g and moisture content of 14%. It is important to highlight that the chosen moisture level in the sample must be maintained for all test analyses. It can be seen that a sample with 14% moisture will present a relatively high percentage of solids. Therefore, it is important to avoid clumping through agitation.
5. If the tower does not lock, the initial test temperature may not be reached. Failure to insert the paddle correctly may cause it to rub or jam during testing. Failure to release the turret immediately when it jams may cause the blade coupling to rub or seize during the test.
6. Always remove the container using a suitable thermal insulation glove, as the container will be hot to the touch.
7. If there is a problem with a test or if it has already produced the desired result, you can cancel the test to save time. This will stop the test and apply any relevant analytics.
8. The recoil value reported by the Thermocline Software means the difference between the final viscosity and the retention force. However, when it comes to rice, the industry generally considers "Setback" to be the difference between final viscosity and peak viscosity. Both of these values are provided in the gluing profile and can be used to calculate rice indentation.
9. If the values are not within the range, repeat the procedure using another packet of starch. If the second set of values is not within the range, notify a supervisor.

10. The project file can contain up to three graphs for each dataset. To highlight a specific chart, place the cursor over a point on that line of the chart. You can also move the cursor slightly to see the data for that point. To display a vertical or horizontal line, place the cursor at the desired position on the horizontal or vertical axis.

5 Conclusion

Different methods are available to determine the gluing properties of starches, the choice naturally depends on the available facilities as well as the objective to be achieved in the experiment. RVA is the most used technique to determine the properties of pastes in several studies. Recently, the use of multivariate data analysis has been seen as a further improvement and refinement of RVA measurements, through the combination of RVA curves or viscograms. Furthermore, a new high-temperature RVA model has been used, which allows the starch suspension to be heated up to 140 °C, providing an advantage over traditional models that do not allow the evaluation of supervisory behaviors at extrusion temperature. However, this method is not completely understood. Finally, it is important to highlight that the gluing property of starch is a very complex process, affected by several factors such as its composition, interaction with other components, and modification, among others. Thus, independent evidence derived from two or more mathematical modeling of the curve obtained by RVA allows cross-comparisons that can provide a more complete understanding of this phenomenon (Table 3).

Acknowledgments

The authors are grateful to the Coordination of Improvement of Higher Education Personnel—Brazil (CAPES); the Minas Gerais Research Foundation (FAPEMIG, Brazil); the National Council for Scientific and Technological Development (CNPq); the Minas Gerais State Research Support Foundation (FAPEMIG); the Minas Gerais Agricultural Research Company—Cândido Tostes Dairy Institute (EPAMIG-ILCT); and Federal University of Viçosa (UFV).

Table 3
Conditions and results of studies using RVA

Starch type	Operating conditions	Results	References
Wheat A	1 min of equilibration at 50 °C, then heated to 95 °C at a rate of 12 °C/min and maintained for 2.5 min at the same temperature (95 °C). Then the temperature was reduced to 50 °C at a rate of 12 °C/min and held for 2 min	The values of PT, PV, BDV, TV, SBV, and FV were 80 °C, 2907 cp, 482 cp, 2405 cp, 1088 cp, and 3493 cp	[39]
Lotus	1 min equilibration at 50 °C, then heated to 95 °C at a rate of 12 °C/min, and held at 95 °C for 2.5 min, then cooled to 50 °C at a rate of 12 °C/min, and kept at 50 °C for 2 min	The values of PT, PV, BDV, TV, SBV, and FV were 60 °C, 2717 cp, 1493 cp, 1224 cp, 512 cp, and 1737 cp	[40]
Potato	1 min of equilibration at 50 °C, then heated to 95 °C for 3.5 min, and maintained for 2.5 min at the same temperature (95 °C). Then the temperature was reduced to 50 °C for 4 min and maintained for 1 min. The rotation speed was maintained at 960 rpm during the first 10 s and maintained at 160 rpm during the remaining process	The values of PT, PV, BDV, TV, SBV, and FV were 77.40 °C, 3343 cp, 1206 cp, 1709 cp, 1516 cp, and 2722 cp	[41]
Kapas	1 min of equilibration at 50 °C, then heated to 95 °C for 3.7 min, and maintained for 2.5 min at the same temperature (95 °C). Then the temperature was reduced to 50 °C for 3.8 min and maintained for 2 min. Initially, after 10 s, the paddle was rotated at 960 rpm to disperse the starch. After that, the blade was rotated at 160 rpm	The values of PT, PV, BDV, TV, SBV, and FV were 79.11 °C, 4037 cp, 3049 cp, 5308 cp, 988 cp, and 2258 cp	[42]
Kepok		The values of PT, PV, BDV, TV, SBV, and FV were 80.70 °C, 3996.5 cp, 2778 cp, 4796 cp, 1218.5 cp, and 2018 cp	
Ambon		The values of PT, PV, BDV, TV, SBV, and FV were 77.27 °C, 4529 cp, 3294.5 cp, 4506.5 cp, 1234.5 cp, and 1212.25 cp.	
Nangka		The values of PT, PV, BDV, TV, SBV, and FV were 76.58 °C, 4535.5 cp, 3409.5 cp, 4995.5 cp, 1126 cp, and 1586 cp	

References

1. Agama-Acevedo E, Flores-Silva PC, Bello-Perez LA (2019) Cereal starch production for food applications. In: Clerici MTPS, Schmiele M (eds) Starches for food application. Academic, London, pp 71–102
2. Seligra PG, Jaramillo CM, Famá L, Goyanes S (2016) Biodegradable and non-retrogradable eco-films based on starch–glycerol with citric acid as crosslinking agent. Carbohydr Polym 138:66–74
3. Schmiele M, Sampaio UM, Clerici MTPS (2019) Basic principles: Composition and properties of starch. In: Clerici MTPS, Schmiele M (eds) Starches for food application. Academic, London, pp 1–22
4. Avérous, L. R., & Halley, P. J. Starch polymers: from the field to industrial products. In: Avérous, L. R., Halley, P. J., Starch polymers: from genetic engineering to green applications. 1st. Elsevier; 2014. p. 3–10
5. Pérez S, Baldwin PM, Gallant DJ (2009) Structural features of starch granules I. In: BeMiller JN, Whistle R (eds) Starch: chemistry and technology, 3rd edn. Academic, New York, pp 149–192
6. BeMiller JN, Huber KC (2010) Carbohydrates. In: Damodaran S, Parkin KL, Fennema OR (eds) Fennema's food chemistry, 4th edn. Artmed, Porto Alegre, pp 75–130
7. BeMiller JN (2019) Carbohydrate chemistry for food scientists. Woodhead Publishing, London, 427 p
8. Schirmer M, Höchstötter A, Jekle M, Arendt E, Becker T (2013) Physicochemical and morphological characterization of different starches with variable amylose/amylopectin ratio. Food Hydrocoll 32(1):52–63
9. Tao K, Li C, Yu W, Gilbert RG, Li E (2019) How amylose molecular fine structure of rice starch affects functional properties. Carbohydr Polym 204:24–31
10. Debet MR, Gidley MJ (2006) Three classes of starch granule swelling: influence of surface proteins and lipids. Carbohydr Polym 64(3): 452–465
11. Li C, Wu A, Yu W, Hu Y, Li E, Zhang C, Liu Q (2020) Parameterizing starch chain-length distributions for structure-property relations. Carbohydr Polym 241:116390
12. Ye X, Zhang Y, Qiu C, Corke H, Sui Z (2019) Extraction and characterization of starch granule-associated proteins from rice that affect in vitro starch digestibility. Food Chem 276:754–760
13. Li H, Dhital S, Slade AJ, Yu W, Gilbert RG, Gidley MJ (2019) Altering starch branching enzymes in wheat generates high-amylose starch with novel molecular structure and functional properties. Food Hydrocoll 92:51–59
14. Regina A, Blazek J, Gilbert E, Flanagan BM, Gidley MJ, Cavanagh C, Ral J, Larroque O, Bird AR, Li Z, Morell MK (2012) Differential effects of genetically distinct mechanisms of elevating amylose on barley starch characteristics. Carbohydr Polym 89(3):979–991
15. Zhu F, Wang S (2014) Physicochemical properties, molecular structure, and uses of sweetpotato starch. Trends Food Sci Technol 36(2): 68–78
16. Waramboi JG, Dennien S, Gidley MJ, Sopade PA (2011) Characterization of sweetpotato from Papua New Guinea and Australia: physicochemical, pasting and gelatinisation properties. Food Chem 126(4):1759–1770
17. Kasturi P, Bordenave N (2013) Oat starch. In: Chu Y (ed) Oats nutrition and technology. Hoboken, Wiley-Blackwell, pp 95–121
18. Cozzolino D (2016) Theuse of the rapid visco analyser (RVA) in breeding and selection of cereals. J Cereal Sci 70:282–290
19. Crosbie, G. B., & Ross, A. S. The RVA handbook. (2007). (St. Paul: AACC International
20. AACC InternationalApproved Methods of Analysis (2000) Determination of the pasting properties of rice with the rapid visco analyser, Method 61-02.01, 11th edn. AACC International, St. Paul
21. Cozzolino D, Allder K, Roumeliotis S, Eglinton J (2012) Feasibility study on the use of multivariate data methods and derivatives to enhance information from the Rapid Visco Analyses. J Cereal Sci 56:610–614
22. Guo J, Lian X, Kang H, Gao K, Li L (2016) Effects of glutenin in wheat gluten on retrogradation of wheat starch. Eur Food Res Technol 242(9):1485–1494
23. Luo X, Li J, Lin X (2012) Effect of gelatinization and additives on morphology and thermal behavior of corn starch/PVA blend films. Carbohydr Polym 90(4):1595–1600
24. Booth R, Bason M (2007) Principles of operation and experimental techniques. In: The RVA handbook, p 152
25. Wani IA, Sogi DS, Hamdani AM, Gani A, Bhat NA, Shah A (2016) Isolation, composition, and physicochemical properties of starch from legumes: a review. Stärke 68(9–10):834–845

26. Vanier NL, Zavareze ER, Pinto Z, Klein B, Botelho FT, Dias ARG, Elias MC (2012) Physicochemical, crystallinity, pasting and morphological properties of bean starch oxidised by different concentrations of sodium hypochlorite. Food Chem 131:1255–1262
27. Avérous LR, Halley PJ (2014) Starch polymers: from the field to industrial products. In: Avérous LR, Halley PJ (eds) Starch polymers: from genetic engineering to green applications, 1st edn. Elsevier, pp 3–10
28. Deffenbaugh LB, Walker CE (1989) Use of the rapid-visco-analyzer to measure starch pasting properties. part i: effect of sugars. Starch – Stärke 41:461–467
29. Zaidul ISM, Yamauchi H, Kim SJ, Hashimoto N, Noda T (2007) RVA study of mixtures of wheat flour and potato starches with different phosphorus contente. Food Chem 102:1105–1111
30. Li J, Shin GH, Lee IW, Chen X, Park HJ (2016) Soluble starch formulated nanocomposite increases water solubility and stability of curcumin. Food Hydrocoll 56:41–49
31. Sun Q, Xing Y, Qiu C, Xiong L (2014) The pasting and gel textural properties of corn starch in glucose, fructose and maltose syrup. PLoS One 9(4):e95862
32. Adebowale KO, Lawal OS (2003) Functional properties and retrogradation behaviour of native and chemically modified starch of mucuna bean (Mucuna pruriens). J Sci Food Agric 83:1541–1546
33. Singh J, Kaur L, McCarthy OJ (2007) Factors influencing the physico-chemical, morphological, thermal and rheological properties of some chemically modified starches for food applications-A review. Food Hydrocoll 31:1–22
34. Chantaro P, Pongsawatmanit R, Nishinari K (2013) Effect of heating–cooling on rheological properties of tapioca starch paste with and without xanthan gum. Food Hydrocoll 31(2): 183–194
35. Nakamura Y, Ono M, Sawada T, Crofts N, Fujita N, Steup M (2017) Characterization of the functional interactions of plastidial starch phosphorylase and starch branching enzymes from rice endosperm during reserve starch biosynthesis. Plant Sci 264:83–95
36. Zhan Q, Ye X, Zhang Y, Kong X, Bao J, Corke H, Sui Z (2020) Starch granule-associated proteins affect the physicochemical properties of rice starch. Food Hydrocoll 101: Article 105504
37. Bangar SP, Kumar M, Whiteside WS (2021) Mango seed starch: a sustainable and eco-friendly alternative to increasing industrial requirement. Int J Biol Macromol 183:1807–1817
38. Zhuang H, Feng T, BeMiller JN (2019) Effect of hydrocolloids in normal maize starch and its distarch phosphates. Food Hydrocoll 96:231–245
39. Chi C, Zou Y, Peng X, Yang Y, Chen B, He Y et al (2023) Medição da gelatinização do amido utilizando um espectrofotômetro. Hidrocolóides Alimentares 108956
40. Yang J, Xu S, Wang W, Ran X, Ching YC, Sui X et al (2023) Preparação e caracterização de bioplásticos a partir de fécula de mandioca sililada e óleos de soja epoxidados. Polímeros de carboidratos 300:120253
41. Akhila PP, Sunooj KV, Aaliya B, Navaf M, Sudheesh C, Yadav DN, Khan MA, Mir SA, George J (2022) Morphological, physicochemical, functional, pasting, thermal properties and digestibility of hausa potato (Plectranthus rotundifolius) flour and starch. Appl Food Res 2(2):100193
42. Marta H, Cahyana Y, Djali M, Arcot J, Tensiska T (2019) A comparative study on the physicochemical and pasting properties of starch and flour from different banana (Musa Spp.) cultivars grown in Indonesia. Int J Food Prop 22: 1562–1575

Chapter 6

Rheological Characterization

Bilge Nazli Altay and Christopher L. Lewis

Abstract

This book chapter examines the range of methods, protocols, and standardization of the rheological characterization of starches, an abundant complex carbohydrate biopolymer found in a variety of plants, including corn, tapioca, wheat, potatoes, pea, and rice. A dry, granular powder form of starch is frequently obtained in modified (acid-thinned, thermally modified, thinned/hydroxy ethylated, etc.) or unmodified (native) forms depending on specific requirements. The unique viscoelastic characteristics of starch have a substantial influence on how well it functions in a given application. Starch is used in diverse industries such as food, paper and packaging, textiles, pharmaceuticals, personal care, construction, and oil and gas industries where starch's binding, thickening, and stabilizing qualities are favorable.

The significance of the key rheological properties of starch, such as the pasting property and viscosity of starch paste and starch gel, is discussed. In addition, the available general standards and specific methods for assessing the rheological characteristics of starches are provided.

Key words Starch, Rheology, Rheometer, Shear rate, Viscosity, Thixotropy, Non-Newtonian, Thermoplastic, Biopolymer, Standards

1 Introduction

Starch, a polysaccharide, holds significant importance in both the food and industrial sectors, encompassing conventional uses. Functioning as a complex carbohydrate, it is a vital energy source for organisms ranging from plants to humans. The exceptional properties of starch render it a versatile ingredient in a wide array of food products, such as baked goods, sauces, and confectionery items. However, its applications extend beyond the food industry. Starch has attracted significant interest and is currently used as a renewable polymer in diverse industrial applications. The attractiveness of starch stems from its renewable characteristics, biodegradability, abundant availability, and ability to form cohesive films, making it suitable for a wide range of applications. As such, starch finds utility in various conventional industrial sectors, including paper and packaging manufacturing, textile production, pharmaceuticals,

Sneh Punia Bangar (ed.), *Standardized Procedures and Protocols for Starch*, Methods and Protocols in Food Science,
https://doi.org/10.1007/978-1-0716-3866-8_6,

personal care products, construction, oil and gas drilling, and drinking water purification [1, 2] Moreover, starch finds usage is diverse applications, including electronics, drug delivery, antimicrobial materials, and structural materials [3].

Understanding the rheological characteristics of starch products is crucial for product development, process design and evaluation, quality control, and consumer satisfaction [4]. Starch serves multiple purposes, primarily functioning as a thickener, colloid stabilizer, gelling, and volume agent, binder and adhesive, moisture retainer, texturizer, and fat substitute in various applications [5]. Its widespread utilization is attributed to its low cost. Additionally, starch is extensively employed as a key raw material for producing biodegradable films, primarily due to its abundant availability and ease of use.

The analysis of starch rheology provides numerous valuable benefits across various fields. In terms of process design and evaluation, rheological data help determine suitable processing conditions, such as temperature, shear rates, and mixing times, to achieve desired product characteristics and ensure efficient production. By analyzing rheological properties for quality control, manufacturers can monitor batch-to-batch consistency, assess product stability, and ensure optimal performance and shelf life. For example, in the baking industry, rheological analysis helps evaluate dough mixing properties and predict final product quality. Rheological analysis aids in the development of new starch-based products or the improvement of existing formulations. By understanding how starch behaves under different conditions, formulators can adjust ingredient compositions and processing parameters to optimize runnability and stability. Moreover, rheology significantly impacts consumer perception and acceptability. Analyzing starch for rheology helps food manufacturers ensure desirable texture attributes such as creaminess, smoothness, viscosity, mouthfeel, and mouth-coating properties. By matching consumer expectations, products can be better tailored to meet market preferences and enhance consumer satisfaction.

The rheological analysis of starch is vital for understanding its flow and deformation behavior, which directly impacts its functional properties in various applications. By conducting comprehensive rheological studies, researchers and industries can optimize process design, ensure quality control, develop new products, and enhance consumer satisfaction. The utilization of various instruments such as rheometers, viscometers, and rapid visco analyzers (RVAs) offers valuable insights into the behavior of starch under diverse conditions. These instruments allow for the comprehensive analysis of starch rheology, providing a deeper understanding of its flow and deformation properties. Understanding the importance of rheology measurement for starch-based products, the following discussion delves into the fundamental principles of rheology and its relationship to the flow and deformation properties exhibited by diverse starch materials. Individuals in the field of rheology are required to familiarize themselves with specific sub-areas of interest, including rheometry (the measurement and analysis of flow and deformation of materials), constitutive equations (mathematical models that describe the relationship between stress and strain), measurement of flow behavior in complex geometries, and the calculation of material behavior in complex flow conditions.

1.1 Rheology Basics and Characterization

Rheology is a field of study that attracts interest from academic researchers and professionals in the industrial sector. The study of rheology often begins by exploring the origins of the term itself. Derived from Greek roots, "rheology" combines "rheo," meaning flow, with "logy," which denotes science. Professor Eugene Bingham of Lafayette College, Indiana, coined the term "rheology" at the suggestion of a colleague, the Professor of Classics. The definition gained acceptance upon the establishment of the American Society of Rheology in 1921 [6].

Rheology, despite being associated with the term "flow," is a field of study that encompasses solids, gases, and liquids, including the different phases and textures of food [7]. Rheology examines how materials respond to external and internal forces while undergoing flow and play a vital role in various industries, including food, paper and packaging, textiles, pharmaceuticals, personal care products, construction, electronics, and oil and gas. Currently, rheology finds more application in the study of various materials such as solutions, emulsions, suspensions, foams, gels, melts, and semisolids like creams and pastes. In addition, powders and granular materials have gained increasing popularity [8]. Rheology is often employed to comprehend various scenarios, including formulation, material behavior, mixing, stability, packaging considerations (such as dispensing products like toothpaste), equipment selection for production, transfer processes like pumping, as well as evaluating the effects of environmental changes. In essence, rheology is a scientific discipline dedicated to investigating how materials deform under stress (shear) and how materials flow. This focus on both deformation and flow has led to its characterization as the science of "deformation and flow." The degree of material deformation depends on factors such as the magnitude, velocity, direction, and viscosity of the applied or encountered stress [9]. Moreover, the inherent properties of the materials themselves, such as weight, can act as stress-inducing factors.

Generally, materials that maintain their shape are commonly associated with solid-like behavior. However, in rheology, a solid material is defined by its ability to exhibit elastic behavior, meaning it can return to its original shape after deformation. As such, the classification of liquids and solids is not a strict binary system but rather represents two extremes along a spectrum of observed behavior. The notion of dynamic behavior, where physical properties can vary depending on the nature of the interaction, presents a challenging aspect to ascribe within the context of the liquid–solid spectrum. In rheology, the deformation of materials described as ideal solids is elastic when a force is applied, and the material can return to its original state when the force is removed [10]. Moreover, ideal solids respond instantaneously (i.e., there is no delay in their mechanical response). In contrast, viscous behavior refers to the phenomenon where the material undergoes a delayed response (e.g., material deformation is out of phase with the applied stress). The ideal liquid material flows with the applied force, and the energy generated is dissipated as heat in the liquid material and cannot be recovered [10]. Materials that exhibit "viscoelastic" behavior display both elastic and viscous behavior. Coatings, inks, paints, and starch are all instances of such materials.

As a useful framework, rheologists commonly categorize liquids as being either structured liquids or elastic liquids. The concept of a structured liquid can be understood as a solid that has the ability to flow. They display characteristics resembling solids when they are at rest but can flow when subjected to applied stress surpassing their yield stress. Examples of structured liquids include suspensions like smoothies and conductive coatings and emulsions such as creams and mayonnaise. When spooned onto a plate, mayonnaise retains its shape, demonstrating solid-like characteristics. However, it can still be easily pushed through with a spoon, exhibiting flow. On the other hand, elastic liquids display a tendency to flow effortlessly, but they exhibit a resilient nature that becomes evident when the rate of deformation becomes high. Hair styling gel, for example, conforms to the shape of its container, resembling a liquid-like behavior. Yet, if the action is performed at a faster timescale, shaking the bottle vigorously, the gel starts to exhibit bouncing, akin to a jelly-like solid characteristic. Other examples of elastic liquids include polymer solutions such as gels and food gums and surfactant solutions such as shower gel and hand soap. This understanding of classification by characterizing products into structured and elastic liquid classifications becomes valuable when embarking on the process of rheology testing. The objective is to roughly categorize materials to provide a general idea of how to approach the initial rheological analysis of the sample under investigation. Rheometers are instrumental in measuring the viscoelastic properties, viscosity, and time-related characteristics of materials. After establishing the definition of rheology, it is crucial to define the variables frequently encountered in the field, starting with viscosity and shear stress.

Viscosity (η) quantifies a material's resistance to flow. It can be calculated by dividing the shear stress (σ) (Eq. 1) by the shear rate ($\dot{\gamma}$) (Eq. 2). In essence, the applied stress can be visualized as the force exerted per unit area, signifying the level of intensity with

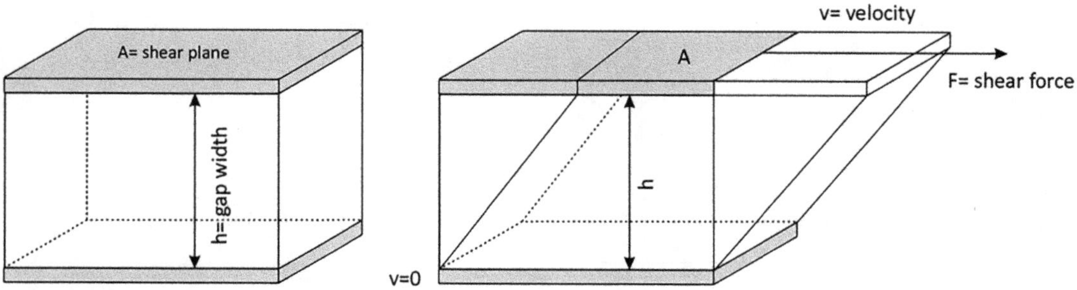

Fig. 1 Illustration of a material that has been displaced by force at a specific velocity

which the material is being stressed. When a substantial force is distributed over a small area, it generates a high shear stress, while a smaller force distributed over a larger area results in a lower shear stress. In the case of a liquid under applied stress, it typically demonstrates a certain level of fluid flow. The shear rate is employed to quantify this flow, representing the ratio of the upper surface's velocity to the sample's thickness (height) and is measured in reciprocal seconds (per second)(/s). It is noteworthy that attaining a high shear rate for a material can be accomplished by not only increasing the speed of the upper surface but also by reducing the gap between the surfaces, even if the speed remains constant. Shear strain (Eq. 3) is often expressed as the displacement ratio that occurs when a material is subjected to shear stress. In the schematic in Fig. 1, the sample of interest is positioned between two surfaces, and stress is applied to the upper surface, causing displacement. Shear strain is calculated by determining the ratio of this displacement to the height of the sample, resulting in a dimensionless value. Alternatively, shear strain can be expressed as a percentage. The unit of viscosity (Eq. 4) is expressed as pascal-second (Pa·s) or millipascal-second (mPa·s). The unit of poise (p) or centipoise (cp) is also common, and the conversion between these units is as follows: 0.001 Pa·s = 1 mPa·s = 1 cp. Within the imperial system, the poiseuille (Pl) serves as the unit of viscosity. In practical terms, the centipoise (cp) or centistokes (cSt) are more prevalent as units of viscosity, even in countries where the imperial system is primarily utilized. Water and honey are frequently employed as examples when discussing viscosity. Honey exhibits higher viscosity (10,000 cp) compared with water (1 cp), indicating a slower flow rate when subjected to identical conditions. This contrast illustrates that viscosity is directly related to a material's resistance to flow, while fluidity is inversely proportional. As a result, when the term "high viscosity" is used, it essentially denotes a condition of low fluidity, whereas "low viscosity" suggests a higher level of fluidity.

$$\text{Shear Stress } (\sigma) = \frac{\text{Force}}{\text{Area}} = \frac{N}{m^2} = Pa \text{ (Pascal)} \tag{1}$$

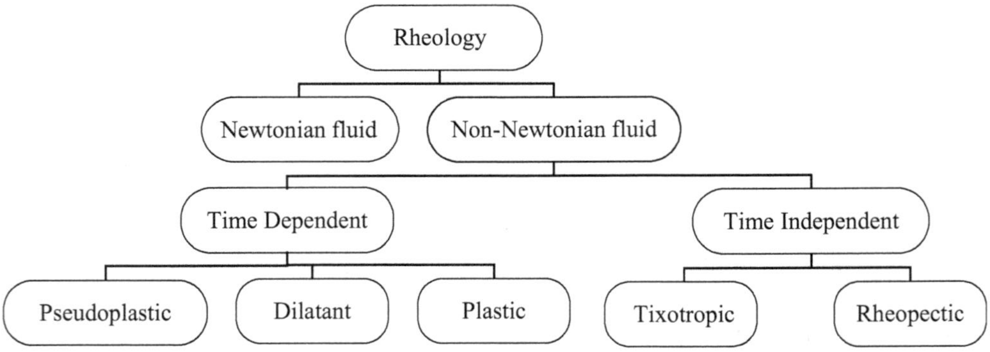

Fig. 2 General classification of materials in rheology science

$$\text{Shear Rate } (\dot{\gamma}) = \frac{\text{Velocity}}{\text{Height}} = \frac{m/s}{m} = s^{-1}\ (1/s) \tag{2}$$

$$\text{Shear Strain } (\gamma) = \frac{\text{Displacement}}{\text{Height}} = \frac{m}{m} = \%\text{or unitless} \tag{3}$$

$$\text{Viscosity } (\eta) = \frac{\sigma}{\dot{\gamma}} = Pa.s \text{ (Pascal second)} \tag{4}$$

In rheology, the flow behavior of materials can be categorized as either Newtonian or non-Newtonian (Fig. 2). Newtonian fluids, such as water, oils, and organic solvents, exhibit a consistent viscosity regardless of variations in shear rate or previous shear history. However, it is important to note that the viscosity of these fluids can be affected by temperature changes [11]. As temperature rises, the viscosity of a fluid decreases. This phenomenon occurs because the individual elements or particles within the fluid tend to attract each other along the flow lines. As the temperature increases, the greater thermal energy leads to an increase in molecular mobility, which counteracts the attractive forces between particles and reduces viscosity.

Materials displaying non-Newtonian flow exhibit either time-dependent or time-independent behavior. Some materials, referred to as "Bingham plastics," deviate from the typical pattern as they require a minimum applied stress to initiate flow. While they fall under non-Newtonian fluids, their behavior closely resembles Newtonian fluids once flow is established, showing a linear relationship between shear stress and shear rate. Mayonnaise and ketchup serve as excellent examples of Bingham plastic rheology.

Rheometers are extensively employed instruments to determine specific rheological behavior exhibited by a material. The flow properties of a sample, for instance, can be assessed by shearing it through rotation between two parallel surfaces. This procedure generates a rheology curve, often known as a rheogram, that provides insights on the flow characteristics (Fig. 3) [9]. The type of non-Newtonian fluid shown in Fig. 3 is characteristic of a

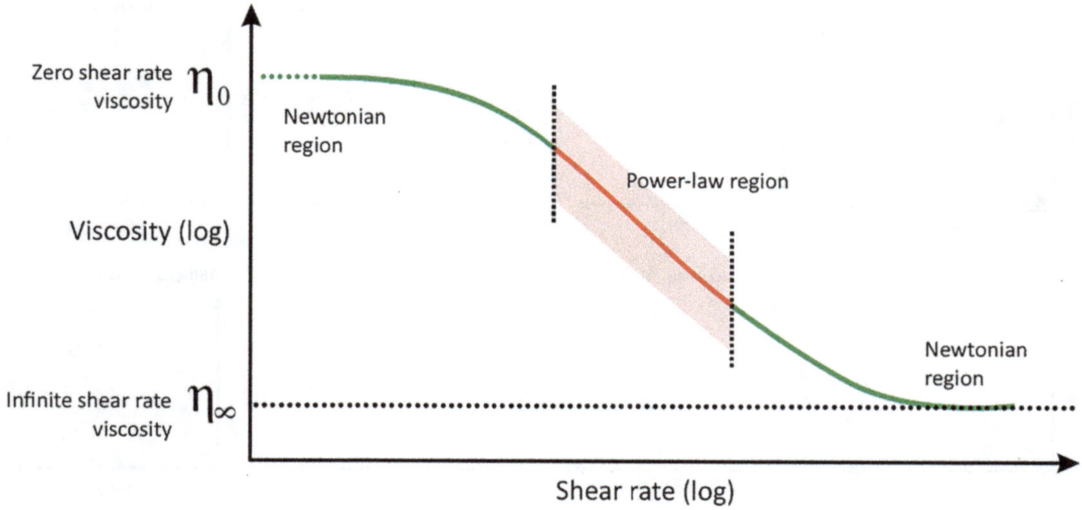

Fig. 3 Depiction of the apparent viscosity behavior for a shear-thinning fluid [9]

pseudoplastic or shear-thinning fluid. As shown, the zero rate viscosity (η_0) and infinite rate viscosity (η_∞) describe low shear and high shear viscosity regions, respectively. Notably, Newtonian-like behavior is observed in these regions where viscosity is independent of shear rate. The zero rate viscosity pertains to the viscosity of a fluid in the absence of shear or flow, indicating the resistance of fluid to flow under static conditions. In contrast, infinite rate viscosity characterizes the viscosity of a fluid when exposed to exceptionally high shear rates; it signifies the resistance of fluids to flow when subjected to intense shear forces.

Viscosity, as a characteristic of materials, has the ability to change in response to applied stress, even when low amounts of stress are present. This change may occur quickly or gradually over a long period of time, and it may take the form of an increase or a decrease. Shear thinning, which denotes a decrease in viscosity as the shear rate increases, is one common example of this non-linear phenomenon. As shown in Fig. 3, the viscosity of the substance decreases as the shear rate increases in the power-law region, making the material less flow-resistant.

There are numerous other types of non-Newtonian fluids, and Fig. 4 presents general rheogram curves of these materials [12]. Whereas shear-thinning materials exhibit a decrease in viscosity as the shear rate increases, dilatant materials demonstrate an increase in viscosity. Inks and shampoos are examples of pseudoplastic materials, while wet sand and starch suspensions exemplify dilatant materials. Thixotropic materials exhibit time-dependent shear thinning followed by a subsequent material recovery. Thixotropic fluids become more fluid with increasing shear rates. When the shearing forces cease, these materials gradually regain their

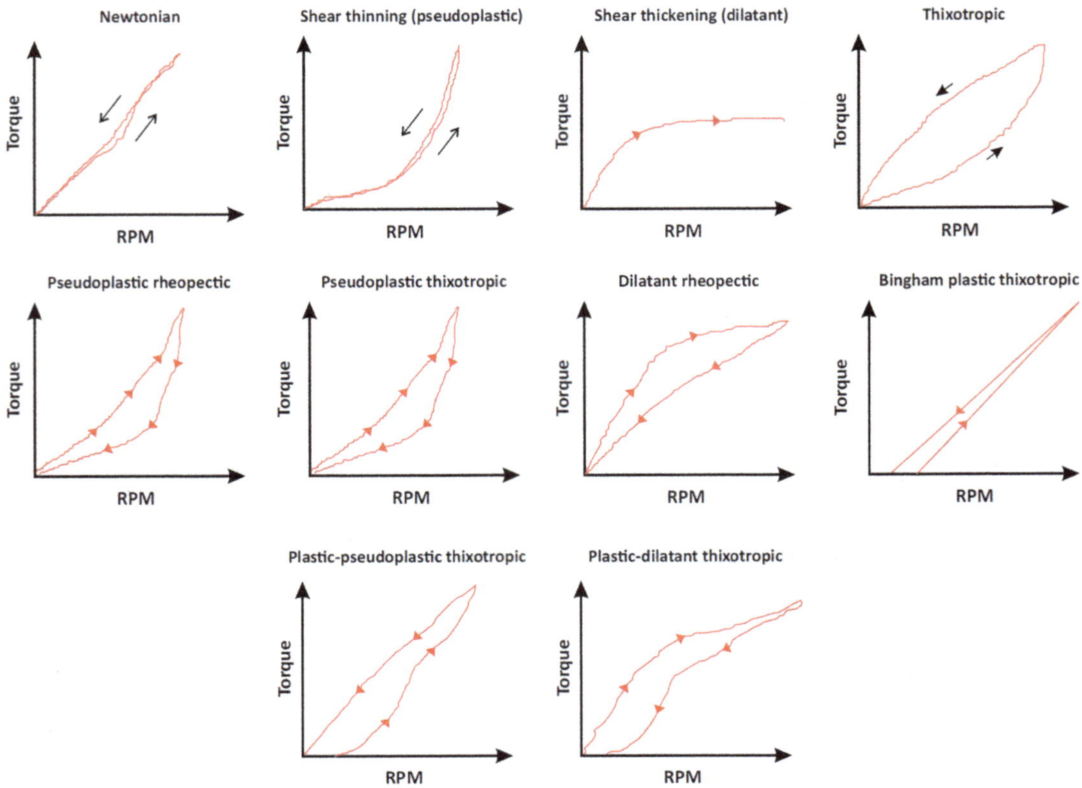

Fig. 4 General rheology curves of materials [12]

original viscosity, reverting to their initial state. This reversible transformation from a higher viscosity to a lower viscosity and back again characterizes the unique quality known as thixotropy. Materials such as offset and screen printing inks, gels, pastes, and certain fluids exhibit this type of viscosity behavior. Unlike thixotropic fluids, rheopectic fluids demonstrate an opposite behavior, wherein their viscosity increases as shear rates rise. When the shear rate is subsequently decreased, the viscosity of rheopectic fluids reverts back to its original value or is higher than the original value. The rheopectic behavior of fluids offers various practical applications, particularly in enhancing the performance of lubricants and suspensions. This viscosity increase contributes to enhanced load-bearing capacity and reduced friction within the system. Moreover, the rise in viscosity also plays a crucial role in preventing sedimentation and maintaining the stability of the suspension over time. There are two common test methods for characterizing thixotropy, the thixotropy index test and the thixotropic loop test. During the thixotropy index test, materials are usually exposed to a controlled shear rate for a specific duration. Then the index is calculated by comparing the viscosity after shearing with the viscosity before shearing, with a value below 1 indicating the presence of thixotropy. On the other hand, the thixotropic loop test involves applying controlled shear stress to the material, allowing it to rest for a duration, and subsequently measuring the recovery of viscosity. This test offers insights into the material's thixotropic extent and its ability to recover viscosity. Numerical integration is used to calculate the size of the loop, which serves as a gauge of the structural breakdown rate and recovery rate for a given ramp time. In general, bigger loops indicate slower breakdown-recovery rates. The Recovery Parameter, another property associated with thixotropy, is found to show a strong correlation with sag resistance and air entrainment. The non-sagging materials are reported to show shorter recovery time and higher final recovered viscosity [13].

Certain materials can behave in intriguing ways, showing traits of both solids and liquids, depending on the rate of the deformation process. Silicone material, for instance, behaves like a slow-flowing liquid when placed in a container and allowed enough time to settle and find its own level. On the other hand, it exhibits brittle fracture under fast and rapid extension, acting like a solid material. This fascinating dual nature demonstrates how, depending on the speed of the deformation process, a material may display properties of both solids and liquids. The concept of time scaling in rheology is accomplished through the utilization of a dimensionless parameter known as the Deborah number (D_e) (Eq. 5), which suggests that if you wait long enough, even mountains have the ability to flow over time [6]:

$$D_e = \tau / t \tag{5}$$

where t represents the specific time scale associated with the observed deformation process, while τ represents a characteristic time scale intrinsic to the material itself. Materials exhibiting high Deborah numbers can be described as solid-like, while those with low Deborah numbers demonstrate characteristics resembling that of a liquid [14]. Therefore, a material can exhibit solid-like behavior due to its significantly long characteristic time or when subjected to a rapid deformation process during the study. This means that even highly mobile liquids with low characteristic times can display elastic solid-like behavior under extremely fast deformation processes. In line with these observations, a solid can be defined as a material that maintains its shape without undergoing continuous deformation when exposed to a particular stress. In other words, for a given stress, there will be a definitive and stable final deformation, which may or may not be reached instantaneously upon the application of the stress. On the other hand, a liquid can be defined as a material that undergoes continuous shape changes, or flows, when subjected to a specific stress, regardless of the magnitude of that stress.

1.1.1 Static Methods

The static methods used for determining linear viscoelastic behavior include creep tests and stress relaxation tests. Creep tests provide insights into the material's deformation behavior over time under constant stress, while relaxation tests examine the relaxation of stress over time under constant strain. These tests offer valuable information regarding the material's capacity to endure prolonged loading, its ability to resist deformation, and its overall viscoelastic properties. Typically, these methods are suitable for materials with a few seconds or longer relaxation times, such as certain gels, elastomers, and complex fluids. The equipment required for static tests is relatively straightforward, involving the application and measurement of the input signal, as well as recording the output. The choice of geometry for static tests depends on the type of material being tested. Solid-like materials can be tested using long slender specimens in tensile or torsional experiments, while liquid-like materials can be assessed using simple shear geometries such as concentric-cylinder or cone-and-plate setups. Commercially available constant-stress rheometers can be used for conducting creep tests in simple shear. The following explanation of the creep test provides two important pieces of information about zero shear viscosity and equilibrium compliance. Zero shear viscosity is discussed earlier and refers to the Newtonian plateau viscosity of the material under extremely low shear conditions. It is a commonly used result obtained from a creep test. Equilibrium compliance refers to the elastic response of the material to strain, which provides insight into its viscoelastic properties. It helps characterize the material's ability to recover from deformation. The creep rheogram in Fig. 5 is explained into three distinct regions: the instantaneous elastic response (1), the viscously retarded elastic deformation (2), and the steady-state viscous flow (3). The instantaneous elastic response (1) represents the immediate and elastic response of the material to the applied stress. In the viscously retarded elastic deformation (2), there is a delayed elastic deformation that occurs after the initial response. For the steady-state viscous flow (3), once

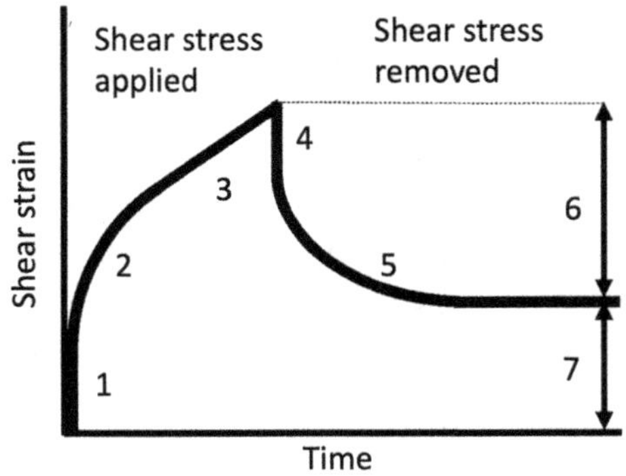

Fig. 5 A representative creep test graph

the material's elastic response reaches its limit, it enters a phase of entirely viscous flow. This part of the graph should appear as a straight line, and its gradient can be used to calculate the zero-shear viscosity of the material. Extrapolating this line back to $t = 0$ allows for the calculation of the equilibrium compliance. The irrecoverable viscous deformation region indicates that the material has undergone deformation that cannot be recovered. In the recoverable elastic deformation region, the material has experienced deformation, but it can recover its original shape once the stress is removed. The second part of the creep curve occurs after the material has achieved steady-state viscous flow and the stress is removed. It consists of two additional regions called the initial recovery step and further recovery. Following the removal of the stress, the initial recovery step corresponds to the instantaneous elastic response (4), where the material begins to recover its deformation and is related to that observed in region (1). Further recovery (5) accounts for the complete recovery of all the elastic deformations in the material and relates to the viscously retarded elastic deformation (2). The recoverable, elastic deformation is denoted by region (6), and the irrecoverable, viscous deformation of the material is represented by region (7) [15].

1.1.2 Dynamic Methods: Oscillatory Testing Methods

Oscillatory measurements are referred to as dynamic measurements because of the continuous variation in either stress or strain [16]. Here a material is exposed to a sinusoidal stress or strain, and the corresponding response in terms of strain or stress is recorded. The behavior of rheological materials can be assessed by measuring their response with respect to various factors such as time, temperature, strain or stress amplitude, and frequency. Oscillatory stress sweeps are a useful method for quantifying modulus and yield stress in a straightforward manner. This test involves applying small, incremental oscillating stresses to a sample, alternating between clockwise and counterclockwise rotations while monitoring the resulting strain or deformation. The modulus, which represents the stiffness of the material, is calculated by dividing the amplitude of the applied stress by the amplitude of the resulting strain. Measurement results can show plateaus in the complex modulus at low stresses, indicating a solid-like behavior in the gels followed by the transitions to yielding as the incrementing applied stress disrupts the internal structure of the gels. One major advantage of oscillatory test methods is their ability to cover a wide frequency range using a single instrument while reporting viscoelastic properties such as tan delta (tan δ, tangent of the phase angle), complex modulus (G^*), storage modulus (G'), and loss modulus (G''). This is particularly useful when studying materials with a broad spectrum of relaxation times. The typical frequency range for oscillatory tests is from 10^{-3} to 10^3 per second,

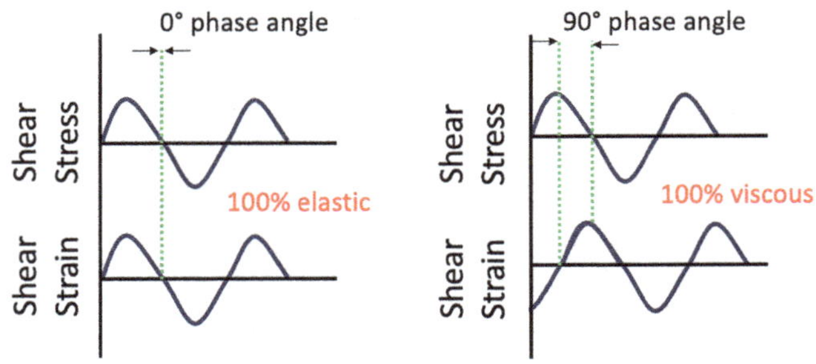

Fig. 6 Oscillation stress/strain waves difference between elastic and viscous materials

enabling the examination of a time spectrum ranging from approximately 10^3 to 10^{-3} s. For longer relaxation times, static tests of longer durations may be necessary, or wave-propagation methods can be employed. To achieve a wide frequency range, forced oscillations are employed. The most common example of forced-oscillation experiments involves the use of a cone and plate geometry, although parallel-plate or concentric-cylinder configurations can also be used.

The oscillation method is a non-destructive technique that involves the application of a sinusoidal stress wave to deform the structure of a material, followed by the measurement of the resulting strain wave. Oscillation rheology testing is typically performed at low applied stresses and strains, often below the sample's yield point, within the linear viscoelastic range [17]. In the case of a purely elastic material, the stress and strain waves are synchronized with a phase angle difference of 0° (stress and strain relation in phase). Conversely, for purely viscous materials, the stress and strain waves are out of phase angle by 90° (stress and strain relation not in phase) (Fig. 6). Most materials exhibit viscoelastic behavior, falling between these two extremes, and the phase difference provides valuable information about their viscous and elastic characteristics. For example, a material is characterized as having equal elasticity and viscosity when there is a 45° phase angle or when the tangent of the phase angle (tan δ) is equal to 1.

The complex modulus (G^*), or so-called combined modulus, measurement is another useful method for evaluating the overall stiffness or rigidity of a sample via oscillatory testing. It represents the resistance to deformation, and rigidity, in a material and can be expressed as the ratio between the amplitude of stress and the amplitude of strain (Eq. 6). A high complex modulus means increased resistance to deformation and greater rigidity in a material. It can endure higher levels of stress and strain before undergoing notable deformation. Materials with high complex moduli often demonstrate solid-like behavior and possess elastic properties. Conversely, a low complex modulus suggests reduced resistance to

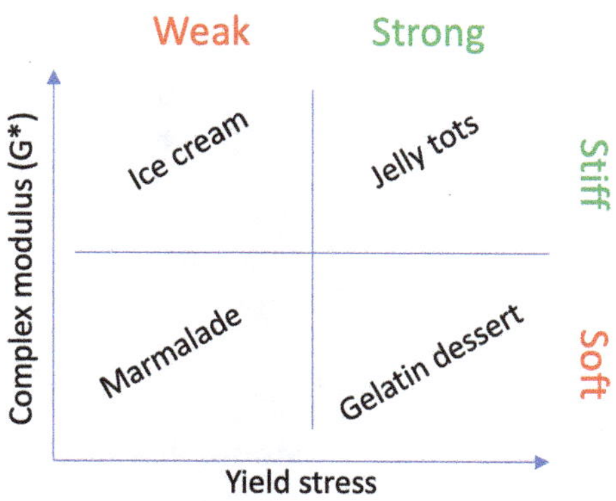

Fig. 7 Complex modulus and yield stress relation observed in everyday food samples

deformation and greater flexibility in a material. It experiences significant deformation under lower stress and strain levels. Materials with low complex moduli typically exhibit more fluid-like or viscoelastic behavior, displaying a tendency for flow and deformation. Figure 7 illustrates the relationship between complex modulus and yield stress observed in everyday food samples containing starch. Rigid structures are evident in products that exhibit stiffness, while soft structures are characterized by a wobbly and gelatinous consistency. Strong structures have the ability to retain a suspension, whereas weak structures pour easily. Within the field of rheology, rigidity is referred to as modulus, and strength is denoted by the term yield stress.

$$\text{Complex Modulus } (G^*) = \frac{\text{Shear amplitude}}{\text{Strain amplitude}} = Pa \quad (6)$$

The complex modulus can be broken into two quantities that represent the solid-like and liquid-like responses of the material (Fig. 8). The storage modulus (G′), also known as the elastic modulus, represents the energy stored in a sample during deformation, reflecting its solid-like resistance to deformation. On the other hand, the viscous modulus (G″), referred to as the loss modulus, represents the energy dissipated during deformation, indicating the liquid-like portion of the material's resistance. It corresponds to the complex modulus portion that signifies the dissipation as heat during deformation. It offers valuable information about a material's capacity to absorb and dissipate energy when exposed to applied stress or strain. Notably, when G′ is greater than G″, the material exhibits more elastic characteristics, while when G′ is smaller than G″, the material displays more viscous characteristics. Conveniently, tangent delta (tan δ) is obtained by dividing G″ by G′ and allows for the determination of the relative energy dissipation in a material (Fig. 8).

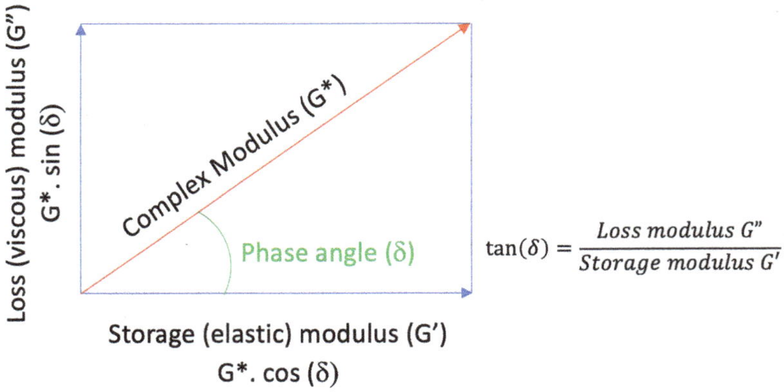

Fig. 8 The relationship between complex, loss, storage moduli, and phase angle

1.1.3 Dynamic Methods: Wave Propagation Method

In wave-propagation methods, waves are generated at the surface of the specimen in contact with the wave generator. To evaluate the viscoelastic functions, it is necessary to measure the velocity and attenuation of the waves as they pass through the specimen. One notable advantage of wave-propagation methods is their adaptability to high-frequency studies. They have been commonly employed in the kilohertz (kHz) range and even higher, reaching frequencies in the hundreds of gigahertz (GHz). Wave propagation tests typically entail applying oscillatory or dynamic forces to induce wave-like disturbances within the material. Several techniques, including ultrasound, shear wave, or surface wave methods, can be utilized to generate and detect these waves. Subsequently, the collected data are analyzed to ascertain the material's mechanical properties, which encompass parameters such as storage modulus, loss modulus, complex modulus, damping coefficient, and phase angle.

1.1.4 Dynamic Methods: Steady Flow Method

The steady flow method enables the measurement of flow properties, including viscosity and shear stress, in a controlled and continuous flow environment. Unlike oscillatory methods that involve unsteady and periodic flow fluctuations, the steady flow method provides a more simplified and stable flow situation for studying the rheological characteristics of materials. Using the steady flow method at a constant rate, time-dependent behavior of materials can be observed. The steady flow method in rheology involves monitoring changes in rheological properties when a system is switched between two steady-state flow conditions. In this method, the initial velocity at a specific point within the system remains constant, allowing flow behavior analysis under steady-state conditions. In order to assess the recovery behavior of a material over a specific time period, a three-step process is commonly employed (Fig. 9). The initial step involves subjecting the material to a low shear rate to establish a baseline measurement. Subsequently, a high shear rate is applied to disrupt any existing structure in the material.

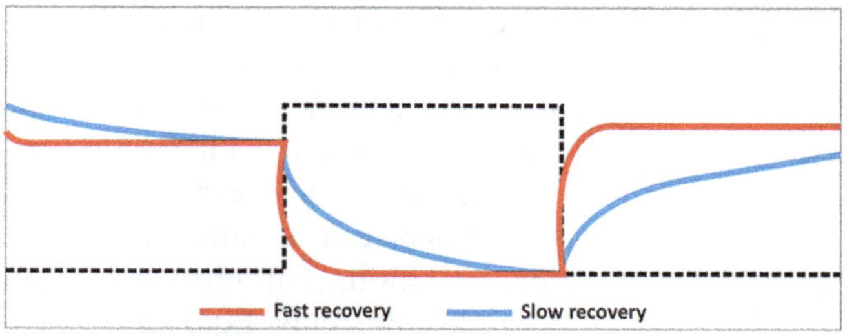

Fig. 9 Detecting time-dependent behavior of materials at two different constant rates using the steady-flow method. Note: Solid line corresponds to shear rate, whereas dashed lines correspond to viscosity

Finally, the shear rate is returned to the initial low level to observe how the material recovers. This approach allows for the detection of rapid or slow recovery, providing insights into the material's resilience and ability to regain its initial state.

1.1.5 Determining Key Measurements

Viscosity is a relevant property for both structured liquids and elastic liquids, although the specific shear rates of importance may vary between the two. However, for structured liquids, the rigidity and strength of the structure become particularly significant. Parameters such as modulus and yield stress, which indicate the force required to induce flow, hold more relevance. Moreover, it is crucial to comprehend the behavior of the structure following disruption. Specifically, determining whether it remains in a broken-down state or has the capability to recover becomes pertinent. If recovery occurs, as indicated by the term thixotropy, understanding whether the recovery happens instantly or requires a significant duration of time is of significance. When dealing with a structured liquid that requires pumping, considerations include the required force for effective flow. The yield stress is not a significant factor for elastic materials. Instead, it is significant to examine the viscoelastic behavior and characteristics, such as the bouncing phenomenon described earlier holds great relevance when studying elastic materials. The extensional behavior of the elastic material, such as observing tailing and stringing during pumping, also becomes important when evaluating their pumpability. In the case of materials intended for spraying, determining the optimal conditions to achieve atomization into droplets becomes important as it is essential to examine for defects, the presence of bubbles, and overall appearance and performance quality. The rheological behavior of both structured and elastic materials is explored via various types of flows, with shear flow being the predominant focus. Shear flow occurs in various practical applications, including coating processes, drainage processes, the application of creams, the spreading of inks and coatings on surfaces, and pumping fluids through pipes. It involves the relative motion of one surface against another. In certain instances, the importance of extensional flow becomes evident. This type of flow is observed predominately in viscoelastic products and manifests as stringing and tailing, as commonly observed in products like liquid soap. Through the manipulation of shear stress and shear strain during the investigation of materials, researchers can acquire data that are subjected to analysis and interpretation. The ratio between these two variables yields the modulus for solids (Eq. 7) and viscosity for liquids, a key parameter in characterizing the material's behavior. When a solid sample undergoes stress, and the resulting deformation is minimal, it signifies a high modulus value. This indicates that the sample is rigid and resistant to deformation. When two materials are subjected to the same stress, and one material undergoes a more substantial deformation compared to the other, it signifies that the former material has a lower modulus, while the latter material possesses a higher modulus:

$$\text{Modulus} = \frac{\text{Shear stress}}{\text{Shear strain}} = Pa \quad (7)$$

When the relationship between the applied shear stress and the resulting shear rate is examined for liquids, the focus is on assessing the speed at which the stress induces flow in the sample. In the case of classical ideal solids and ideal liquid materials, a linear relationship can be observed by calculating the ratio between these two variables (Fig. 10). Viscosity can be considered a single value that represents the fluid's resistance to flow for Newtonian liquids. In the majority of cases, particularly for non-Newtonian liquids, the viscosity of a fluid is affected by the specific conditions it experiences, leading to variations in its viscosity (Fig. 3).

Two primary approaches for evaluating viscoelastic behavior include static and dynamic methods. In static methods, a step change in stress or strain is applied, and the subsequent development of strain or stress over time is observed. Dynamic tests involve the application of a sinusoidally varying strain. It is important to note that regardless of the chosen method, it is essential to ensure that measurements are conducted within the linear range. Otherwise, the results may become dependent on experimental details

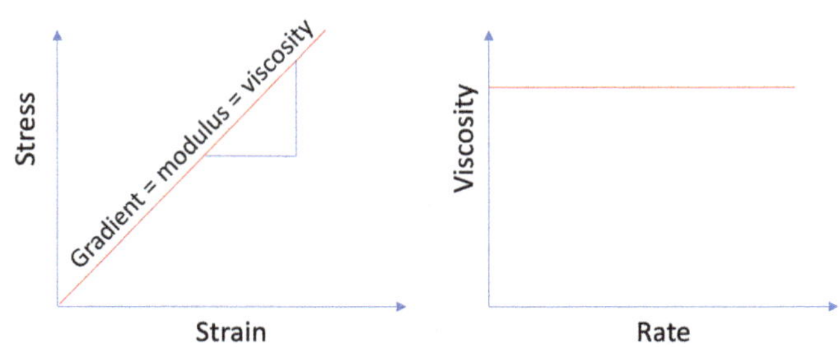

Fig. 10 Linear relationship for ideal solids and liquids

Table 1
Rheological components of various processes

Process	Deformation type or attribute	Rheological property
Roll coating	Squeezing flow Stretching flow Shearing flow High strain rates Large accelerations Large decelerations Surface area transients	Biaxial extensional viscosity Uniaxial extensional viscosity Shear viscosity High-shear viscosity Elasticity (G′) G′ recovery Dynamic surface tension
Spray	High shear rate Large accelerations High-strain-rate extensional deformation Surface area transients	High-shear viscosity Elasticity (G′) Extensional viscosity/elasticity Dynamic surface tension
Brushing/rolling	Medium shear rate Stretching flow Surface area transients	Shear viscosity Extensional viscosity/elasticity Dynamic surface tension
Leveling/sagging	Slow shear flow Surface-stress driven Transient (nonequilibrium process)	Low shear stress viscosity Dynamic surface tension Structure recovery kinetics
Curtain coating	Stretching flow Surface area transients Shear (pumping, extrusion)	Extensional viscosity/elasticity Dynamic surface tension Shear viscosity

and will not be representative of the material's unique characteristics. The linearity of the measurements can be assessed by verifying that the computed viscoelastic functions remain unaffected by the magnitude of the applied stresses and strains. The most significant deformation types are simple shear and extension, although the deformation history of a material can be complex during the lifetime of materials (Table 1) [13].

1.1.6 Rheology Fitting Models

When flow curves of materials are obtained, it becomes feasible to acquire a more comprehensive understanding of the material behavior by employing diverse mathematical fitting models. The models act as a link between the measurement, prediction, and comparison of materials in response to shear, time, and temperature. Table 2 provides an overview of widely used models that assist in understanding the behavior of materials in this context.

The *Power Law* model is the simplest and most commonly employed form of the general viscous constitutive relation applied to steady simple shear flows [18]. Empirical evidence has indicated that the shear stress of such fluids can typically be correlated with the shear rate using a Power-Law function. In this two-parameter model, the viscosity is defined by the consistency coefficient (K) and the flow behavior index (n) (Eq. 8). While the Newtonian

Table 2
Commonly used rheology models

Flow behavior	Model
Pseudoplastic	Power Law (Ostwald de Waele)
Dilatant	Cross
Bingham	Sisko
Casson	Williamson
Herschel-Bulkley	Carreau

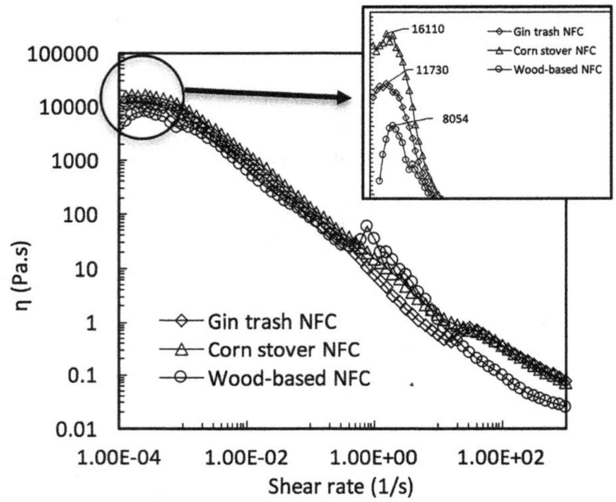

Fig. 11 Viscosity-shear rate rheograms of 1% nano-fibrillated suspensions

model assumes a flow behavior index of = 1, the Power-Law model, as indicated by its name, can account for non-linearity in the curve through the value of n. The model is capable of representing both shear thinning (pseudoplastic) and shear thickening (dilatant) behavior by employing a flow behavior index greater than 1 for shear thickening or less than 1 for shear thinning. A clear drawback of the Power-Law model is its inability to accurately describe the low shear rate region [18]:

$$\eta = K.\dot{\gamma}^{n-1} \tag{8}$$

Controlled stress rheometers showed that the yield stress is an apparent phenomenon and that the measuring method has an impact on its amplitude. An example in Fig. 11 shows that the yield stress of corn stover, gin trash, and wood-based nano-fibrillated suspensions at maximum are 16,110, 11,730, and 8054 Pa·s, respectively. Below the yield stress point, the suspensions present no real macroscopic flow [19]. It is now understood that the majority of materials possess a high but finite viscosity, referred to as zero shear viscosity, rather than an infinite viscosity below a certain yield stress threshold. Three models incorporate a yield stress component and vary in complexity. The *Bingham* model (Eq. 9) is a linear model that includes an additional yield stress term, is the most complex among the shear stress/shear rate models, and offers greater flexibility in fitting data. The *Casson* model (Eq. 10) is the second model that considers a linear relationship taken to the square root. Lastly, the *Herschel-Bulkley* model (Eq. 11) combines the Power Law with an added yield stress term. Modern rheometer software suggests visually comparing the experimental data curve of a material with the curve shapes of "standard" models. This helps identify which model is likely to provide the most accurate fit. The proposed model is accompanied by standard error and regression coefficient metrics, where a smaller value for the standard error indicates a better fit. Typically, a standard error value below 10 indicates a suitable fit. If the standard error falls below 15, it is considered acceptable for the purpose of model fitting. However, if the data are observed to be noisy, more lenient standards should be considered, ranging from 10 to 30 [20]. The regression coefficient should ideally be close to 1, indicating a strong correlation between the fitted model and the observed data.

$$\sigma = \sigma_\gamma + \eta_p \gamma \quad (9)$$

$$\sigma^{1/2} = \sigma_o^{1/2} + \eta^{1/2} \gamma_c^{1/2} \quad (10)$$

$$\sigma = \sigma_y + K\gamma^n \quad (11)$$

The *Cross* model categorizes materials based on the logarithmic relationship between viscosity and shear rate and is extensively employed in various industries. Figure 12 illustrates a typical

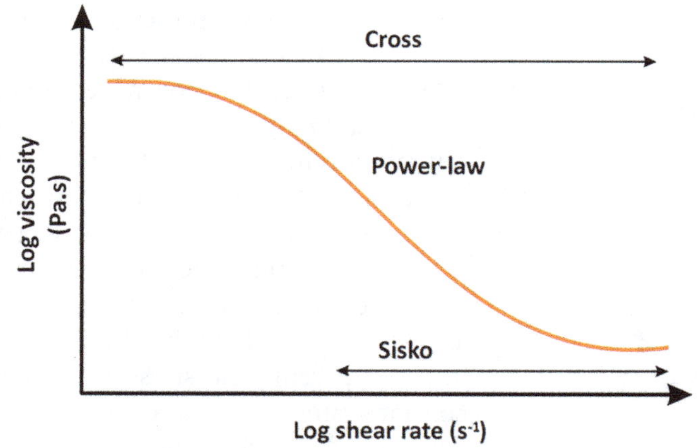

Fig. 12 Defining the regions of applicability of Cross, Sisko, and Power Law models to a pseudoplastic curve [21]

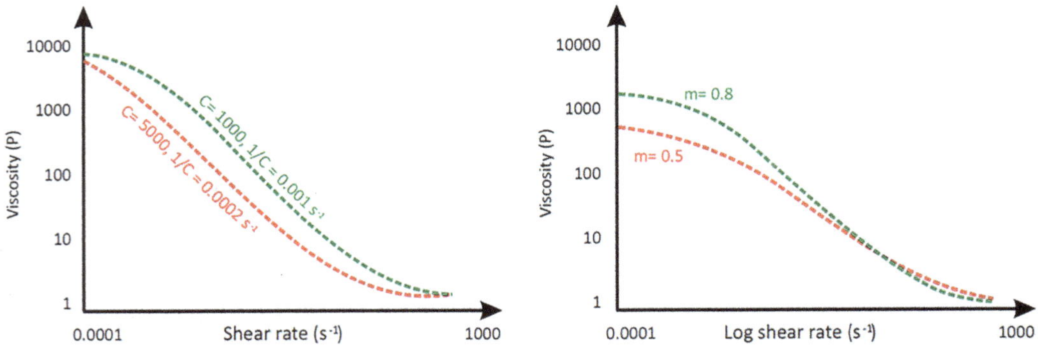

Fig. 13 Viscosity curves of materials with different C and m values

viscosity curve conforming to the Cross model utilizing a viscosity formula that is dependent on the shear rate, as represented by Eq. 12:

$$\eta = \eta_\infty + \frac{\eta_0 - \eta_\infty}{1 + (C\dot{\gamma})^m} \tag{12}$$

The parameter η_0 represents the zero shear viscosity of the material at a steady state. The steady-state viscosity is valuable for assessing the stability of materials and monitoring the impact of changes in polymer molecular weights and formulations. The symbol η_∞ denotes the value of the infinite viscosity. The parameter m is known as the Cross speed coefficient, and its value depends on the shear rate in the region where the viscosity begins to decrease. A zero value for m indicates Newtonian behavior, while values approaching zero indicate shear thinning. C is called the Cross time constant (consistency). $1/C$ value gives information about where shear thinning starts. A comparison of materials with varying $1/C$ and m values is presented in Fig. 13.

As variants of the Cross model, other models, such as the *Carreau* model, have been developed. The selection of the most appropriate model is not guided by extensive theoretical principles but rather is frequently determined using trial and error. TA Instruments, a prominent rheological instrument manufacturer, provides some guidelines for fitting flow data and describes several available models. Generally, the choice of model depends on the type of experiment conducted. For simple linear flow curves, models such as Newtonian to Herschel-Bulkley are recommended. If determining an apparent yield stress is necessary, it is advised to maintain a constant ramp rate. For logarithmic data sets—a type of non-linear data—models such as the Cross, Sisko/Williamson, or Carreau are suitable. In the end, quantitative measurements such as regression coefficients and standard errors are used to evaluate the model's efficacy.

1.1.7 Pasting Properties and Gel Strength

For starch, gel strength is often evaluated to understand starch's ability to form a gel-like structure that is influenced by the starch type, concentration, temperature, and processing conditions. It is important across various applications in food and industrial applications. Pasting is a basic procedure in many industries where starch granules are exposed to heat and water and become a thick paste. In this setting, the subtle interactions between starch properties—such as gel strength—become evident as they have a substantial impact on pasting behavior and, by extension, the general usefulness of starch in a variety of applications. Torque rheometers are typically employed to characterize pasting properties and the gelatinization process of starch dispersions. Instruments like the viscoamylograph comprise a temperature-controlled mixing chamber where the torque applied to the material by the mixing element is used to measure viscosity. Devices such as the RVA (rapid viscosity analyzer) and Brabender Viskograph-E are specially designed mixing instruments used to assess peak viscosity, gelatinization, and retrogradation of both native and modified starch products [22]. However, the dynamic rheology approach offers more in-depth insights into the starch system compared to viscoamylography and is relatively straightforward to implement. The cost of an appropriate stress-controlled rheometer is reported to be comparable to that of a viscoamylograph like the RVA [23].

1.2 Rheometry Experimental Methods

Rheometers are designed to manipulate and measure materials within a precisely regulated geometric structure. These devices allow researchers to exert control over materials, allowing them to apply and quantify various stress, strain, and strain rate parameters under various situations, ensuring precise and accurate experimental setups for comprehensive analysis. Rheometers enable the examination of materials under various simple flow conditions, such as steady simple-shear flow, small-amplitude oscillatory-shear flow, and extensional flow. These flow conditions serve as common scenarios in which the properties of materials are studied and analyzed. The classification of measurement methods is determined by the specific flow geometry involved in the measurement process.

In the case of non-Newtonian materials, their viscosity can exhibit variability depending on factors such as flow rate and duration. This means that the fluid can undergo changes, becoming either more or less viscous due to various factors, highlighting the dynamic nature of the material under investigation. Consequently, relying solely on viscosity measurements conducted at a constant shear rate might not provide a comprehensive understanding of the behavior exhibited by non-Newtonian fluids. To accurately comprehend the intricate characteristics of these fluids, it becomes imperative to perform viscosity measurements at different shear rates. For this purpose, rheometers prove to be highly suitable as they enable viscosity measurements at distinct shear rates and varying time intervals. Such measurements provide valuable insights into the viscosity profile and time-dependent behavior of the fluid material. The rheogram obtained by running the instrument from zero to maximum shear rate and then back to zero enhances the understanding of unique rheological properties.

In order to measure the rheological properties of materials, it is necessary to determine the numerical values of constants used in rheological equations. Depending on the chosen equation, experiments are conducted to examine the influence of these constants on the results. Two key characteristic functions of materials—viscosity as a function of shear rate and viscoelastic characteristics, such dynamic moduli, as a function of frequency—are always central, regardless of the model for rheological behavior that is chosen. Experimental methods for measuring rheological properties encompass rheometry. The more specific method, called viscometry, is commonly employed for viscosity measurements. These methods can be categorized into two approaches: absolute and relative measurements. In the absolute methods, viscosity is computed as a ratio of the shear stress to the shear rate. The primary objective of absolute methods used for viscosity measurement is to establish a connection between pairs of values that depend on the flow geometry, as defined by the design of the measuring instrument. These pairs of values include the relationship between force (or torque) and stress, as well as the connection between flow velocity and strain rate. The quantities of force, pressure, torque, and others may be used in the experiment to determine the dynamic parameters. Relative methods of viscosity measurement involve comparing the properties of a fluid with those of a model fluid (reference fluid) with known properties. There may be instances where it is not crucial to determine the absolute viscosity of a fluid. Rather, it may be satisfactory to compare particular characteristics of different fluids in order to obtain the required information or insights. Typically, this involves determining the ratio between the dynamic viscosity of the unknown sample and that of the reference fluid to determine relative viscosity characteristics, such as flow time.

The development of a rotating rheometer has resulted in considerable advances in the field of materials testing. One significant advancement is the possibility of using a separated actuator (motor) and detector (torque transducer) arrangement, which provides tighter strain control and smaller inertial (torque) correction. While this arrangement is more expensive, it provides researchers with more precision and accuracy in their measurements. A further option is the combined actuator and detector design, in which the lower plate stays fixed and the motor handles both torque and strain measurement. This arrangement is more cost-effective and allows for the use of smaller torques. Furthermore, it simplifies testing such as creep, yield, and recovery while allowing for the management of large inertial corrections as needed. Finally, these

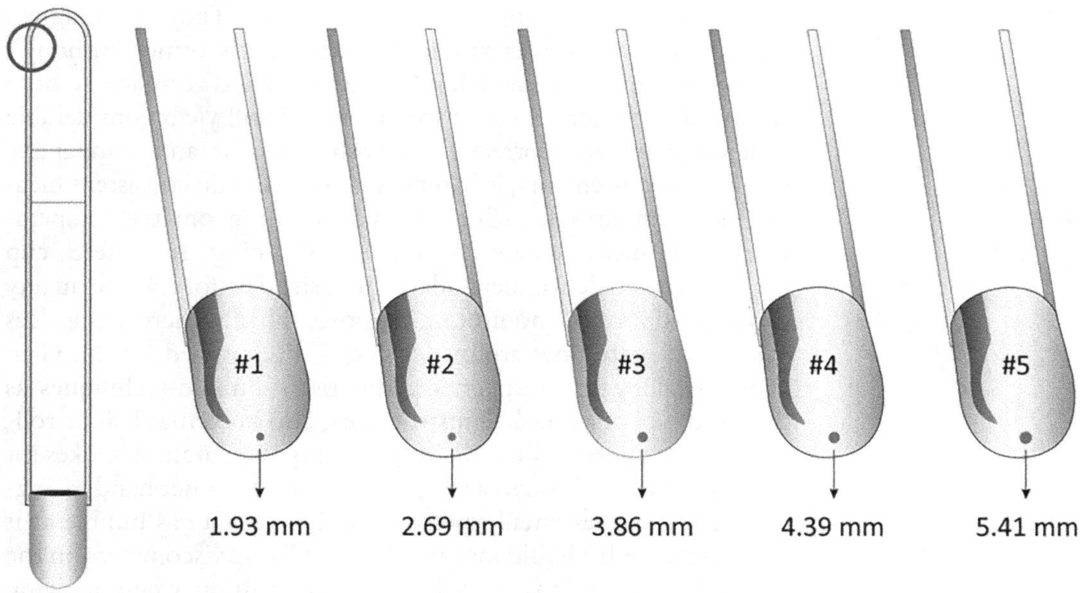

Fig. 14 Handheld production floor viscometers [25]

various designs provide diverse benefits, allowing scientists to adjust their rheological experiments to individual needs and budgetary constraints.

1.2.1 Flow Viscometers used in Production Floor in Industrial Settings

Three main types of industrial viscometers are used directly on the production floor in industrial settings. The first type consists of rotational devices, such as the Brookfield viscometer, offering some potential for consistent data interpretation. For this type of viscometer, the torque is measured while the sample is subjected to constant rotation using a spindle of specified geometry. With adherence to appropriate calibration and operational protocols, the Brookfield viscometer can deliver consistent and reliable viscosity data across diverse applications and industries. The second type involves "flow through constrictions" and is exemplified by arrangements such as the Ford cup and Zahn cup viscometers (Fig. 14). These handheld flow cup (efflux cup) viscometers often exhibit complex flow and stress fields despite their simple operation. These instruments are designed to allow quick and efficient monitoring and control of various processes to make informed decisions regarding process optimization, troubleshooting, and quality control, especially in the printing industry [24]. The test is simple and relies on the time for a specified volume of fluid to flow (under gravity) through a fixed orifice at the bottom of a shaped cup. While these instruments can be easy using standard liquids for Newtonian fluids, the measurement of viscosity becomes more challenging when dealing with non-Newtonian fluids. Theoretical derivations become intricate, and in certain cases, even complex mathematical approaches may fail to generate consistent viscosity data. They are designed particularly for materials with low to medium viscosities. Although they do not offer the same level of precision and accuracy as more advanced viscometers, cup viscometers can still yield considerable data when utilized correctly. To ensure reliable and comparable viscosity measurements, it is important to maintain consistent measurement conditions, including uniform cup geometry, temperature, and measurement protocol. By doing so, these cup viscometers can deliver dependable viscosity data for routine quality control and process monitoring purposes. Finally, there are devices that incorporate flow around obstacles, exemplified by the Glen Creston falling-ball instrument, and rising-bubble techniques as the third type. Falling-ball instruments, also known as ball or rolling ball viscometers, function by measuring the duration it takes for a ball to descend through a sample fluid. On the other hand, rising-bubble techniques entail assessing the speed of a gas bubble as it ascends through a liquid sample. When utilizing viscometers on the production floor, it is crucial to exercise caution when applying formulas designed for Newtonian liquids to non-Newtonian fluids. The complexity of non-Newtonian behavior requires careful consideration to ensure accurate viscosity measurements and interpretations.

1.2.2 Rotational Instrument Methods

Various types of viscometers use rotational motion to generate a simple shearing flow. There are two approaches employed by these instruments to generate the flow. It either involves driving a component and measuring the resulting couple, while the other involves applying a couple and measuring the subsequent rotation rate. In terms of applying rotation and measuring the resulting torque, two methods can be employed. One is driving and measuring the couple on the same component, or the second is driving and measuring the couple on separate components. In modern viscometers, the first method is utilized in instruments such as Brookfield, while the second method is employed in viscometers like Weissenberg and Rheometrics rheogoniometers [6]. For couple-driven instruments, the couple is applied to one member, and its rate of rotation is measured. The commercially available constant stress instruments can apply couples within the range of 10^{-6}–10^{-2} Nm, and the measurable shear rates range from 10^{-6} to 10^{3}/s. The specific dimensions of the instruments and the viscosity of the material determine the shear rates achievable. Similar to all viscometers, regular calibration and zeroing of viscometers using calibrated Newtonian fluids within the viscosity range of the measured materials are essential. The rapid viscosity analyzer (RVA) is another optimized rotational viscometry and mixing device for determining gelatinization and retrogradation of native and modified starch products while employing controlled heating and cooling cycles

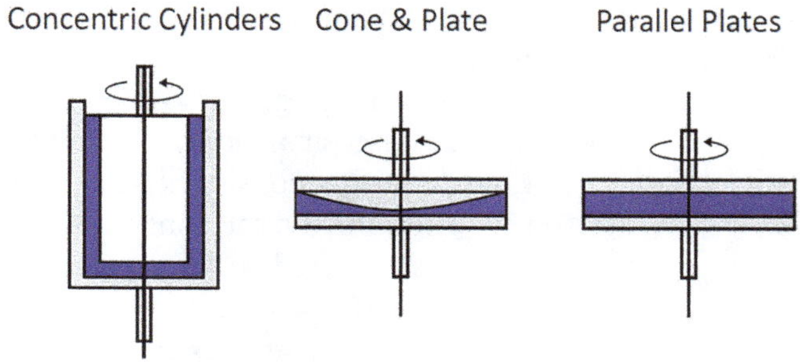

Fig. 15 Rotational geometries

to analyze samples. The alterations in torque offer valuable information regarding the viscosity of samples and how they behave during the pasting process. The viscosity is calculated as the "Newtonian equivalent" based on a calibration with a standard material [26].

The design of a rotational rheometer incorporates three primary geometries: concentric cylinders, cone and plate, and parallel plates (Fig. 15). These geometries offer several key advantages including: (1) circular closed streamlines allow for the generation of large steady-state strains; (2) the rotation can be precisely controlled, measured, and converted into strain; (3) the torque can be accurately controlled, measured, and converted into stress; and (4) a single instrument can accommodate all three geometries, providing versatility and convenience [27].

In standards such as ISO 3219-2 and comparable documents, measuring geometries are categorized into two groups: absolute measuring geometries and relative measuring geometries. Absolute geometries consist of measuring configurations such as coaxial cylinders based on Couette and Searle principles, double gap geometry, and cone-plate geometry. Conversely, relative geometries encompass plate–plate, coaxial relative cylinder measuring geometries with a radius ratio >1.1, as well as geometries featuring surface modification and special designs like rotors adhering to ISO 2555 and ASTM D562 guidelines.

Concentric Cylinder

Concentric cylinder systems, also known as coaxial, Couette, cup, and bob systems, are commonly employed for samples with lower viscosities that may not be effectively contained within the gap of cone and plate or parallel plate system. There are various types of concentric cylinder systems, including recessed end, conical end, vane, and double concentric designs. Concentric cylinders are classified into two based on the gap between the spindle: narrow gap and wide gap. When the gap between two concentric cylinders is sufficiently small and they undergo relative rotation, the liquid being tested within the gap undergoes an almost uniform shear rate. The shear rate in narrow gap viscometers is mostly governed by the rotating speed or angular velocity of the viscometer, rather than by the particular characteristics of the test liquid. This property of narrow gap viscometers allows for more regulated and uniform viscosity readings, making them suited for specific applications requiring consistent shear rates. To consider the gap as narrow, the ratio of the inner cylinder radius to the outer cylinder radius is reported to be exceeding 0.97. End effects are likely to occur if due consideration is not given to the different shearing conditions that may exist in any liquid covering the ends of the cylinders. The concept of end effect refers to the influence or change in the flow behavior that occurs near the boundaries or ends of a sample or measurement system. This effect arises due to the interaction between the sample and the measuring apparatus, particularly at the edges or surfaces. The presence of end effects can lead to non-uniform flow patterns, alterations in shear rates, or localized variations in viscosity close to the boundaries. Consequently, the accuracy and reliability of viscosity measurements can be impacted. It is essential to appropriately address and mitigate end effects to ensure meaningful and dependable rheological data acquisition. Fortunately, modern viscometer manufacturers ensure that the ratio of the liquid depth to the gap between the cylinders is designed so that end correction is considered negligible. Otherwise, to ensure accurate measurements with narrow-geometry concentric cylinder viscometers, it is recommended to vary the immersed length of the materials being tested while maintaining a constant rotational rate. The challenges of achieving parallel alignment and handling suspensions with large particles in narrow-gap concentric-cylinder viscometers led to the development of wide-gap viscometers as a solution to these limitations. In wide-gap geometry, the shear rate is determined by the properties of the test liquid, unlike in narrow-gap instruments, where it is independent of the liquid properties. Thus, the shear rate varies with the liquid's viscosity and flow behavior. The selection of either wide gap or narrow gap is contingent upon the specific needs of the viscosity measurement, the nature of the sample under examination, and the desired level of precision and manipulation of shear rates. Although the ranges can vary depending on the specific design and specifications, narrow gaps typically operate within a shear rate range of 10^0–10^5/s, while wide gaps cover a wider shear rate range of 10^{-2}–10^3/s. The typical sample size for concentric cylinder geometries is approximately 10 mL.

Cone and Plate

Cone and plate geometries are commonly employed for single-phase homogeneous samples or samples that contain submicron particles. However, particulate matter samples are generally unsuitable for cone and plate geometries. This is because the particles tend to migrate toward the apex of the cone and can become trapped in the truncation area, leading to inaccurate data. When dealing with liquids that have low viscosity, it is often necessary to employ high rotational speeds to generate measurable torques accurately. However, in such cases, secondary flow can occur, which absorbs additional energy and may increase the measured couple. This can be mistakenly interpreted as shear thickening. An empirical formula (Eq. 13) has been proposed that offers some assistance in addressing this issue [28], where T is actual torque (dyn cm), T_0 is theoretical torque assuming primary flow (dyn cm), $B(\theta, B)$ is a dimensionless function, Re is Reynolds number (dimensionless), and β is cone angle (radian). The study lists the values of the function B for $(Re/\beta)^2$ to be less than about 0.01, both in theory and experiment showing that the effect of the secondary flow is negligible.

$$\frac{T}{T_0} = 1 + B(\theta, B)\left(\frac{Re}{\beta}\right)^2 \tag{13}$$

In cone-and-plate instruments, it is crucial to reset the cone and plate properly so that the tip of the cone aligns with the surface of the plate. In the case of a 1° gap angle and a cone radius of 50 mm, a deviation of 10 μm in the axial separation results in an additional 1% error in the shear rate [6]. The cone and plate geometry is well-suited for materials that exhibit a strong dependence on shear rate. Typical sample sizes for cone geometries are approximately between 0.03 and 4.8 mL depending on the geometry diameter and angle.

Parallel Plate

The parallel plate technology can successfully measure particle-containing materials. One advantage of this method is the flexibility to change the gap distance, which eliminates particle size concerns. As a general rule, the gap size should be at least 10 times greater than the greatest particle size for particulate materials. For example, if the maximum particle size is 100 μm, then the optimum gap size should be set to 1000 μm. The non-uniform distribution of stress throughout the full diameter is a significant disadvantage of the parallel plate design. However, the software utilized by the rheometer providers typically compensates for this shortcoming. Otherwise, a simple correction method, known as the single-point method, can be employed, as described in the previous study [29]. In this study, researchers proposed a single-point correction technique that provides approximate values for material functions without the need for numerical differentiation. The single-point correction method offers the advantage of producing more accurate results compared to the apparent Newtonian values. Due to the nonhomogeneous nature of flow in the parallel plate geometry, it becomes necessary to differentiate the measured quantities (torque and normal force) with respect to the shear rate at the rim of the parallel plate. In sinusoidal oscillation tests using the parallel plate geometry, there is no need for correction. When these tests are conducted within the linear region, the results obtained from the parallel plate geometry are considered to be exact. Typical sample sizes for plate geometries are also approximately between 0.03 and 4.8 mL depending on the geometry diameter.

Selecting Geometry

Geometries for rheometers are typically fabricated using stainless steel, aluminum, or plastic materials. It is important for the geometry material to be chemically compatible with the test sample to prevent any potential corrosion issues. Despite its relatively high weight, stainless steel offers the advantage of having a low coefficient of thermal expansion. It is compatible with most test materials and possesses robustness that allows it to endure heavy usage. On the other hand, aluminum, while lighter, has a higher coefficient of thermal expansion and is more limited in terms of chemical compatibility. This lower weight, however, results in reduced inertial effects. Plastic geometries such as acrylic, polycarbonate, and rigid PVC are typically transparent, enabling visual observation of the sample's behavior. Plastic geometries are notably lighter compared to metallic ones. Acrylic and polycarbonate, being relatively lightweight, present fewer inertial challenges but have some limitations in terms of chemical compatibility. It is important to avoid using plastic geometries above temperatures of 40 °C [30]. The cone and plate and parallel plate geometries have numerous benefits in common, including the need for minimal sample quantities, ease of cleaning, low inertia, and the potential to attain high shear rates. Regarding the diameters of cone and plate as well as parallel plate geometries, they commonly exist in sizes ranging from ~8 to 60 mm. The smaller geometry is generally appropriate for handling stiffer and high-viscosity materials, while the larger geometry is better suited for managing materials with low viscosity. A smaller angle (in the case of a cone and plate) or a narrower gap (in the case of a parallel plate geometry) allows for higher maximum shear rates. As the diameter of both cone and plate or parallel plate geometry decreases, the shear stress increases. The cone and plate combination has the extra benefit of ensuring equal shear rate distribution across the sample, whereas the parallel plate design can accept bigger particles. For both geometries, it is important to properly load the sample (Fig. 16a) and exercise caution to avoid underfilling or overfilling the geometry to prevent edge fractures [27]. The geometry is subsequently lowered until it reaches the gap, where the sample is expected to exhibit a slight bulge, signifying the proper fill (Fig. 16b). In flow tests, the occurrence of edge fracture (Fig. 16c) or instability can lead to misleading data that appears to exhibit shear thinning behavior. However, the actual cause of this phenomenon is the sample escaping from the edges, rather than genuine shear thinning properties. As the shear rate, viscosity, or normal stress variations increase, edge fracture

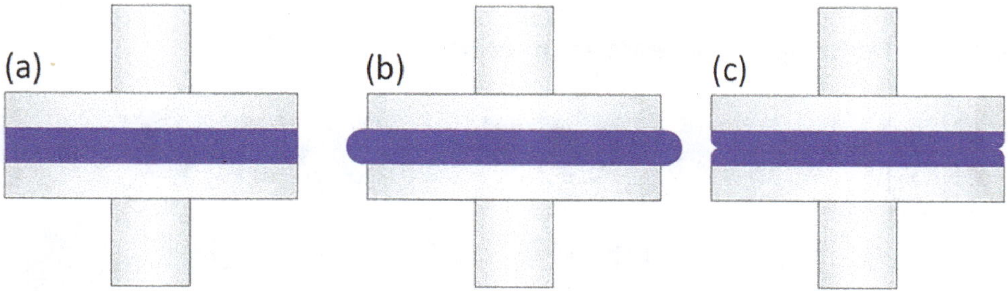

Fig. 16 (**a**) The optimal correct loading that has axially straight edges, (**b**) bulging edge, and (**c**) edge fracture

may occur. It may also occur at low rates, frequently below 1/s [27]. In cases where the gap is overfilled, there is a possibility of the excess sample migrating and accumulating on the surface of the geometry. However, this occurrence is less likely when dealing with low-viscosity samples. On the other hand, if the gap is underfilled, it leads to a modification in the diameter of the geometry. This alteration unavoidably introduces significant errors, making it crucial to avoid such a situation.

Rheometry Calibration and Considerations

It is worth mentioning that ensuring precise measurements is a significant concern in rheometry. To verify the proper functioning of a rheometer, one can measure the viscosity of a certified standard Newtonian oil as a monthly practice. Examples of such oils include Cannon S600, which has a nominal viscosity of 1.4 Pa·s at 25 °C, or PTB 1000A, which has a nominal viscosity of 1 Pa·s at 20 °C. These oils are chemically stable and low in volatility. An experimental procedure recommended by TA Instruments entails utilizing a 60-mm 2° stainless steel cone attachment on the rheometer to measure the oil and the rheometer should be set to have a zero gap. A 4-min flow test, employing a continuous ramp and controlling the stress range from 0 to 88 Pa at 20 °C, should be conducted to determine the Newtonian viscosity after ensuring accurate filling of the sample. If the Newtonian viscosity deviates by more than 4% from the certified value, it indicates the potential need for attention from the service provider for the rheometer. It is important to highlight that there are various potential sources of operator error that can lead to inaccurate results. However, these errors do not necessarily indicate a malfunctioning instrument. They can arise from factors such as improper gap setting, incorrect temperature settings, or over- or under-filling of the gap. When choosing an appropriate measurement geometry, it is crucial to consider the type of experiment to be conducted, the nature of the sample, the presence of particles, and the real-world scenario intended to be replicated. For the latter, there are certain general recommendations that might help such as commonly used shear rates in materials and processes presented in Table 3 [18]. If a

Table 3
Commonly used shear rates in materials and processes

Situation	Shear rate range	Application
Sedimentation powders in suspending liquid	10^{-6}–10^{-4}	Medicines, paints
Leveling	10^{-2}–10^{-1}	Printing inks, paints
Draining under gravity	10^{-1}–10^{1}	Coating, painting, bleaches
Extruders	10^{0}–10^{2}	Polymers
Chewing and swallowing	10^{1}–10^{2}	Foods
Dip coating	10^{1}–10^{2}	Confectionary, paints
Mixing and stirring	10^{1}–10^{3}	Manufacturing liquids
Pipe flow	10^{0}–10^{3}	Pumping, blood flow
Spraying and brushing	10^{3}–10^{4}	Spray-drying, painting, fuel atomization
Rubbing	10^{4}–10^{5}	Creams and lotion application to the skin
Milling pigments in fluid bases	10^{3}–10^{5}	Printing inks, paints
High-speed coating	10^{5}–10^{6}	Paper
Lubrication	10^{3}–10^{7}	Gasoline engines

constant shear rate is desired, suitable options include the narrow-gap concentric cylinder or the small-angled cone and plate viscometers. When the shear stress varies inversely with distance, the wide-gap concentric cylinder can be selected as a viable option. For situations where there is a linear change in shear rate or shear stress, the parallel plate, capillary, or slit viscometers are appropriate choices.

1.2.3 Capillary and Slit Viscometers

Although both capillary and slit viscometers are used to measure viscosity, their construction and operation are different. In a capillary viscometer, the material flows through a thin tube, or capillary, of a defined length and diameter. Here material flow occurs principally as a result of the pressure differential and resistance. The flow is restricted to the inside of the capillary; hence, capillary viscometers are generally cylindrical or tapered in shape. Capillary viscometers are frequently employed particularly in lab settings to measure the viscosity of liquids and solutions. Slit viscometers feature two fixed, parallel walls with a small opening between them. Slit viscometers analyze the flow between the parallel sides while applying a pressure gradient to determine the viscosity. The sample flows in a flat, constrained channel between the parallel walls of slit viscometers with a planar shape and therefore is two-dimensional. The rheological characteristics of complicated fluids, such as polymer melts, liquid crystals, and suspensions, are frequently studied with slit viscometers. Capillary rheometry is the predominant method employed to investigate the rheological properties of thermoplastic starch (TPS). The starch polymer melt exhibits some characteristics similar to those of conventional synthetic polymers, and therefore, it is commonly referred to as TPS in the literature.

1.2.4 Rheoscope

A rheoscope is a high-tech rheometer that combines controlled rate/controlled stress capabilities with an integrated high-resolution video camera with a microscope lens and light source. It functions as a cone plate viscometer, applies controlled shear forces for the precise management of the shear rate, and is capable of observing the response to shear using either a microscope or video. The camera collects one image for each data point in the rheological measurement at a rate of 15 or 30 per second depending on the specifications of the rheoscope and the computer used. This novel approach enables the measurement of viscosity under shear stress while also seeing variations in starch particles.

1.2.5 Systemic Rheology

Systemic rheology involves transforming conventional geometries of rheometers into stirred microreactors, allowing certain steps of a process to be reproduced to some extent or even closely mimicked to investigate rheological behavior under conditions closely resembling real-world processing [31]. The approach involves utilizing the rheo-reactor, which necessitates replacing standard geometries with stirred microvases operating in batch or semi-batch mode. The setup allows for performing mixing procedures and in-situ rheological monitoring that can be challenging to achieve with complex fluids (*see* Subheading 2.1.4). Moreover, complex materials can have a microstructure that matches the size of the gaps present in conventional geometries (cone-plate, parallel plates, and Couette). Such samples frequently face difficult experimental hurdles during conditioning and analysis, such as material structural changes, slippage, sedimentation, or characterization during an uncontrolled pre-strain state, which are all resolved by the rheo-reactor design. The systemic rheology method requires using a software with special geometry option that involves computing the stresses and deformations in a Couette system.

2 Materials

2.1 Starch Overview

Starch is a complex matrix with structural and chemical complexities, commonly found in a semi-crystalline state, whether in the form of granules or retrograded starch [32]. Starch granules serve as the primary storage structure for carbohydrates in plants. Green plants primarily store carbohydrates in the form of starch, which is predominantly found in seeds (corn, wheat, rice, barley), roots (potato, tapioca), tubers (arrowroot), stems, leaves, and fruits. From a chemical standpoint, starches consist of two distinct molecules, namely, amylose, which makes up approximately 20–30% of the starch, and amylopectin, which accounts for around 70–80% of the starch composition. Both amylose and amylopectin influence the rheology and structure of starch [33]. It is worth noting that plants also incorporate certain proteins within their starch structure. The impact of water on the amylose and amylopectin molecules within a starch granule differs depending on its temperature. While cold water has minimal effect on these molecules, hot water significantly alters them.

Starch is typically obtained as a dry, granular powder, available in either a modified or unmodified (natural) form. Unmodified starch is frequently unsuitable for practical applications without undergoing appropriate modifications. Although it is economically attractive, its high molecular weight (>300 million g/mol) needs to be reduced by cooking depending on the application [34]. Therefore, the use of modified starches with enhanced functional properties is more common. Modification processes, including mild degradation, crosslinking of chains, acid-thinning, thermal modification, thinning/hydroxy ethylation, derivatization with phosphate or esters, and pre-gelatinization can be applied based on specific needs [31]. The modification methods can be of a physical, chemical, or biochemical nature and can be performed in an aqueous suspension solution or through semi-dry processes like extrusion [1]. Starch, being composed of poly-glucose, contains numerous hydroxyl groups that are highly amenable to chemical substitution. This allows for the introduction of side groups along the chain structures of starch and improves the clarity and stability of the gel during cooking, mixing, and freezing. Table 4 presents some of the application areas for modified starch [1, 3, 5].

The solid or semi-solid phase typically consists of swollen granules or fragments ranging from 300 μm to less than 1 μm in diameter, depending on the source of the starch [35]. When starch-containing items are cooked, an irreversible process known as gelatinization occurs and leads to significant changes in texture and properties. Again, the characteristics of starch gels are influenced by various factors, including the source of the raw starch, environmental conditions such as seasons, and any modifications applied to the starch. At temperatures over 60 °C (140 °F) and beyond, starch undergoes a structural transformation. The organized crystalline layers of starch granules start to experience a process of melting with the heat applied. The granules absorb water, causing the starch to swell and rupture. This causes the starch molecules, both amylose and amylopectin, to be released and dispersed in the surrounding liquid. As a result, the mixture thickens and forms a gel-like consistency. Due to its unique

Table 4
Modified starch for different application areas

Modification	Application
Viscosity adjustment	Wet-end additive
Stability improvement	Coating and surface sizing
Adhesion and binding improvement	Ceramics
Change in solution consistency	Fabric printing
Adaptation in ionic character	Finishing additive
Improvement of paste clarity	Convenience products
Introduction of specific properties	Confectionery
– Rate of dissolving – Fermentation resistance	Extruded products Dairy products
– Increased salt tolerance	Infant foods
– Chemical reactivity	Bakery products
– Thermal stability	Preserved foods
– Hydrophobicity	Oil ang gas drilling
Building industry	Water purification
Mining operations	Agglomeration
Soil erosion	Dry batteries
Biodegradable plastics	Personal care
Self-healing material	Antimicrobial agent
Feedstock	Porous foam structures
Pharmaceuticals	Drug delivery
Moisture barrier	Packaging, papermaking, printing

viscoelastic properties, starch plays a significant role as a vital component in numerous industries benefitting from the binding, thickening, and stabilizing properties offered by starch, which is commonly subjected to modifications with the objective of improving its functional properties [36].

The utilization of starch can present challenges during the cooking process, including issues such as gelling, incomplete starch granule expansion, increased viscosity upon cooling, bacterial growth in cooked starch, and difficulties in controlling viscosity during storage. To address these challenges, a novel type of nanoparticle starch has been introduced, offering various advantages [34]. The literature describes the method for producing starch nanoparticles by employing shear force and cross-linking agents simultaneously in a high-temperature extruder during the

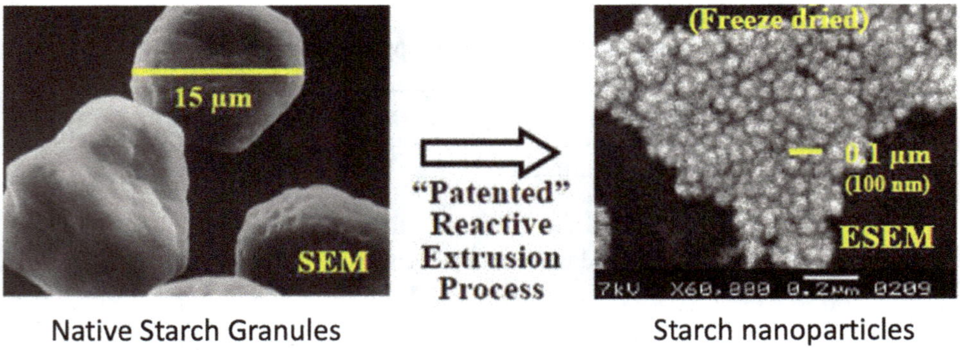

Fig. 17 Particle size comparison of natural starch and starch nanoparticles [34]

manufacturing process [37–39]. The average size of starch-based biopolymer nanoparticles, which can be produced at an average of one metric ton per hour in the extruder, is between 20 and 100 nm. The starch nanoparticles are provided in a dry state and can be readily dispersed in water without the requirement of cooking. Their utilization brings about advancements in quality, cost reduction in manufacturing, improved processability, and the opportunity for developing formulations with higher solids content. The performance of starch nanoparticles surpasses that of conventional cooked coating starches [40] (Fig. 17).

2.1.1 Starch Sources and Rheological Behaviors

Rheology finds more appropriate applications in various materials, including solutions, emulsions, suspensions, foams, gels, melts, and semisolids such as creams and pastes. Putting aside technical terminology, the study of starch rheology primarily entails analyzing starch in different states such as air-dry starch, starch suspensions, starch paste, cooked pastes, cold pastes, pastes and gels of pre-cooked starch, hydrolyzed and oxidized starches, dextrins, and syrups, as extensively detailed in previous literature [35]. Overall, rheological testing methods offer a distinctive advantage as they not only assess viscosity but also directly measure properties such as strength, rigidity, and brittleness. Different rheological techniques can be employed to analyze the properties of starch, with the most sensitive methods including oscillatory testing for determining the critical gel point, swelling behavior in relation to equilibrium modulus, and creep testing at constant stress to observe the rate of deformation [41]. The degree of gel swelling is connected to the gel's equilibrium modulus, with more rigid gels exhibiting less swelling compared to more pliable gels [41].

The pioneering work of Wolfgang Oswald involved the initial measurement of starch viscosity during the pasting process where he defined the pasting temperature as the specific temperature at which the viscosity starts to increase [31]. As the temperature rises, the granules undergo rupture, leading to the release of amylose into the solution, followed by amylopectin. Both granule rupture and polymer alignment caused by mechanical shear contribute to a reduction in viscosity. This subsequent process, occurring after gelatinization, is commonly referred to as pasting. Upon cooling, the starch molecules undergo reassociation, resulting in the formation of a gel. This gel formation typically leads to an increase in viscosity, ultimately reaching a final viscosity. During this stage of the pasting curve, the starch molecules undergo retrogradation or reordering. The final viscosity, which represents the viscosity at the end of this process, is widely utilized as a key parameter for assessing the quality of a specific starch sample.

Cooked starch is commonly characterized as non-Newtonian, showing substantial shear thinning in its rheological behavior. The literature indicates that uncooked starch slurries, comprising 40–50% starch by weight, exhibit dilatant behavior [42]. Furthermore, many starches exhibit thixotropy, a phenomenon where their viscosity decreases over time with variations in shear rate. Consequently, there is no universally fixed measurement for starch viscosity. The final viscosity is influenced by factors such as heating and cooling rates, duration of holding, and shear rate. The rheological characteristics of starch are governed by numerous other factors, each playing a significant role in shaping its behavior. These factors include the chemical structure of starch, concentration, pasting, and storage conditions [33]. The chemical structure of starch itself is an important variable because it affects how starch molecules interact with one another and how they are arranged in a paste or gel matrix. The rheological properties of starch are also influenced by the concentration of starch, commonly referred to as solid content, employed in the formulation. Higher concentrations of starch tend to result in a thicker consistency and an overall increase in viscosity. Pasting conditions, including temperature, shear rate, and heating rate, have a significant impact on the rheological properties of starch-based systems. Temperature plays a crucial role in the swelling and gelatinization of starch granules, influencing the viscosity and flow behavior of the resulting paste or gel. Shear rate, on the other hand, influences the degree of structural breakdown and alignment of starch molecules, affecting the overall viscosity and texture. Furthermore, the heating rate used during the pasting process can influence the gelatinization of starch and subsequent retrogradation, leading to alterations in its viscoelastic properties [36]. Storage conditions, including temperature and time, are important factors to consider when dealing with starch-based pastes and gels. Over time, these materials can undergo various changes, such as retrogradation and syneresis, which can significantly affect their rheological behavior.

Temperature fluctuations during storage can have an impact on the stability and viscosity of starch systems. Therefore, it is important to pay close attention to the storage conditions to guarantee

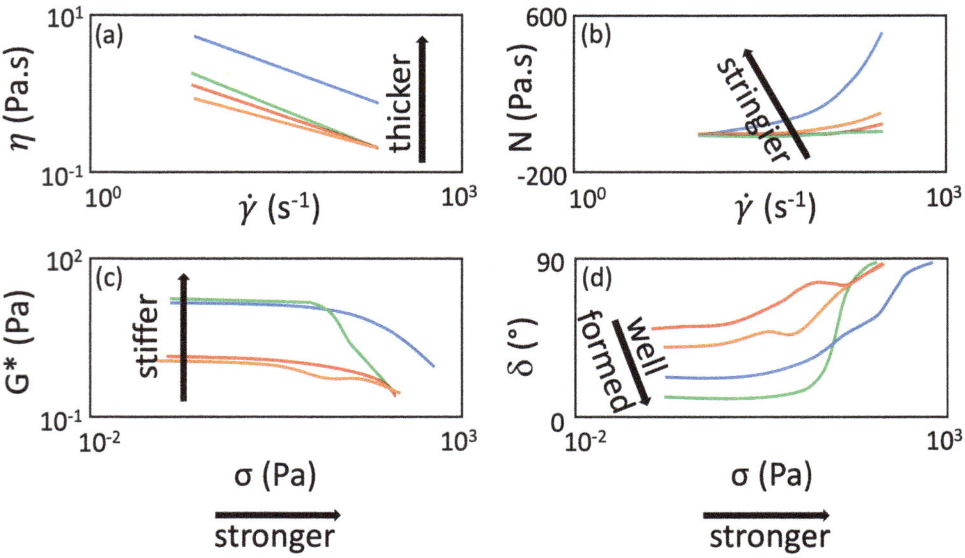

Fig. 18 5% w/v gelatinized starch (• potato, • dent maize, • waxy maize, and • tapioca)

the stability and required rheological characteristics of goods made from starch. Given the multitude of factors that influence starch rheology, it is crucial to maintain consistency and comparability in test results. This can be achieved by adhering to standardized test procedures when comparing different starch samples. An example showcased in Fig. 18 by the Centre for Industrial Rheology lab demonstrates the wealth of data that can be obtained solely from an oscillation stress sweep test conducted on various starches gelatinized at 60 °C. In the analysis, viscosity was observed in relation to shear rate across a wide range of shear rates (Fig. 18a). The starch from the potato sample stood out with consistently higher viscosity compared to the other samples. The other starches exhibited similar viscosities at higher shear rates. At lower shear rates, there were noticeable differentiations between the samples, suggesting that these differences would likely become more pronounced at even lower shear rates. In the normal stress data (Fig. 18b), it was found that the potato starch sample stood out and continued differentiation between the samples that exhibited similar viscosities. Further exploration of the oscillation stress view data (Fig. 18c, d), particularly the complex modulus and phase angle data, revealed differences in rigidity and structural levels among the samples. Of particular interest was the fact that despite having distinct groupings of complex modulus data representing different levels of stiffness, the yielding behaviors of the samples varied significantly. This was exemplified by the dent maize and potato starch, which both yielded at a similar point but in distinct ways, highlighting the complex and dynamic nature of the yielding process.

Researchers have taken varied approaches, utilizing distinct methodologies as outlined in Table 5. In order to gain a comprehensive understanding of the viscoelastic properties of starch, researchers commonly employ a range of rheological testing techniques. These include utilizing dynamic rheometers [33, 43] to assess behaviors through oscillatory [44], controlled stress/controlled rate [45], strain-controlled [46], and concentric cylinder [35] methods and capillary rheometers [47]. Dynamic rheometers allow for the comprehensive study of the response of starch pastes and gels to dynamic forces, enabling the characterization of their elastic and viscous behavior. By subjecting the sample to controlled deformation and analyzing the resulting stress and strain responses, a dynamic rheometer provides valuable data on the mechanical properties, viscoelasticity, and flow behavior of starch-based systems. The capillary method has been used to investigate the rheological properties of starch nanoparticles [40], TPS [47], and maize starch [48]. The studies utilized a wide range of starch sources, including oat [44], corn [22], banana [49], wheat [50], corn starch nanoparticles [40], rice [43], mungbean [51], wheat/barley [52], mango kernel [53], pearl millet [54], black rice [55], Pongamia pinnata seeds [56], avocado seed starch [57], and tapioca [58].

In numerous studies, when subjecting starch to higher shear rates, the shear stress exhibited an initial rapid increase followed by reaching a plateau [43–45]. This behavior is typical of shear thinning, a common trait observed in pseudoplastic fluids. The mechanism behind this phenomenon involves the disruption during the shearing process of the network formed by interlocked polysaccharide molecules. With increasing shear rate, the disruption of these intermolecular entanglements outpaces their reformation, resulting in a decrease in the apparent viscosity of the fluid. It is also reported that the high shear thinning behavior of all starch samples could be associated with the greater destruction of the intra- and intermolecular bonding systems in the starch network due to shearing at higher shear rates [46]. In one study, concentrated suspensions of starch exhibited dilatancy. Starch milk with a dry matter content of 40% or higher posed challenges for pumping as they became difficult to handle and demonstrated irregular flow, even at low velocities [35].

Most studies conducted a comprehensive analysis of the rheological properties of pastes using the Herschel-Bulkley model within a shear rate range of 0–1000/s. In one study, a high coefficient of determination ($R^2 = 0.99$) was found, indicating a good fit of the model to the experimental data [33]. In the same study, the flow behavior index (n) that quantifies the extent of shear thinning as it deviates from 1, ranged from 0.41 to 0.50 and 0.59 to 0.94, indicating a substantial degree of shear thinning in the starch pastes. The consistency index (K), a rheological parameter

Table 5
Oscillatory and steady shear procedures adapted in the literature

Starch source	Solid content (%)	Method	Geometry	Gap (μm)	Strain frequency	Temp (°C)	Frequency sweep	Steady shear (/s)	Model	References
Oat	15 w/w	Oscillatory Steady shear	Parallel plate 4 cm diameter	1000	2%, 10 Hz	45–95	0.1–100 rad/ s at 25 °C	1–1000	Herschel-Bulkley	[44]
Potato	3 w/w	Steady shearretrogra dation and freeze-thaw property strain sweep test	Parallel plate 50 mm diameter	500	2%, 0.63 rad/s	4	0.63–63 rad/ s at 4 °C	0.01–1000	Power-Law Casson	[46]
Wheat	10 w/w	Oscillatory Steady shear	Parallel plate 4 cm diameter	1000	2%, 10 rad/s	45–95	0.1–100 rad/ s at 25 °C	1–500	Herschel-Bulkley	[50]
Starch NP	14, 20, 23, 28, 33	Stress rheometer High shear viscometer Ultra-high shear capillary And slit viscometer				24		1–1000 10,000–100,000 10^5–10^6 and 1–3 million		[40]
Corn	10 w/w	Oscillatory	Couette C25 bob		0.7 pa	95 95 to 5	0.01–20 Hz			[22]
Rice	20 w/w	Oscillatory	Parallel plate 4 cm diameter	1000	0.5% 1 Hz	45–85				[43]
Mungbean	10 w/w	Oscillatory	Parallel plate 4 cm diameter	1000	2%, 10 rad/s	45–95	0.1–100 rad/ s 25 °C 2%			[51]
Wheat/ barley	10 w/w	Oscillatory Steady shear	Parallel plate 4 cm diameter	1000	2%, 10 rad/s	45–95	0.1–100 rad/ s 25 °C 2%	1–500	Herschel-Bulkley	[52]
Mango kernel	15 w/w	Oscillatory Steady shear	Parallel plate 4 cm diameter	1000	2%, 10 rad/s	45–95	0.1–100 rad/ s 25 °C 2%	1–1000	Herschel-Bulkley	[53]

Pearl millet	8 w/w	Oscillatory Steady shear	Parallel plate 4 cm diameter	1000	2%, 10 rad/s	45–95	0.1–100 rad/s 25 °C 2%	1–500	Herschel-Bulkley	[54]
Black rice	10	Oscillatory Steady shear	Parallel plate 4 cm diameter	1000			0.1–100 rad/s 25 °C	1–1000	Herschel-Bulkley	[55]
Pongamia pinnata	6 w/w	Oscillatory Steady shear	Parallel plate 4 cm diameter	1000	2%, 10 rads/s		0.1–100 rad/s 25 °C	1–1000	Herschel-Bulkley	[56]
Tapioca	5 w/w	Oscillatory Steady shear	Parallel plate 60 mm diameter and starch pasting cell	1000	2%	30–95 95 to 30	0.01–10 Hz, 0.5% at 25 °C	0.1–100 (2°, 60 mm parallel)	Power-Law	[59]
Corn	7.5	Oscillatory Steady shear	Starch pasting cell		30 rads/s	25–120 120 to 25	0.01–10 Hz, 0.5%	0.1–100	Power-Law Herschel-Bulkley	[60]
Tapioca	6	Oscillatory	Cone and plate 1°, 50 mm diameter	50	10 rads/s	25	0.5% strain			[61]
Wheat	8–15	Steady shear	MV cup and MV-II bob			23 and 60		1–1000	Power-Law	[62]
Banana	4 w/v	Oscillatory Steady shear	Cone and plate 4°, 60 mm diameter	100,000	10%	60 (steady) 25 (strain)	0.01–100 rad/s	1–300		[49]
Tapioca (cold soluble)	6 w/w	Oscillatory Steady shear	Cone and plate Parallel plate Vane	1000 150	0.01, 1 Hz	25–95 95 to 25	0.01–10 Hz	0.001–1000	Power-Law	[16]

reflecting the viscosity values, varied between 0.06 and 3.09 Pa·s. A higher value of K signifies greater structural strength, resulting in less pronounced thixotropic behavior. The storage modulus (G′) in general is consistently higher than the loss modulus (G″) in all starches, indicating a more elastic response compared to a viscous behavior [33]. Similar to the Herschel-Bulkley model, in another study the Casson model was also found to exhibit high accuracy ($R^2 = 0.99$) [46]. Overall, in studies on starch rheology, the Power Law, Casson, and Herschel-Bulkley models are commonly used for best fitting [53–55]. Specific to TPS, the rheological behavior is adequately described by the Power-Law model, primarily attributed to the progressive reduction of intermolecular bonds within the starch structure during shearing [47]. Comprehensive information on the rheology and rheometry of starch polymer melts can be found in the review paper by Xie et al. [59]

For starch nanoparticles, rheological characterizations are carried out using a capillary viscometer, stress rheometer, high-shear rheometer, and an ultra-high capillary and slit rheometer [40]. The stress rheometer was used to measure the low shear viscosities of pure starch nanoparticles and cooked starch solutions at solid contents of 14.3%, 20.1%, 23.8%, 28.3%, and 33.2%. The starch nanoparticles in the study had levels of low, medium, and relative crosslink density. Among the measured solids, the starch nanoparticles and the conventional cooked starch binder all exhibited shear-thinning behavior. On the other hand, the medium crosslinked starch nanoparticles demonstrated intermediate viscosity relative to the cooked starch solution at the same solid concentrations. Once again, when the samples with the same solid content were evaluated using the high shear rheometer, both the starch nanoparticle dispersions and cooked starch solutions displayed shear thinning behavior. Lastly, 25% and 35% starch nanoparticle dispersions were measured at shear rates exceeding 105/s using a slit or capillary configuration on an ACA viscometer (ACAV). The ACAV was reported to comprise a movable cylinder capable of operating under pressures up to 400 bar. Samples within the cylinder are forced through a small capillary or slit, allowing for the attainment of viscosities and shear rates ranging from 105 to 106/s, depending on the flow rate and the dimensions of the slit or capillary gap. The shear rates used in the study were reported to be commonly encountered in industrial coating processes utilizing rod or blade coaters. The capillary dimensions used were 0.5 mm in diameter and 50 mm in length, while the slit dimensions were 10 mm in height, 0.095 mm in width, and 0.5 mm in length. Due to the narrower gap in the slit, higher shear rates could be achieved for the respective samples. The study highlighted that the distinct geometries of the capillary and slit configurations result in different rheological behaviors, with the slit configuration being more suitable as a model for blade coating. The outcomes of the capillary tests conducted on the 25% and 35% starch nanoparticle dispersions revealed that the starch nanoparticles displayed shear-thinning characteristics throughout the measured shear range. A clear correlation was observed between the starch swell ratio and viscosity in the samples: Starch nanoparticle samples with higher swell ratios exhibited higher viscosity, whereas those with lower swell ratios displayed lower viscosity. Raising the solids content of the starch nanoparticles was found to result in increased viscosities. Nevertheless, the consistent trend persisted, showing that a higher swell ratio was associated with higher viscosity. The rheological properties were measured at higher shear rates (1–3 million/s) using a slit attachment. Similar trends to those observed with the capillary results were apparent, with the starch nanoparticle samples exhibiting higher viscosity as the swell ratio increased. Despite the internally cross-linked starch nanoparticle samples having equal or greater effective volumes compared to the 50% XSB latex samples, the bio-samples exhibit shear thinning behavior, in contrast to their XSB counterparts, which display shear thickening behavior. The overall results presented suggest that starch nanoparticles have the potential to surpass the fundamental rheological properties and practical high-speed coating runnability performance of conventional cooked coating starches [40].

2.1.2 Investigating Phase Transition in Starch

A novel approach using a modified Haake rheometer with a twin-roll mixer was introduced in the previous literature [63]. The method involved measuring torque variations with time and temperature to investigate both the gelatinization process and rheological properties of starch under shear stress conditions. The study observed two distinct torque peaks for pure starch. The initial peak, which was sharp, was attributed to the loading process. Subsequently, the torque increased with time, indicating the phase transition of starch granules under shear stress. The second measured torque was found to be influenced by factors such as water content, mixing speed, and initial temperature. Another study used a rheoscope to investigate the gelatinization process of corn starch samples with different amylose contents. This was the first instance where viscosity measurements and microstructural variations were recorded simultaneously [64]. The study established a direct relationship between viscosity and changes in starch granules during gelatinization under shear stress. It was observed that starches rich in amylopectin gelatinized at lower temperatures, leading to complete destruction of the starch particles. The gelatinization process under shear stress conditions involved two steps: swelling of the starch granules and gradual dissolution of the outer granule layer, followed by rapid complete particle dissolution. Currently, there are other implementations, like Modular Microscope Accessory

(MMA) from TA Instruments specifically created for the Discovery Hybrid Rheometer allowing for the simultaneous execution of precise flow control and measurement, in conjunction with high-resolution optical microscopy [65].

2.1.3 Materials to Minimize Evaporation Losses

In situations where the samples are water-based or contain volatile solvents, the evaporation that occurs during measurements can present difficulties. To address this issue, a solvent trap cover is utilized, positioned above the geometry while ensuring no direct contact is made with the sample. The trap allows adding a small amount of the suitable solvent, enabling the surrounding space near the sample to become saturated with solvent vapor. This saturation effectively inhibits evaporation from taking place. If the solvent trap system is planned to be utilized during an experiment, it is generally recommended to execute inertia correction with solvent present in the trap, but without any sample loaded.

Other methods found in the literature regarding starch measurement include covering the sample with a thin layer of low-density silicone oil [43, 44, 50, 53] and light paraffin oil (to prevent evaporation during measurements) [16]. However, certain standards advise against the utilization of silicone oils due to their tendency to contaminate tools, containers, and other equipment [66, 67]. Further highlighting the need to avoid their usage is the possibility that silicone oils may display shear-thinning behavior at high shear rates [67].

2.1.4 Advanced Equipment

The viscoelastic properties of starch have been characterized using oscillatory, controlled stress/controlled rate, strain-controlled, and concentric cylinder rheometers. Recent developments however suggest that traditional rotational rheometers with viscometry test geometries are not suitable for testing starch suspensions due to their specific nature requiring continuous mixing during measurement to prevent sedimentation [31]. Additionally, due to the non-Newtonian nature of starch dispersions, conventional viscosity calibration methods are considered insufficient for conducting more comprehensive tests. Furthermore, controlling the water content during the heating/cooling cycle poses a challenge in the analysis of starch properties. A mixing chamber called a starch pasting cell (SPC) that minimizes water losses during the cooking cycle has been incorporated into rheometers in order to analyze and characterize starch in a more comprehensive manner (Fig. 19) [31]. The SPC employs an innovative impeller design including heating and cooling rates of up to 30 °C/min to facilitate mixing, minimize water loss, and regulate sedimentation throughout the testing process. With the utilization of the SPC, the conventional rheological methods involving steady, transient, and oscillatory measurements remain accessible for evaluating starch characteristics throughout

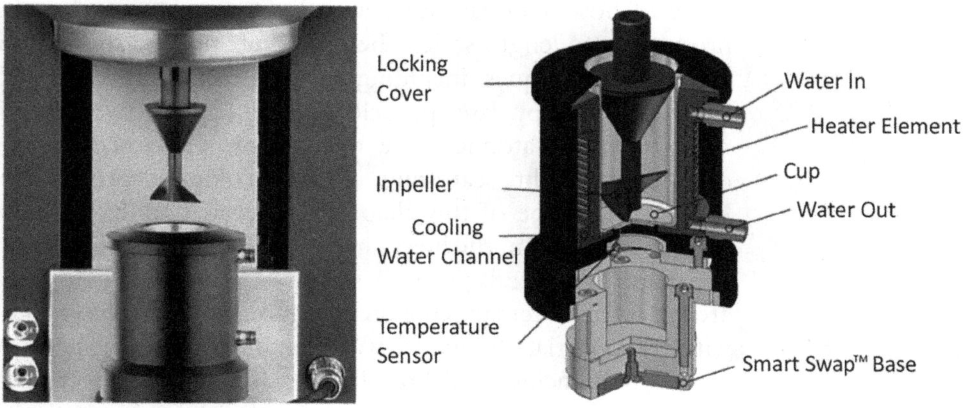

Fig. 19 The SPC system offered by TA Instruments

the gelatinization process of unprocessed and modified starch products, along with examining the attributes of the resulting starch gels.

By integrating the SPC cell with a sensitive rheometer, it becomes possible to conduct not only standard heating/cooling cycles but also advanced rheological characterization of starch gels using oscillatory testing methods. This allows for the evaluation of strain (rate) and stress coefficients, providing an analysis that considers the appropriate shear rate and stress. Furthermore, the use of a mixing element to measure torque, rather than an ideal "viscometric flow" geometry typical in rotational rheometers, enhances the accuracy of measurements. Conventional SPCs are limited to temperatures up to 95 °C. However, the latest pressure cell designs offer enhanced sensitivity at low torques, enabling measurements at temperatures from −5 to 105 °C. A recent research article covers pressurized SPC design (up to 5 bar) and evaluates the starch dispersion rheology at 120 °C, where the starch exhibited shear-thinning and significant thixotropic behavior [60].

New approaches such as microrheology for characterizing materials at the microstructural level have also evolved as a result of equipment developments. These techniques enable a better understanding of complicated gel structures and dynamics by considering numerous lengths and timeframes. Traditional techniques in rotational or capillary viscometers primarily assess the relative movement of entire macromolecular chains and layers within the medium. This characterization referred to as macrorheology or macroviscosity focuses on the overall motion of the system. In microrheology, the speed of diffusion for small molecules (probes) within gel-like media is frequently influenced by both the characteristics of the solvent within the gel matrix and the segmental motion of macromolecules [68]. Viscosity determined from the diffusion rate of probe molecules in a medium containing biopolymer is commonly termed microviscosity. Microrheology is the study of how materials store and dissipate mechanical energy in proportion to length scale. Microrheology may be studied using a variety of techniques, including particle tracking rheology, diffuse wave spectroscopy, laser particle tracking (optical tweezers), magnetic tweezers, atomic force microscopy, piezorheometers, and quasi-elastic light scattering. Microrheological procedures are beyond the scope of this chapter; however, they have been adequately discussed in the literature [41, 69]. This chapter focuses on the material's overall properties, known as macrorheology, which are expected to remain consistent regardless of the geometry used within appropriate limits. However, in microrheology, geometry is crucial, influencing spatial resolution, confinement effects, boundary conditions, material response, and the ability to capture accurate data at small scales. Researchers must thoughtfully select the geometry based on the materials being studied to derive meaningful conclusions in microrheological studies.

3 Methods

3.1 Standard Procedures and Protocols

While the specified global standards might not be specifically tailored for starch, they still provide consistency and comparability and facilitate inter-laboratory comparisons and proficiency testing.

3.1.1 Rheological Properties of Non-Newtonian Materials by Rotational Viscometer

ASTM D2196–20, approved in 2020, outlines the measurement of apparent viscosity, shear thinning, and thixotropic properties of non-Newtonian materials within a shear rate range of 0.1–50/s at 25 ± 0.5 °C. The measurements are performed using a rotational viscometer by measuring the torque exerted on a rotating spindle at a constant speed within a 600 mL low-form Griffin beaker. The standard in Method A provides information about the apparatus, reference oils for calibration, calibration verification, preparation of specimens, testing procedure, calculation of apparent viscosity, reporting, precision, and bias. In Method B, the procedure for viscosity under changing speed conditions, degree of shear thinning and thixotropy, calculations and interpretation of results, reporting, precision, and bias are provided. In Method C, the viscosity and shear thinning of a sheared material, preparation of the specimen, testing procedure, calculations and interpretation of results, reporting, precision, and bias are provided for high-speed measurements of at least 200 rad/s (2000 r/min) equipped with a 50-mm (2-in.) diameter circular dispersion blade. The widely used thixotropic index test determines the ratio of viscosities measured at arbitrary high- and low-shear conditions. However, it is important to note that since the test solely measures equilibrium viscosities, it does not provide any information about time dependence. The test should be referred to as the "shear-thinning index" as indicated by ASTM D2196 to accurately represent the essence [13].

3.1.2 Cone/Plate Viscosity at a 500/s Shear Rate

The standard test method ASTM D7395–18 outlines the procedure for measuring the viscosity of paints and similar products. It utilizes a cone/plate viscometer set at a shear rate of 500/s to determine the viscosity of the sample at a closely controlled temperature of 25 ± 0.1 °C. The method is designed to be used as an alternative to No. 4 Ford cup measurements. The standard proposes that if viscosity measurement at a higher shear rate is required, the Test Method D4287 can be utilized, which covers viscosity measurement at shear rates ranging from 10,000 to 12,000/s. The standard provides information about apparatus, reagents and materials, sampling procedure, preparing the apparatus, testing procedure, precision, and bias. The standard recommends reporting the results by including this test method to reference to the test procedure, the specific equipment used, product identification, the type of cone used, the rotational speed in revolutions per minute (rpm), and the test result in millipascal seconds (mPa·s) or centipoise (cp) rounded to the nearest 0.1 mPa·s (cp).

3.1.3 High-Shear Viscosity Using a Cone/Plate and Capillary Viscometer

ASTM D4287–00, reapproved in 2019, outlines the procedure for measuring the viscosity of paints, varnishes, and related products at a rate of shear of 12,000/s. The flow characteristics under high-shear conditions in this method closely resemble the conditions encountered during various application methods like brushing, spraying, electrostatic disk coating, or roll coating. The standard provides information about apparatus, reagents and materials, sampling procedure, preparing the apparatus, testing procedure, precision, and bias. The standard recommends reporting the results by including this test method to reference to the test procedure, the specific equipment used, product identification, the type of cone used, the rate of shear, the temperature, the specifics of reporting units, any deviation, whether agreed upon or otherwise, from the specified test procedures outlined in the test method, and the date on which the test was conducted.

ASTM D446 and D2162 are the preferred methods for measuring viscosity using a capillary viscometer. They are suitable for measuring rheological properties at very high shear rates. The standard outlines viscometer calibration in detail and encompasses various commonly employed viscometers suitable for Test Method D445.

3.1.4 General Principles of Rotational and Oscillatory Rheometry

The ISO 3219, approved in 2021, is the replacement of ISO 3219: 1993.

Plastics—polymers/resins in the liquid state or as emulsions or dispersions—determination of viscosity using a rotational viscometer with a defined shear rate. It outlines the general specifications for rotational viscometers for the determination of the viscosity of polymers in the liquid or liquid-like state. In comparison to the previous standard, the updated version has introduced several modifications, including the addition of plate–plate measuring geometry, relative measuring geometries, and oscillatory rheometry. In the current standard, Part 2 outlines the overall principles governing rotational and oscillatory rheometry from various aspects, including terminology, measurement techniques, data interpretation, and instrument calibration. Part 1 covers the terminology.

3.1.5 Dynamic Mechanical Properties Melt Rheology

This standard method ASTM D4440–15, approved in 2015, describes how to determine and report the rheological characteristics of thermoplastic resins and other types of molten polymers using dynamic mechanical instruments. It may be used to calculate the complex viscosity and other important viscoelastic properties of such materials as a function of frequency, strain amplitude, temperature, and time. Fillers and other additions can have an impact on these qualities. The standard can be relevant where the conversion of starch to a thermoplastic is considered [47].

3.1.6 Powder Rheology

The standard method ISO 8130-15, published in 2023, outlines techniques for determining rheological properties of coating powders. Although the standard covers powder in the molten state, in a specific coating system, the procedures for evaluating the particle state can be adopted to measure starch powder (granules) to ensure consistency and comparability. The standard recommends using a rheometer in accordance with ISO 3219-2 to determine the melt and polymerization characteristics and outlines which geometry specifications, such as diameter, height, and type, prevent electrostatic charging during measurement. The sample for measurement is recommended to be taken according to ISO 15528. The fluidization and cohesion strength measurements, test specifications, and data interpretations are covered in the standard. The special terms used in the field of coating powders are covered in the same standard, ISO 8130 Part 14.

3.1.7 Dough Rheology

Well-established empirical rheological techniques are employed to measure dough properties. The commonly utilized instruments include the farinograph, mixograph, extensograph, and alveograph. Criticism is often directed toward these methods due to the interpretation of results in non-SI units, substantial sample requirements, and the inability to define fundamental rheological parameters such as stress, strain, modulus, or viscosity [70]. The Association of Official Agricultural Chemists (AOAC), an organization involved in creating analytical methods, has developed a method called AOAC Official Method 926.08, which details how to determine the rheological properties of dough using a Brabender

Farinograph. The food industry frequently uses the Brabender Farinograph to assess the rheological properties of dough, such as its water absorption, dough development time, and stability. CEN/ISO/TS 17780:2018 is another standard for cereal and cereal products guidelines on the determination of the water absorption properties of dough using a farinograph.

4 Notes

Various standards are dedicated to addressing the field of rheology and establishing standardized terminology and principles within it. Having distinct specialized rheology standards, defined nomenclature, and principles is critical for the progress of the profession, promoting collaboration, and ensuring the quality and dependability of rheological data in many sectors and applications. The specific standards below are the common ones used.

ISO 3219-1:2021 Rheology—Part 1: Vocabulary and symbols for rotational and oscillatory rheometry: The standard outlines commonly used terms, definitions, and symbols related to rotational and oscillatory rheometry.

ISO 3219-2:2021 Rheology—Part 2: General principles of rotational and oscillatory rheometry: This standard specifies the general principle related to rotational and oscillatory rheometry. In addition to comprehensive symbols and units, the rotational and oscillatory rheometry, common geometries, formulas, and calculation methods related to each geometry are explained in detail. The standard offers advantages and disadvantages of the geometries and includes utilization of cover with or without a solvent trap.

ASTM E473—22d Standard Terminology Relating to Thermal Analysis and Rheology: This compilation of terminology comprises definitions of terms utilized in ASTM documents concerning thermal analysis and rheology, not intended to encompass all possible terms associated.

ASTM E2161—23b Standard Terminology Relating to Performance Validation in Thermal Analysis and Rheology: The standard presents terminology concerning the validation of performance in thermal analysis and rheology methods and instrumentation. It excludes terms that are widely understood or adequately defined in other readily accessible sources.

ASTM E1953—20 Standard Practice for Description of Thermal Analysis and Rheology Apparatus: This practice gives broad explanations of the equipment used in thermal analysis or rheometry measurements. The goal is to provide uniformity in the representation of thermal analysis, rheometry, and viscometer instruments in standard test techniques. The descriptions are intended to be used as templates for any test procedure that makes use of the thermal analysis apparatus described in this practice.

5 Conclusion

Starch is typically described as non-Newtonian, displaying significant shear-thinning rheological behavior. Uncooked starch slurries, on the other hand, demonstrate dilatant behavior. Moreover, starch exhibits thixotropy, wherein its viscosity reduces over time with changes in shear rate. In terms of cold-water-soluble starch nanoparticles, these demonstrate shear-thinning behavior at ultra-high shear rates, indicating enhanced runnability. The results presented in the literature suggest that the starch nanoparticles have the potential to surpass the fundamental rheological properties and practical high-speed coating runnability performance of conventional cooked coating starches.

Given the various factors that impact the rheology of starch, it is imperative to maintain consistency and comparability in test results. This can be accomplished by utilizing standardized test protocols when evaluating different starch samples. By following established procedures, the variability introduced by environmental conditions, raw material sources, and modifications can be minimized, enabling more accurate and meaningful comparisons between starches. Graphing low-shear or high-shear data is a common approach for assessing materials. It should be noted that the research findings emphasize relying solely on low-shear data can lead to inaccurate predictions of material performance on high-speed metered size press, rod, and blade coaters. To obtain a more accurate understanding of rheological performance under commercial conditions, specialized ultra-high shear assessment may be required. While materials may display similar behavior at moderate shear stresses, which is comparable to the mixing process, this behavior may not accurately represent how the product behaves during storage. To ensure accurate and meaningful comparisons between different starch samples, the performance of the product under storage conditions is equally important to understand, as it accounts for the majority of its lifespan when it remains static. It is important to consider the storage behavior in addition to the behavior during application or mixing.

Acknowledgments

The authors would like to thank Neil Cunningham from the Centre for Industrial Rheology and Waters Technologies Corporation for their generous collaboration in providing valuable images for incorporation into this book chapter. We also thank Çiğdem Kıncak Yaşar for sketching figures.

References

1. Kraak A (1992) Industrial applications of potato starch products. Ind Crop Prod 1(2-4):107–112
2. Altay BN, Klass C, Chen T, Fleck A, Aydemir C, Karademir A, Fleming PD (2021) The effect of green biobased binder on structural, mechanical, liquid absorption and wetting properties of coated papers. Clean Eng Technol 5:100274. https://doi.org/10.1016/j.clet.2021.100274
3. Ogunsona E, Ojogbo E, Mekonnen T (2018) Advanced material applications of starch and its derivatives. Eur Polym J 108:570–581. https://doi.org/10.1016/j.eurpolymj.2018.09.039
4. Rao MA (1977) Rheology of liquid foods – a review. J Texture Stud 8, 2:135–168
5. Gamage A, Liyanapathiranage A, Manamperi A, Gunathilake C, Mani S, Merah O, Madhujith T (2022) Applications of starch biopolymers for a sustainable modern agriculture. Sustainability 14(10):6085. https://doi.org/10.3390/su14106085
6. Barnes HA, Hutton JF, Walters K (1998) An introduction to rheology. Elsevier Science B. V, Amsterdam
7. Gibson M, Newsham P (2018) Rheology. In: Food science and the culinary arts. Elsevier, pp 89–103. https://doi.org/10.1016/B978-0-12-811816-0.00007-5
8. McKenna BM, Lyng JG (2003) Introduction to food rheology and its measurement. Texture Food 1:130–160. https://doi.org/10.1533/9781855737082.1.130
9. Ionescu CM, Birs IR, Copot D, Muresan CI, Caponetto R (2020) Mathematical modelling with experimental validation of viscoelastic properties in non-Newtonian fluids. Philos Trans R Soc A 378:20190284. https://doi.org/10.1098/rsta.2019.0284
10. Goodwin JW, Hughes RW (2000) Rheology for chemists, an introduction. Royal Society of Chemist, Cambridge
11. Fogel'son RL, Likhachev ER (2001) Temperature dependence of viscosity. Tech Phys 46(8):1056–1059. https://doi.org/10.1134/1.1395131
12. Triantafillopoulos N (1998) Measurement of fluid rheology and interpretation of reograms. Kaltec Scientific Supplies and Instruments LLC, Livonia
13. Eley RR (2012) Rheology and viscometry. In: Koleske JV (ed) Paint and coating testing manual: fifteenth edition of the Gardner-Sward handbook. ASTM International, West Conshohocken, p 2012
14. Bertolo MRV, Rodrigues MÁV, Horn MM, de Oliveira Filho JG, Marangon CA, Ferreira MD, de Guzzi Plepis AM, Junior SB (2023) Chapter 7 – Rheology of nanoscale polymer-based coatings. In: Polymer-based nanoscale materials for surface coatings, pp 131–149. https://doi.org/10.1016/B978-0-32-390778-1.00014-1
15. Centre for Industrial Rheology (2023) Creep test for liquids and semisolids. Accessed in July 2023 on https://www.rheologylab.com/articles/creep-test/
16. Ruß N (2016) Rheology and thermodynamics of starch-based hydrogel-mixtures, Doctoral Dissertation, Johannes Gutenberg-Universität, Mainz, Germany
17. Kulkarni VS, Shaw C (2016) Rheological studies. In: Essential chemistry for formulators of semisolid and liquid dosages, pp 145–182. https://doi.org/10.1016/B978-0-12-801024-2.00009-1
18. Macosko CW (1994) Rheology principles, measurements, and applications. Wiley-VCH, New York
19. Agoda-Tandjawa G, Durand S, Berot S, Blassel C, Gaillard C, Garnier C, Doublier J-L (2010) Rheological characterization of microfibrillated cellulose suspensions after freezing. Carbohydr Polym 80(2010):677–686
20. TA Instruments (2023) Rheology software models. Accessed in July 2023 from https://www.tainstruments.com/pdf/literature/RN9.pdf

21. Malvern (2016) A basic introduction to rheology. Accessed in November 2023 from https://cdn.technologynetworks.com/TN/Resources/PDF/WP160620BasicIntroRheology.pdf
22. Wang Y-J, Truong V-D, Wang L (2003) Structures and rheological properties of corn starch as affected by acid hydrolysis. Carbohydr Polym 52(3):327–333. https://doi.org/10.1016/S0144-8617(02)00323-5
23. Corke H, Zhu F (2018) Chapter 5. Starch properties and modification in grains and grain products. In: Beta T, Camire ME (eds) Food chemistry, function and analysis. Royal Society of Chemistry, Cambridge, pp 85–102. https://doi.org/10.1039/9781788012799-00085
24. Altay BN, Pekarovicova A, Fleming PD (2017) Printed dot quality in response to doctor blade angle in gravure printing. Hilltop Rev 10(1): article 12
25. e-Bay (2023) Viscosity Cup Zahn Cup #1 to #5 Dip Type 44ml. Accessed in Nov 2023 from https://www.ebay.com/sch/i.html?_from=R40&_trksid=p2332490.m570.l1313&_nkw=Viscosity+Cup+Zahn+Cup+%231+to+%235+Dip+Type+44ml&_sacat=0
26. Franck A (2023) Systemic rheology, TA Instruments AN035e. Accessed in July 2023 from https://www.tainstruments.com/pdf/literature/AAN035e_Systemic_Rheology.pdf
27. Giles D, Macosko C (2017) Shear rheometry. Rheology short course notes, May 16–18, Irvine
28. Cheng DC-H (1968) The effect of secondary flow on the viscosity measurement using the cone-and-plate viscometer. Chem Eng Sci 23: 895–899. https://doi.org/10.1016/0009-2509(68)80023-5
29. Carvalho MS, Padmanabhan M, Macosko CW (1994) Single-point correction for parallel disks rheometry. J Rheol 38(6):1925–1936. https://doi.org/10.1122/1.550532
30. TA Instruments (2007) AR 2000 rheometer, rheometric series operator's manual, PN 500106.002 Rev. L, TA Instruments, New Castle
31. Franck AJ (2023) New approach to characterize starch dispersions. TA Instruments AN112. Accessed in July 2023 from https://www.tainstruments.com/pdf/literature/AAN112e_New_approach_to_characterize_starch_dispersions.pdf
32. Peris-Tortajada M (2018) Measuring starch in food, in starch in food, structure, function and applications, 2nd edn. Woodhead Publishing Series in Food Science, Technology and Nutrition, pp 255–281. https://doi.org/10.1016/B978-0-08-100868-3.00006-8
33. Punia S, Dhull SB, Sandhu KS, Kaur M, Purewal SS (2020) Kidney bean (Phaseolus vulgaris) starch: a review. Legume Sci 2(3):e52. https://doi.org/10.1002/leg3.52
34. Altay BN, Bloembergen S, Aydemir C, Karademir A, Fleming PD (2017) Use of nanoparticle binders for paper coatings: a review. J Graph Eng Des 8:39. https://doi.org/10.24867/JGED-2017-1-039
35. De Willigen AHA (1976) The rheology of starch. In: Radley JA (ed) Examination and analysis of starch and starch products. Springer Netherlands, Dordrecht, pp 61–90. https://doi.org/10.1007/978-94-010-1332-1_3
36. Ai Y, Jane J-L (2015) Gelatinization and rheological properties of starch. Starch/Stärke 67: 213–224. https://doi.org/10.1002/star.201400201
37. Giezen FE, Jongboom ROJ, Feil H, Gotlieb KF, Boersma A (2004) Biopolymer nanoparticles. US6677386 B1 (Patent)
38. Van Soest JJG, Stappers FHM, Van Schijndel RJG, Gotlieb KF, Feil H (2004) Method for the preparation of starch particles. US6755915 B1 (Patent)
39. Bloembergen S, Kappen F, Beelen B (2005) Environmentally friendly biopolymer adhesives and applications based thereon. US6921430 B2 (Patent)
40. Shin JY, Jones N, Lee DI, Fleming PD, Joyce MK, DeJong R, Bloembergen S (2012) Rheological properties of starch latex dispersions and starch latex-containing coating colors, conference proceedings of PaperCon 2012, New Orleans, LA
41. Nazir A, Asghar A, Aslam Maan A (2017) Food gels: gelling process and new applications. In: Advances in food rheology and its applications. Elsevier, pp 335–353. https://doi.org/10.1016/B978-0-08-100431-9.00013-9
42. Fischer EK, Lindsley CH (1948) Rheological measurements with the rotational viscometer. Tex Res J 18(6):325–337. https://doi.org/10.1177/004051754801800601
43. Sodhi NS, Singh N (2003) Morphological, thermal and rheological properties of starches separated from rice cultivars grown in India. Food Chem 80(1):99–108. https://doi.org/10.1016/S0308-8146(02)00246-7
44. Singh S, Kaur M (2017) Steady and dynamic shear rheology of starches from different oat cultivars in relation to their physicochemical and structural properties. Int J Food Properties

20(12):3282–3294. https://doi.org/10.1080/10942912.2017.1286504
45. Park S, Chung M-G, Yoo B (2004) Effect of octenylsuccinylation on rheological properties of corn starch pastes. Starch/Stärke 56:399–406. https://doi.org/10.1002/star.200300274
46. Won C, Jin YI, Chang D-C, Kim M, Lee Y, Ganesan P, Lee Y-K, Chang YH (2017) Rheological, pasting, thermal and retrogradation properties of octenyl succinic anhydride modified potato starch. Food Sci Technol Campinas 37(2):321–327. https://doi.org/10.1590/1678-457X.23616
47. Liu H, Xie F, Yu L, Chen L, Li L (2009) Thermal processing of starch-based polymers. Prog Polym Sci 34(12):1348–1368. https://doi.org/10.1016/j.progpolymsci.2009.07.001
48. Brouillet-Fourmann S, Carrot C, Mignard N, Prochazka F (2002) On the use of an internal mixer for the rheological characterization of maize starch. Appl Rheol 12:192–199. https://doi.org/10.1515/arh-2002-0011
49. Agama-Acevedo E, Nuñez-Santiago MC, Alvarez-Ramirez J, Bello-Pérez LA (2015) Physicochemical, digestibility and structural characteristics of starch isolated from banana cultivars. Carbohydr Polym 124:17–24. https://doi.org/10.1016/j.carbpol.2015.02.003
50. Punia S, Sandhu KS, Dhull SB, Kaur M (2019a) Dynamic, shear and pasting behaviour of native and octenyl succinic anhydride (OSA) modified wheat starch and their utilization in preparation of edible films. Int J Biol Macromol 133:110–116. https://doi.org/10.1016/j.ijbiomac.2019.04.089
51. Punia S, Siroha AK, Sandhu KS, Kaur M (2019b) Rheological and pasting behavior of OSA modified mungbean starches and its utilization in cake formulation as fat replacer. Int J Biol Macromol 28:230–236. https://doi.org/10.1016/j.ijbiomac.2019.01.107
52. Punia S, Siroha A, K., Sandhu, K. S., Kaur, M. (2019c) Rheological behavior of wheat starch and barley resistant starch (Type IV) blends and their starch noodles making potential. Int J Biol Macromol 130:595–604. https://doi.org/10.1016/j.ijbiomac.2019.03.009
53. Kaur M, Punia S, Sandhu KS, Ahmed J (2019) Impact of high pressure processing on the rheological, thermal and morphological characteristics of mango kernel starch. Int J Biol Macromol 140:149–155. https://doi.org/10.1016/j.ijbiomac.2019.08.132
54. Siroha AK, Punia S, Sandhu KS, Karwasra BL (2020b) Physicochemical, pasting, and rheological properties of pearl millet starches from different cultivars and their relations. Acta Alimentaria 49(1):49–59. https://doi.org/10.1556/066.2020.49.1.7
55. Dhull SB, Punia S, Kumar M, Singh S, Singh P (2021) Effect of different modifications (physical and chemical) on morphological, pasting, and rheological properties of black rice (Oryza Sativa L. Indica) starch: a comparative study. Starch/Stärke 73:1–2. https://doi.org/10.1002/star.202000098
56. Siroha AK, Punia S, Kaur M, Sandhu KS (2020a) A novel starch from pongamia pinnata seeds: comparison of its thermal, morphological and rheological behaviour with starches from other botanical sources. Int J Biol Macromol 143:984–990. https://doi.org/10.1016/j.ijbiomac.2019.10.033
57. Bangar SP, Dunno K, Dhull SB, Siroha AK, Changan S, Maqsood S, Rusu AV (2022) Avocado seed discoveries: chemical composition, biological properties and industrial food applications. Food Chem X 16:100507. https://doi.org/10.1016/j.fochx.2022.100507
58. Singh A, Geveke DJ, Yadav MP (2017) Improvement of rheological, thermal and functional properties of tapioca stach by using gum arabic. LWT Food Sci Technol 80:155–162. https://doi.org/10.1016/j.lwt.2016.07.059
59. Xie F, Halley PJ, Avérous L (2012) Rheology to understand and optimize processibility, structures and properties of starch polymeric materials. Prog Polym Sci 37(4):595–623. https://doi.org/10.1016/j.progpolymsci.2011.07.002
60. Ahuja A, Lee R, Latshaw A, Foster P (2020) Rheology of starch dispersions at high temperatures. J Texture Stud 51:575–584. https://doi.org/10.1111/jtxs.12517
61. Chaisawang M, Suphantharika M (2006) Pasting and rheological properties of native and anionic tapioca starches as modified by Guar Gum and Xanthan Gum. Food Hydrocoll 20(5):641–649. https://doi.org/10.1016/j.foodhyd.2005.06.003
62. Bagley EB, Christianson BB (1983) Yield stresses in cooked wheat starch dispersions. Starch Stärke 35(3):81–86. https://doi.org/10.1002/star.19830350304
63. Xue T, Yu L, Xie F, Chen L, Li L (2008) Rheological properties and phase transitions of starch under shear stress. Food Hydrocoll 22:973–978. https://doi.org/10.1016/j.foodhyd.2007.05.008

64. Yu L, Kealy T, Chen P (2006) Study of starch gelatinization in a flow field using simultaneous rheometric data collection and microscopic observation. Int Polym Proc 21:283–289. https://doi.org/10.3139/217.0098
65. TA Instruments (2023b) Rheo-microscope accessory. Accessed in Nov 2023 from https://www.tainstruments.com/modular-microscope/
66. ASTM Standard D7395 (2023) Standard test method for cone/plate viscosity at a 500 s-1 shear rate. ASTM International. www.astm.org
67. ASTM Standard D4287 (2019) Standard test method for high-shear viscosity using a cone/plate viscometer. ASTM International. www.astm.org
68. Gulnov DV, Nemtseva EV, Kratasyuk VA (2016) Contrasting relationship between macro- and microviscosity of the gelatin- and starch-based suspensions and gels. Polym Bull 73:3421–3435. https://doi.org/10.1007/s00289-016-1664-9
69. Waigh TA (2005) Microrheology of complex fluids. Rep Prog Phys 68:685
70. Ronda F, Pérez-Quirce S, Villanueva M (2017) Rheological properties of gluten-free bread doughs: Relationship with bread quality. In: Advances in food rheology and its applications. Elsevier, pp 297–334. https://doi.org/10.1016/B978-0-08-100431-9.00012-7

Chapter 7

X-Ray Diffraction Analysis for Starch

Parveen Kumari, Ankur Luthra, Prerna Sethi, and Sonika Banyal

Abstract

Starch is a polysaccharide generated by plants as a form of energy storage. The combination of the two macromolecules amylose and amylopectin, in particular, results in the complicated structure of starch granules. The X-ray diffraction (XRD) analysis is one of the powerful nondestructive technologies utilized to characterize crystalline material and is helpful to explain the preferred crystalline orientation, phases, structure, and structural parameters like crystal defects, strain, grain size, and crystallinity. Using this technique, the starch's long-range crystalline order in relation to the arrangement of the double helices is determined. In X-ray diffraction, native starch exhibits three different crystalline polymorph types (A-, B-, and C-type), with C-type being a combination of A and B types. This chapter includes an overview of the diffraction theory of X-rays by crystals, structural characterization of different starches using X-ray diffraction, and sample preparation protocols utilized in X-ray diffraction analysis for various starches.

Key words Starch, X-ray diffraction, Protocols, Characterization, Structure

1 Introduction

Starch is a polysaccharide that plants produce as a means of storing energy. It consists of two major macromolecules, amylose and amylopectin, contributing to the starch granules' complex structure [1, 2]. These granules exhibit a concentric arrangement of amorphous & crystalline layers. The amylopectin side chains (A- or B-chains) form double helices inside the crystalline regions. Additionally, certain longer amylose chains might interact with amylopectin by means of double-helix interactions [3]. X-ray diffraction (XRD) is the technique most frequently used to identify the kind and degree of native starches'crystallinity [4, 5]. It is an effective nondestructive technique that measures additional structural characteristics like crystallinity, strain, average grain size, and crystal defects. Moreover, it enables the assessment of the starch's long-range crystalline order with respect to the organization of the double helices. Typically, a two-phase model is employed to predict the crystalline structure [6, 7], where the X-ray scattering

Sneh Punia Bangar (ed.), *Standardized Procedures and Protocols for Starch*, Methods and Protocols in Food Science,
https://doi.org/10.1007/978-1-0716-3866-8_7,

contribution from the amorphous regions is subtracted from the scattering of the crystalline domains. The degree of starch crystallinity can be obtained using the ratio of the two phases. However, some differences exist between the calculated starch granule crystallinity and the positions of the fitting peaks [8]. In X-ray diffraction, native starch exhibits three different crystalline polymorph types (A-, B-, and C-type), with C-type being a combination of A and B types. In general, cereals containing amylopectin of significantly shorter chain length, such as waxy maize, rice, and wheat, exhibit A-type patterns, whereas tubers (such as potatoes) and high amylose maize (>40%) starches exhibit B-type patterns. Legumes primarily display the C-type pattern [9, 10]. In A-type starches, water molecules are enclosed within the compact packing of each double helix structure. On the other hand, in B-type polymorphs, a core cavity containing water molecules is surrounded by six double helices, resulting in a more open arrangement [3]. In terms of X-ray diffraction patterns, A-type starch exhibits two prominent peaks at approximately 15° and 23° 2θ, along with a doublet at approximately 17° and 18° 2θ, while B-type starch displays a prominent peak at approximately 17° 2θ and smaller peaks near 15°, 22°, and 24° 2θ and a characteristic peak at about 5.6° 2θ. The crystallinity of C-type starch is a combination of A- and B-types. Based on the proportion of A- and B-type polymorphs, it can be further divided into three distinct subtypes: CA-type (nearer to A-type), CC-type (characteristic C-type), and CB-type (nearer to B-type). The diffraction patterns of CA- and CB-type starches share similarities with CC-type starches, but they exhibit certain different characteristics. CA-type starch exhibits a prominent singlet at 23° 2θ and a shoulder peak at about 18° 2θ. In contrast to this, CB-type starch has two shoulder peaks at roughly 22° and 24° 2θ. The XRD pattern of CC-type starch exhibits a singular peak at around 17° and 23° 2θ and a few smaller peaks at roughly 5.6° and 15° 2θ [5, 11–13].

The packing of the "A"-type structure is monoclinic, having four water molecules enclosed in each unit cell. The "B"-type structure, on the other hand, has a hexagonal packing that creates a space for an internal channel that can accommodate up to 36 water molecules per unit cell [14]. The distance between amylopectin molecule branches defines the crystal structure of the starch granule [15, 16]. The granule will exhibit an "A"-type pattern if the chain length falls less than 19.7 anhydroglucose units. On the other hand, when the length is expanded to 21.6 units, the most typical type "B" pattern will be visible. The granule will, however, display the type "C" pattern for values that are intermediary. Finally, a different crystalline structure related to the amylose percentage is visible. Pattern V, which represents a single left-handed amylose helix, is typically found in high amylose starches. The V_h type pattern is typically formed due to the formation of the amylose-lipid complex following severe heat treatment [17].

Table 1
Crystallinity index for different starch polymorphs

Type of polymorph	Starch source	Degree of crystallinity (%)	Reference
A-type	Wheat Maize Rice Corn Finger millet	27–36 ~ 27 ~ 38 14–39 ~ 30	[20, 21]
B-type	Potato Banana	23–25 18–22	[20, 22, 23]
C-type	Soybean Tapioca Green gram Cassava	27–36 35–38 ~ 32 ~ 13	[19, 20, 22]
V-type	Parboiled rice	3–12	[24]

1.1 Application of XRD in the Characterization of Different Starches

It has been demonstrated that XRD, a fundamental tool from the domain of material sciences, can evaluate the material quality and pertinent variations resulting from various operations and material composition. But in the last few years, XRD has been widely adopted by the community of food scientists and is increasingly finding use in the assessment of food quality [19]. XRD is frequently utilized in examining alterations in the morphology and crystallinity of grain starches after undergoing different processing methods. Wide-angle X-ray scattering exposes the molecular arrangement of the nanoparticles (between 1 and 10 nm), suggestive of the liquid/solid state, whereas small angle X-ray scattering reveals details regarding the size of the recurring lamellar structure (10–50 nm). The crystallinity index for different starch polymorphs is given in Table 1.

1.1.1 Cereal Starches

XRD has been used to determine the effectiveness of parboiling on both normal and fungal-infested rice grains in non-issuable rice from a storage facility. At diffraction angles of 20°, 15°, and 18.4°, prominent peaks were seen in the diffractogram of parboiled sound rice whereas similar peaks were less developed in parboiled fungal-infested grains. As a result, the study found that fungus-infested grains had a considerably lower proportion of crystallinity (8.69%) than sound grains (16.81%) [25]. Buckwheat starch exhibited a decrease in crystallinity as a consequence of pressure-induced gelatinization when subjected to high-pressure processing [26]. A unique quantitative technique based on XRD has been developed to assess samples with high levels of moisture, such as cooked rice, with an emphasis on the diffractogram close to a reflection angle of 5.5°. The method has been effectively utilized for studying cooked rice retrogradation. The abovementioned method eradicates the time-consuming sample preparation protocols for XRD requiring drying and milling of the samples, as well as post-processing analyses including computational peak splitting procedures for crystalline structure measurement [27].

Dome et al. [18] examined the effect of mechanical process on the amorphous phase formation in A-type maize and C-type tapioca starches with different crystalline structures. The application of mechanical force reduced the degree of crystallinity in a planetary ball mill in direct proportion to the length of pretreatment. Maize and tapioca starch granules undergo total breakdown and aggregation after undergoing mechanical processing for 600 s. The highest level of crystallinity was identified in C-type tapioca starch. Maize starch exhibits two prominent peaks at approximately 15.00° and 22.93° 2θ, along with a doublet at approximately 17.02° and 17.92° 2θ, which are characteristic of the A-type crystalline structure (Fig. 1). In contrast, tapioca starch displays prominent peaks at 15.15° and 22.78° 2θ, along with a doublet at 17.12° and 18.19° 2θ which corresponds to the C-type crystalline structure.

1.1.2 Tuber Starches

Jyothi et al. [28] reported the impact of hydrothermal treatments on the crystallinity index of tropical tuber starches viz., cassava, sweet potato, and arrowroot. In contrast to sweet potato and arrowroot starches, which displayed a rise in crystallinity following heat–moisture treatment at 120 °C for 14 h with 20% moisture, cassava starch demonstrated a slight reduction in crystallinity,

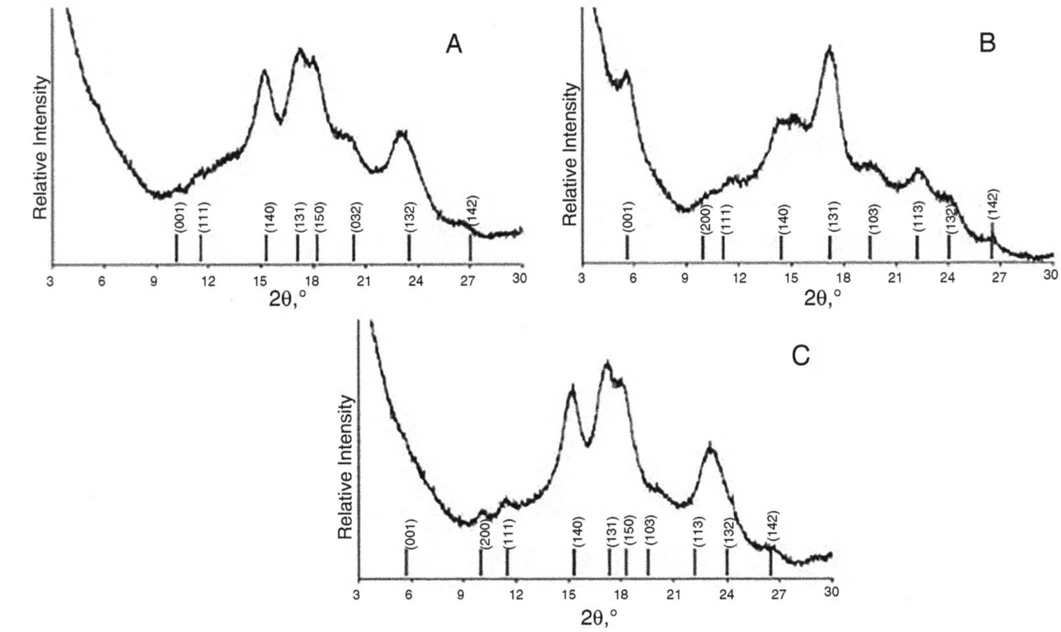

Fig. 1 XRD pattern for corn (**a**), potato (**b**), and tapioca (**c**) starch with labeled ideal reflections [18]

according to X-ray diffraction measurements. The most prominent peaks for native cassava starch were found at diffraction angles 15.047°, 16.8°, 17.95°, and 22.817° 2θ. Peaks were detected at 15.07°, 17.01°, 20.04°, and 23.447° 2θ for the native starch in the case of sweet potatoes, while the native starch of arrowroot displayed strong peaks at diffraction angles 15.01°, 17.35°, 20.06°, and 23.327° 2θ. Cassava starch samples treated with heat–moisture exhibited significant decline in peak intensity and crystallinity while an increment in peak intensity and crystallinity was observed in both heat–moisture-treated sweet potato and arrowroot starches.

The effect of microwave heating on the crystallinity of potato starch was studied in water suspension at different temperatures (50 °C and 60 °C) by Nadiah et al. [29]. The diffraction patterns of native potato starch and heat-treated potato starch differ significantly. The native potato starch's B-type X-ray pattern, which had a prominent peak at 17° and 18° 2θ, changed to an A-type pattern with a prominent peak at 15°, 17°, 18°, and 23° 2θ. The difference in the results was caused by the heat–moisture treatment of potato starch, which altered its crystalline structure. Additionally, the shifting of double helices in the starch granule's core chain and the enhanced packing after heating were the causes of the shift in the X-ray pattern.

1.1.3 Legume Starches

Bashir et al. [30] reported the effect of gamma irradiation on chickpea starch crystalline structure. The XRD examination revealed a C-type pattern (a combination of both A and B type), which is typical of legume starch. Peaks were observed in the starch granules at around 15°, 17°, 18°, and 23.3°. The peak intensities decreased with increasing irradiation dose, demonstrating a change in the crystalline structure of the irradiated starch. The disordered double helices of the starch granule and the loss of the organized structure (amylopectin) are responsible for this alteration.

The effects of annealing and heat–moisture treatments on the crystallinity of sword bean starches were investigated by Olu-Owolabi et al. [31]. Heat–moisture treatment at 25% and 30% moisture levels changed the starch's B-type X-ray diffraction pattern to C-type. Heat–moisture treated sword bean starch at 20% moisture level and native sword bean starch both displayed the B-type X-ray pattern with prominent peaks at 17.2 and 17.1 Å, respectively. The B-type X-ray pattern of the annealed sword bean starch also showed prominent peaks at 15.1, 17.15, and 23.35 Å. sword bean starch at 25% moisture level displayed a C-type X-ray pattern with prominent peaks at 15.35, 17.3 and 23.50 Å and smaller peak at 10.05 and 20.25 Å. Similarly, sword bean starch at 30% moisture level had typical C-type pattern with prominent peaks at 15.4, 17.3, and 23.50 Å and weak peaks at 10.00 and 20.40 Å. Various XRD conditions and results of starches from different sources are presented in Table 2.

Table 2
XRD conditions and results of starches from different sources

Type of starch	Operating conditions	Results/interpretations	Reference
Irradiated arrowroot starch	XRD measurements were taken with a range of 2θ angles from 5° to 37°, applying a step size of 0.04 and a measurement duration of 10 s per step. The receiving slit was configured at 0.4 mm, while the divergence and soller slits were both adjusted at 1.0 mm.	Starch maintained its A-type crystalline structure, as indicated by the prominent peaks observed at approximately 15°, 17°, 18°, and 23° (2θ) in the XRD pattern. However, there was no notable increase in the relative crystallinity as a result of the irradiation treatment.	[50]
High hydrostatic pressure treated rice starch	Analysis was conducted by scanning diffraction angle from 4° to 40°. Each measurement was taken in steps of 0.05 with a 2 s duration between each step.	Native starch displayed A-type X-ray patterns, characterized by strong peaks at approximately 15.04°, 16.84°, 17.96°, and 23.02° 2θ. However, following HHP treatment at 600 MPa for 30 min, a significant transformation from the A-type to the B-type pattern was observed.	[51]
Irradiated aria' (*Goeppertia allouia*) starch	Analysis was done with a range of 2θ angles from 5° to 100° applying a step size of 0.01313° and a measurement duration of 60 s per step.	It was shown that when irradiation dosages increased from 1 to 50 kGy, the amorphous component gradually increased and the mean crystallite sizes significantly decreased.	[52]
Hydrothermally modified *Dioscorea hispida* Dennst (Thai yam) starch	Angular range was covered from 5° to 45° (2θ), using a step angle of 0.04° between each measurement.	Native starch initially exhibited a type B pattern, which transformed into a type C pattern when modified at 90 °C. Subsequently, when modified at temperatures ranging from 100 to 130 °C, the starch adopted a type A crystallographic pattern.	[53]
Native and resistant chickpea starch	Analysis covered a 2θ range of 5°–37° with a step size of 0.02°. At each step, data was collected for a duration of 6.7 s.	Both native and resistant starches exhibited C-type crystallinity, as indicated by the presence of a weak peak at around 6° (typical of B-type polymorphs) and strong peaks at 17° and 23° (typical of A-type polymorphs).	[33]

(continued)

Table 2 (continued)

Type of starch	Operating conditions	Results/interpretations	Reference
Sweet potato starches (C-type)	Angular range was covered from 3° to 40° 2θ using a step angle of 0.02° between each measurement	All seven types of sweet potato showed C-type XRD patterns in their starches, with distinct diffraction peaks at 5.6°, 15°, 17°, and 23° 2θ	[54]
Debranched cassava starch	Analysis was covering a range of 5°–37.5° 2θ, with a step size of 0.01°. At each step, data was collected for a duration of 6.5 s.	Materials are predominantly made up of different ratios of A- and B-type polymorphs, with only a small proportion of V-type structures. Heat–moisture treated samples display a distinct A-type polymorph signature, characterized by a doublet peak at 2θ values of 17.1° and 18.1°. In contrast, other treated samples exhibit a mixture of A-polymorphs and B-polymorphs in their XRD patterns.	[55]
Irradiated rice starch	Scanning was performed at a scanning speed of 2°/min, using X-ray settings of 40 kV and 30 mA. The scanning covered an angular range of 2θ from 4° to 40°.	Control starches from both cultivars exhibited an A-type X-ray pattern. This pattern was characterized by a weak intensity peak at a diffraction angle of 2θ 20.0°, a medium-strong intensity peak at 15.1°, a strong intensity doublet at 17.1°/18.0°, and a strong intensity peak at 23.0°. After the irradiation treatment, there were no significant changes observed in the X-ray diffraction pattern of the starch samples.	[56]
Ultrasonically modified corn and cassava starch	Covered an angular range of 10°–50° for 2θ. The scanning rate during the analysis was 0.02° per min.	Both native cassava and corn starch showed a characteristic A-type crystalline pattern, with prominent peaks observed at approximately 15°, 17°, 18°, and 23° in the X-ray diffraction (XRD) pattern. However, after sonication treatment, both corn and cassava starches exhibited a decrease in relative crystallinity.	[57]

(continued)

Table 2 (continued)

Type of starch	Operating conditions	Results/interpretations	Reference
Horse chestnut starch	Sample was subjected to X-ray diffraction (XRD) analysis with a scanning range from $2\theta = 5°$ to 50°. The scanning speed used for the analysis was 2° per min.	X-ray diffraction (XRD) analysis of hydrogen cyanide (HCN) starch revealed a typical C-type crystalline pattern, as indicated by characteristic peaks at 5.7°, 15°, 17.3°, and 22.3° in the XRD pattern. The percentage crystallinity of HCN starch was determined to be 32.78%.	[58]

1.2 XRD Principle

X-rays are widely utilized through imaging and scattering modes for the quality determination of foods. However, their diffraction mode has gained even more popularity these days. The basic principle behind X-ray diffraction lies in the generation of X-rays [32]. Further, it is the ability of a substance to diffract these rays in a manner that is significant for a food component like starch. These are explained in further sections of this chapter.

1.2.1 X-Ray Generation and Imaging

The X-ray emission is understood on the basis of atomic spectra and energy levels. Laboratory equipment could be used as popular radiation sources like anodes and sealed tubes [32]. The tungsten filament acts as cathode while the target food material acts as anode. Electrons, being high-energy components, are subjected to high potential field of about 40–160 kV within the sealed tube. This excites the photons, which, in return, emit radiations falling within the wavelength range of 10^{-3}–10^{-9} m. These X-rays are therefore produced as a result of two major effects. The first is the creation of energy gaps, which results in the formation of photons, and the second is the excitation of such photons within a high-voltage tube [19]. X-rays are highly penetrating, due to which they are used in several industries including medical, pharmaceutical, and food industries. The growing interest in food items could be attributed to its ability to highlight the crystalline aspects of some important components. The X-ray pattern for food as a target material is shown in Fig. 2.

XRD has proved to be one of the most effective tools for evaluating the proportionality of starch polymorphs. Besides this, the imaging technique is functional for determining the crystalline structures of starch. The starch granules are considered semicrystalline whose polymorphic structures could be observed by the application of a detector. Not all polymorphs are found in one type of starch. For instance, a study by Sun et al. [33] mentioned that A-, B-, and C-type polymorphs are most commonly found in

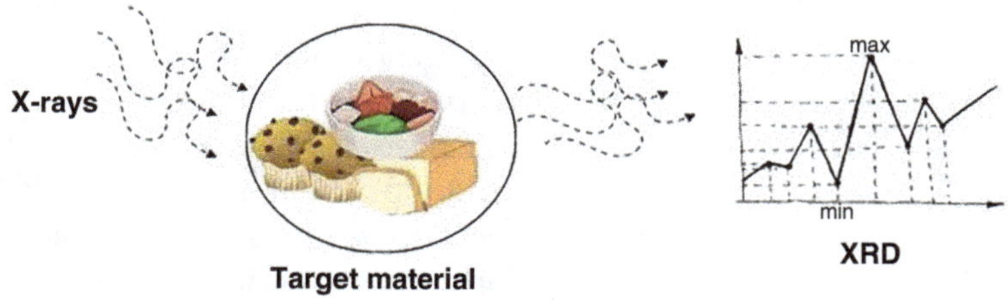

Fig. 2 XRD pattern for food as target material

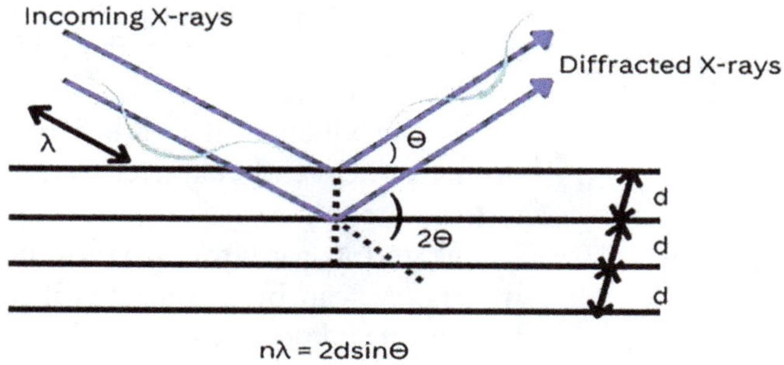

Fig. 3 Diffraction principle based on Bragg's law

cereals, tubers, and pulse starches respectively. The same study also highlighted that a differential scanning calorimeter (DSC) could study the thermal properties of A- and B-type polymorphs; however, XRD is used for analyzing the individual proportions in C-type polymorphs.

X-ray diffractometers are a combination of three major elements, namely, sample holder, cathode ray tube, and detector for image analysis. As discussed in previous sections, X-rays are produced by heating a filament where electrons are bombarded to the target material. Each material has its own characteristic diffraction peak, which makes it different from others. The terms "crystallinity" and "crystalline" polymorphs are evaluated by studying separate diffraction patterns. Purohit et al. [19] mentioned that the crystallinity could be judged by considering the total area under the crystalline peak. In contrast, the crystalline polymorphs are studied at a characteristic diffraction angle of 2θ. The X-rays are detected and the data can be analyzed to figure out the desired crystalline structure of starch. The detection of the diffracted ray is based on the principle of Bragg's law (Fig. 3) as indicated below:

$$n\lambda = 2d\sin\theta \quad (1)$$

where n represents a positive integer, λ represents wavelength for X-rays, d is the lattice parameter, and θ indicates the diffraction angle.

The powdered materials like starch consist of different substances with random orientations. Based on Eq. 1, the source (cathode tube) and the detector scan the material to obtain their entire diffraction band. As stated earlier, this could be done at a characteristic diffraction angle range of 2θ. Several factors are of great concern to obtain a better resolution of the material. These include the scan time, scan size, and signal-to-noise ratio [34]. In other words, an increased scan time and signal-to-noise ratio, along with a decrease in the scan size (also known as the increment size), results in data with better resolution. However, this might take a longer time than expected, due to which it is advised to obtain data at lower resolution during the initial phase followed by increased resolutions later.

The production of X-rays from sealed tubes and anodes is based on the principle of characteristic spectra, which is significant for a target material.

K_α and K_β form the two most notable components of the X-ray spectra. K_α can be further classified into $K_{\alpha 1}$ and $K_{\alpha 2}$ where the former one is known to have a shorter wavelength. These radiations are considered during XRD, and the rest of the radiations are removed by utilizing several number of filters and monochromators. This facilitates the production of a monochromatic radiation, which is detected to give XRD spectra. However, the case is slightly different when X-rays are produced from synchrotrons [32]. The same study also mentioned the occurrence of various scattering effects in food on exposure to X-rays. These are collectively called Rayleigh scattering, which occurs due to reaction between the electrons and photons. The scattering could be constructive or destructive depending upon the nature of the crystalline target material. The interpretation of the constructive radiations as perceived by the XRD phenomenon could be seen by the geometrical demonstration as shown in Fig. 4.

2 Materials

2.1 Sample

Starch is extracted from the plant source depending upon the study. The finely powdered starch material is used for further analysis. The test is conducted using a diffractometer (mostly powder diffraction), and no preprocessing conditions are required unless specified. Further, water or alcohol is used for wet grinding of samples.

2.2 Instrumentation

- Mortar pestle (Manual grinding of sample).
- Mc Crone mill (Mechanical grinding of sample).
- X-ray diffractometer: It is a combination of various components (Fig. 4) including:

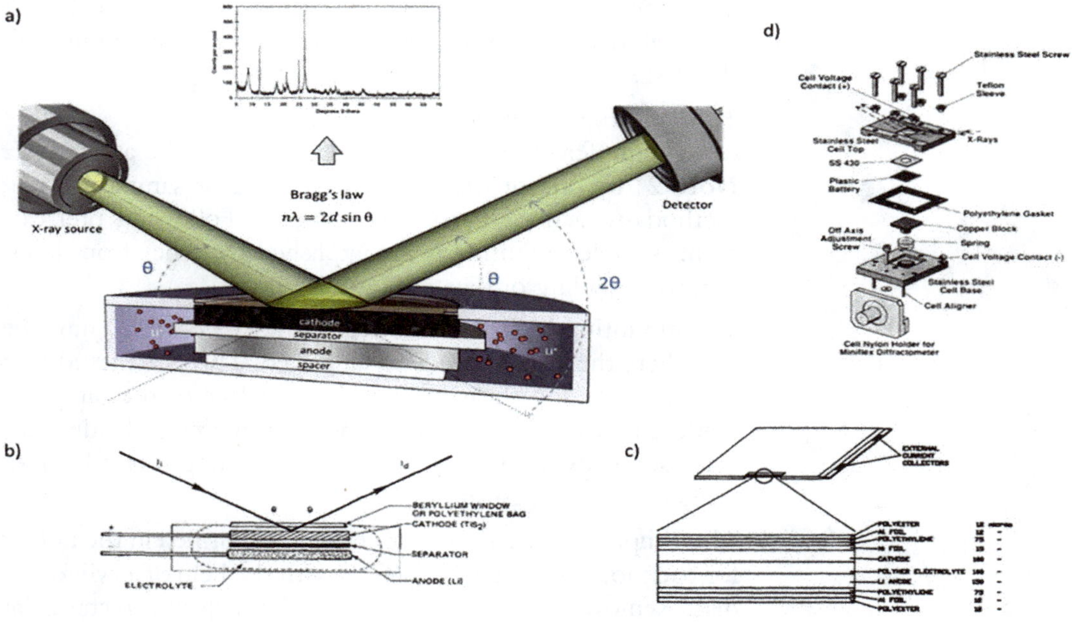

Fig. 4 Components of XRD phenomenon

- X-ray tube.
- X-ray filter (like Nickel).
- Sample holder.
- X-ray detector.
- Soller and receiving and divergence slits.
- Circular photographic film.
- Monitor for data interpretation.

2.3 Sample Preparation

The sensitivity of the XRD technique makes it important for powdered samples like starch to be prepared well. The analysis is based on the pore size, orientation, thickness, and the removal of unwanted substances from the test sample [35]. Several research papers and books have highlighted the proper sample preparation for XRD [35, 36]. These are broadly categorized into three parameters of sample preparation [37]:

1. *Grinding or Milling of the Sample.*

   Dry grinding could be carried out by either mechanical or manual means. Ideally, a particle size of 1–5 μm is required. Besides this, particles with less than 20 μm of pore size are acceptable (*see* **Note 1**). Additionally, the spinning of the samples improve accuracy. On the other hand, *wet grinding* is done using mechanical mills in the presence of water or alcohol. A raw material of particle size less than 0.4 mm is required in this case.

2. *Filling or Formation of Sample Mount.*

   The preferred pretreatment method could be applied (*see* **Note 2**) by one of the several mounting or sample loading methods to achieve particle orientation. Following pretreatment, sample mounting is accomplished by either front loading, back loading or slice methods.

3. The mounting surface for front-loading (*see* **Note 3**) must be flat where the sample needs to be placed. This is achieved by a normal force and shear motion. The surface of the sample is made uniform and rough with the help of a razor blade. Further, some extra amount of powdered sample is added to facilitate randomness in peaks.

4. The sample is poured over the central hole drilled in the holder for back-loading. Press the sample with the help of a cylindrical disc. Remove the extra sample with the help of a rectangular object. Sprinkle some additional samples over the top.

5. Side-loading consists of a side-opened sample holder. It can be used for normal powders as well.

6. *Preservation of the Hydrated Sample.*

   The mounting of the hydrated samples must be done within 15 min. Liquid samples require pretreatment where these could be preserved as freeze-dried or spray-dried samples before diffraction analysis (*see* **Note 4**).

Refer **Note 5** for understanding the preferred orientation mounts. There are several factors that must be kept in mind before sample preparation for analysis [35]. These include:

1. The surface of the sample highly impacts the peak intensity obtained during analysis. The change in peak intensities could be minimized by preparing samples in viscous solutions followed by drying to obtain random orientations.

2. Besides orientation, the shape of the granule also forms a significant parameter. The anisotropic shape of the powdered sample could be reduced by grinding it. However, it must be ensured that the level of grinding does not exceed the extent which facilitates lattice dislocation.

3. Sample placement at the sample holder is crucial for estimating the peak positions. The height of the holder and thus the sample placement at too high or too low levels relatively affect the peak positions.

3 Methods

Distinct methods have been widely studied to develop a suitable procedure for evaluating starch crystallinity. As a result, the protocol could be considered a two-phase system for starch. However, the advances in understanding the molecular basis including orientation, lattice dislocation, folding, and crystalline growths led to the development of a procedure encompassing the crystal defects [8]. Powder XRD is one of the most widely used methods for starch analysis.

3.1 For Powder XRD Analysis (Fig. 5)

1. Starches are isolated from various sources like cereals, tubers, pulses, and other starch-rich commodities.
2. It is important to consider a powdered crystalline material as the sample.
3. The finely powdered sample is held inside a thin-walled capillary tube.
4. The monochromatic X-rays are filtered to pass them at a specific wavelength (generally Cu–K_α radiation at wavelength 1.540 A°) (*see* **Note 6**).
5. These pass through a collimator and form a narrow beam that falls over the sample through a hole in the photographic film.
6. A scan angle ranging between 0° and 40° is used for powdered starch samples.
7. A scan speed range of 1°–3°/min is used.
8. The minute crystals of the sample present in the random orientations will diffract the X-rays, thus satisfying Bragg's Eq. 1.
9. The diffraction peaks for the sample are noted based on the principle as shown below.
10. Since the entire procedure is based on Bragg's law, the detected intensities are evaluated using Bragg's Eq. 1, and the peaks are identified based on constructive interference.

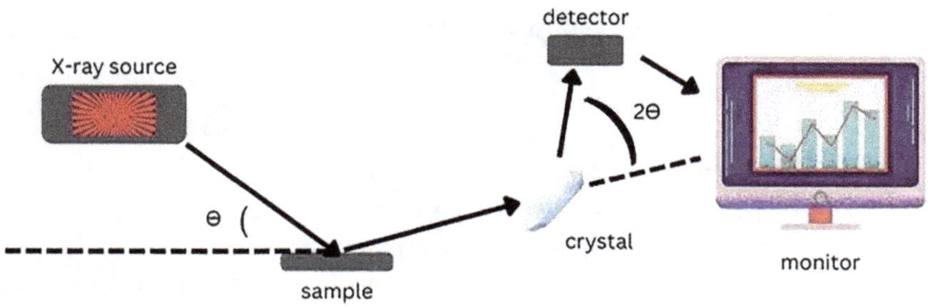

Fig. 5 Basic diffractometer system

Table 3
The structure and crystal parameters for A- and B-type polymorphs [18]

System	Unit cell characteristics	Polymorph
Monoclinic	$a = 11.90$ A° $b = 17.70$ A° $c = 10.52$ A° $\alpha = \beta = \Upsilon = 90°$	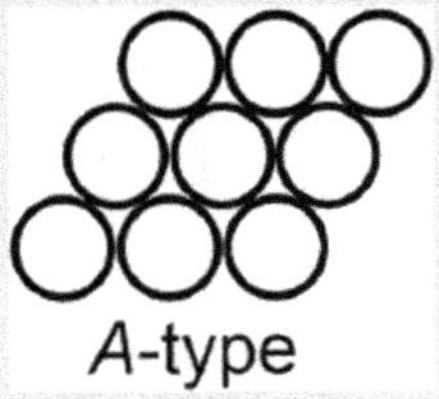
Hexagonal	$a = b = 18.50$ A° $c = 10.40$ A° $\alpha = \beta = 90°$ $\Upsilon = 120°$	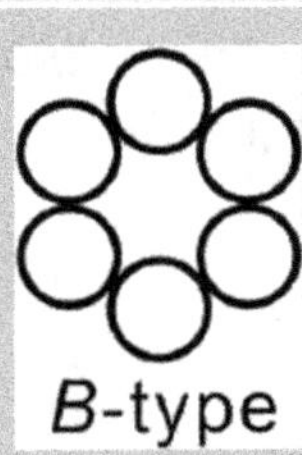

11. Now, the peak signals are detected with the help of an X-ray detector as shown in Fig. 3.
12. The undeviated X-rays will leave, whereas the diffracted rays will provide a pattern that is interpreted to analyze the results.
13. The detector is often connected to a computer, where the signals are converted into count rates, and the analysis could be done based on the output displayed on the monitor.

The protocol is applied to study starch characteristics like polymorph proportion, crystallinity, gelatinization, and retrogradation. For example, Lopez-Rubio et al. [8] conducted a study using a peak fitting procedure (discussed in further sections) to determine starch crystallinity. Several starch sources used for this study include maize, rice, potato, barley, and wheat varieties.

3.1.1 Relative Crystallinity of Starch Polymorphs

The A-, B-, and C- type polymorphs are widely studied to investigate the crystallinity of starch. Following Eq. 2, X-ray diffractograms can be used to determine the crystallinity index of starches.

$$\text{Crystallinity Index } (\%) = \frac{Ia - Ib}{Ia} \times 100 \qquad (2)$$

where I_a refers to the intensity of the lattice (002) diffraction and I_b refers to the diffraction in arbitrary units (same as I_a) at 2θ value representing amorphous material.

Even though both A- and B-type polymorphs possess double-helix structure, the former is known to have a compact structure compared with the open structure of the latter polymorph fraction.

XRD interpretation becomes easier with a brief understanding of the structure and lattice parameters for these polymorphs. The crystal parameters are shown in Table 3 [18, 38].

When it comes to C-type polymorph, the study of crystalline proportionality becomes important. The A-type is known to surround the B-type polymorph which is present at the center of the starch. Several studies like Bultosa and Taylor [39] and Noosuk et al. [40] reported very strong intensities for tapioca and sago starch at 2θ values of 16.8^0 and 17.1^0 alongside *d*-values of 5.28 A^0 and 5.19 A^0, respectively.

For descriptive evaluation, relative crystallinity is observed using one of the three methods, namely Fourier transfer infrared spectroscopy (FT-IR), differential scanning calorimeter (DSC), and X-ray diffraction (XRD). DSC is considered as a less reliable method due to the occurrence of crystallization during the heating process. This is overcome by FT-IR as it is a rapid and nondestructive technique compared to DSC. Further, it requires less sample preparation time [41]. The same research mentioned the absorption bands of 1047/cm and 1022/cm for crystalline and amorphous regions of starch, respectively. It could be noted that the results of FT-IR are in accordance with the XRD analysis. However, to get a more credible result, XRD analysis for the samples is conducted for starch crystallinity. The relative crystallinity for starch samples can be calculated using the following formula:

$$\text{Relative crystallinity (RC, \%)} = \frac{Ac\ X\ 100}{(Ac + Aa)} \quad (3)$$

where A_c represents the crystalline area and A_a represents the amorphous region on the X-ray diffractogram.

It must be noted that the relative crystallinity of starch is often inversely proportional to the amylose content [42]. Additionally, Sun et al. [33] used two methods to quantify the relative crystallinity of A- and B-type polymorphs in chickpea starch. The first one focused on a deconvolution method which was in accordance with that performed by Lopez-Rubio et al. [8]. The deconvolution method or the peak fitting procedure for estimating structural change in starch is usually performed by following the steps given below:

3.1.2 Selection of the Peaks for the Fitting Procedure

The diffraction pattern for the starch samples is obtained as per the XRD analysis performed. Several studies have been conducted to obtain the reference powder X-ray diffraction peaks. The selection of the peaks is also dependent upon the quality of the pattern obtained for starch samples. Further, it must be noted that the better the resolution of the peaks, the more convenient it is to initiate the fitting procedure without affecting the data analysis.

- The attempt to fit the peaks is initiated by comparing the reference peaks with the pattern observed. However, there are a wide number of uncertainties that may arise as a result of various fitting parameters.
- A specific peak is considered to fit the crystalline phase of the sample. The diffraction peak for A-type starch lies at 15° and 23° 2θ whereas B-type starch shows a peak at 2θ value of 17° [43] (*see* **Note 7**).
- The typical C-type polymorph shows a single peak at 2θ values of 17° and 23° (*see* **Note 8**).
- Besides these double-helix conformations, various amylose complexes are formed with lipids that are noted as single-helix conformations. These represent V-type crystallinity in starch. These reflections at 2θ values of 7°, 13°, and 20° (20° being the most common) are also considered for the peak fitting procedure.

3.1.3 Selection of the Most Suitable Fitting Function

- Gaussian is one of the most suitable functions used for this purpose. Lorentz and Voigt functions are also used; however, Gaussian lines provide the most desirable results [8].
- The approximate values are obtained by manual determination. These might not be completely accurate; however, they provide a vague approach to fitting the scattering profile of the sample.
- The goal of the fitting procedure is to obtain the least standard deviation values, thereby reducing the errors in the analysis. Attention to fitting parameters is crucial.
- The combined studies of XRD with other methods like FT-IR [44] also propose the use of fitting functions. The fitting of the holocrystalline peak is carried out using Gaussian lines within an FT-IR range of 1800–1000/cm [41].
- The Gaussian lines are used as a reference band for chickpeas to estimate the relative crystallinity of starch obtained.

3.1.4 Estimating Starch Crystallinity

- Starch, as we know, is considered a two-phase system consisting of both crystalline and amorphous regions. The relative crystallinity is calculated using the area under the diffractogram which is in accordance with Eq. 2.
- Consider the relative crystallinity of A-type and B-type starch as RC_A and RC_B, respectively where

$$RCA = \frac{\text{sum area of peaks constituting A} - \text{type polymorph}}{\text{area of total XRD pattern}}$$

$$RCB = \frac{\text{sum area of peaks constituting B} - \text{type polymorph}}{\text{area of total XRD pattern}}$$

3.1.5 For Investigating the Gelatinization of Starch

- The process involves the heating of starch in the presence of water. Further, starch is generally cooled before conducting diffraction analysis [45].
- For maize starch gelatinization, a diamond culet is used along with the gasket consisting of a 0.2 mm hole where the sample (starch) is being held.
- The screws are tightened, and the pressure is generally measured using ruby balls. Further, in situ X-ray diffraction is performed at room temperature.
- Unlike the procedure mentioned for starch crystallinity, a double-crystal monochromator is used for starch gelatinization. An X-ray beam of energy 18 keV is applied along with the wavelength of 0.68 Å. The diffraction patterns are recorded for a period of 10 min [46] (*see* **Note 9**).

3.1.6 For Retrogradation of Starch [47]

- Generally, a water addition method is applied.
- The X-ray diffraction pattern is attained using a diffractometer with an X-ray source of Cu–K_α radiation. A voltage of 45 kV is applied along with a scan speed of 3° per min.
- The peaks are generally not clear at 2θ angle less than 10°, due to which the scan axis of 10°–37° is used for the estimation of starch retrogradation (*see* **Note 10**).

3.2 Interpretation of XRD Data

To interpret the XRD data, it is of utmost importance to understand the structure and parameters of starch polymorphs (Table 1). However, the basic procedure to interpret the obtained XRD data is as follows:

- The data for X-ray diffraction were obtained in the form of an Excel worksheet.
- The interpretation lies with the conversion of this Excel data to a publishable graph to observe peaks.
- This can be achieved by one of the most commonly used software, Origin Pro 2018.
- The data is imported, followed by its labeling. The *x*-axis and *y*-axis are labeled as 2θ vs. intensity of the sample, respectively.
- If the peaks are not clear, the data is plotted as *y*-offset data, and the peaks are evaluated for crystallinity.
- The crystallinity index for the starch is calculated by either using Eq. 2 or the given formula:

$$\mathrm{CI}\ (\%) = \frac{At - Aa}{At} \times 100$$

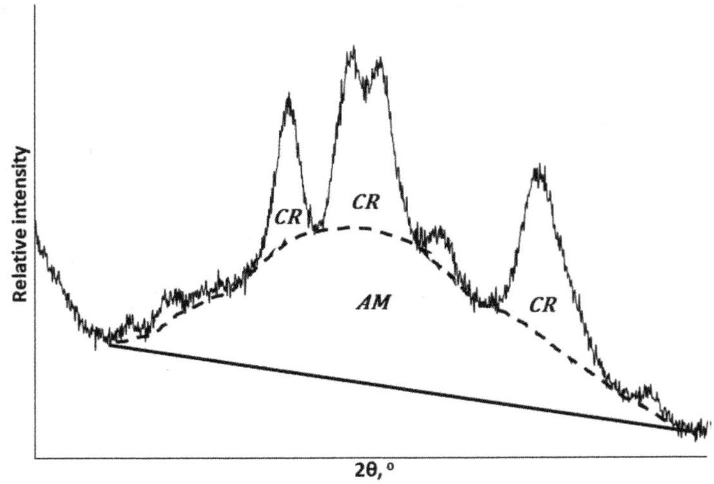

Fig. 6 Graphical representation to estimate the degree of crystallinity of native starch [18]

where A_t refers to the total area of the diffractogram, that is, which contains both amorphous and crystalline regions and A_a represents the area under the curve for amorphous regions.

A graphical representation of this formula is seen in Fig. 6.

Here, AM represents the amorphous region and CR represents the crystalline region. These regions are separated by a distinct smoothed line drawn.

Depending upon the starch source, the degree of crystallinity could be obtained for a specific polymorph. For instance, the A- and B-type structures are distinct for cereal and tuber starches, respectively. Similarly, studies have been conducted on tapioca starch to investigate the peaks for C-type polymorph [18]. The standard starch sources (Table 3) could be used to evaluate the XRD data for distinct polymorphs.

For evaluating A-, B-, and C-type polymorphs, corn, potato, and tapioca starches are used as sources. The characteristic peaks for A-type are observed at 2θ values of 15° and 22.93° and a doublet around the value of 17°. For B-type, peak values are generally observed around 5.5°, 15°, 17.2°, 19.6°, 22°, 24°, and 26.3°. On the other hand, peak values of 15° and 22.7° and doublet peaks at 17.12° and 18.19° are observed in tapioca starch for C-type polymorph (Fig. 1).

3.2.1 Millet Starch Representation

Millets have gained global importance due to their nutritional benefits and XRD could be a technique to evaluate the crystallinity of these underutilized crops [48]. Even though the starch polymorphs are of four types, that is, A-, B-, C-, and V-type, millets are known to exhibit A-type polymorphs as suggested for cereal starches.

Bangar et al. [48] highlighted a procedure to obtain XRD patterns for a variety of millets, which is summarized as follows:

- X-ray diffractograms are obtained using diffractometers.
- An ambient temperature of 25°C is used.
- The 2θ angle range of 4°–40° is used along with a step size of 0.02.
- A scan speed of 10 s is generally employed.
- Origin Pro 8E software is used for plotting the graph and the relative crystallinity is calculated using Eq. 3.
- A range of 24.73–31.62% is considered to be a significant relative crystallinity range for millet starches.
- When it comes to crystalline peaks, millets are known to exhibit "A-type" pattern with 2θ peak values at 15°, 17°, 18°, and 23.1°.

The same study also mentioned that the relative crystallinity was found to be highest for Proso millet starch as compared to other subjects. The general XRD representation for various millet starches can be seen in Fig. 7.

Besides the basic protocol for polymorphs, starch fractions like amylose and amylopectin could be studied by XRD. Subroto et al. [49] explained amylopectin analysis using XRD as:

1. Amylopectin is first separated from other components by soaking operations at temperature of 80 °C for 2 h.

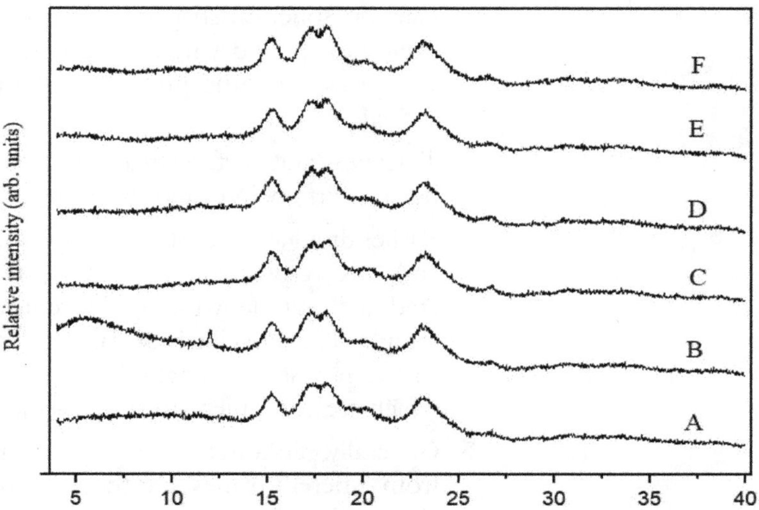

Fig. 7 XRD patterns for millet starches representing barnyard millet (A), finger millet (B), foxtail millet (C), kodo millet (D), little millet (E), and proso millet (F) [48]

2. Once the dried powder is obtained, it can be used to identify the crystalline structure of amylopectin.
3. Wide-angle X-ray diffractometer is used to identify the reflector angle.
4. The rays are passed using Ni filter.
5. The scan angle is generally in the range of 4°–33° 2θ.
6. A capture time of 2 s per step is employed.
7. A silicon-based X-ray detector, and a characteristic X-ray of Co–Kα radiation with a wavelength with 1.79 A° are used.
8. The result is interpreted using software like Origin and the peaks are analyzed.

4 Notes

1. Particle statistics are evaluated by the degree of fineness of the starch sample. For the purpose of quality evaluation, the samples are often grounded to achieve a uniform particle exposure. However, the extent of grinding is crucial in determining the deformation of structural aspects of the crystal that might occur. A characteristic particle size is preferable because large pore sizes produce various unknown contents/peaks in diffraction patterns. Also, wet grinding methods are preferable over dry grinding ones due to the formation of amorphous regions in the latter method of grinding.
2. Possible pretreatment methods involve the formation of irregular or spherical-shaped aggregates via employing freeze-drying or spray-drying procedures. Depending on the sample and procedure, the pretreatment method could be skipped as well.
3. Thickness and surface area must be kept in mind to avoid any beam overflow in case of front loading.
4. Earlier dried and wet starch samples were evaluated using XRD and DSC respectively. Nowadays, starch samples can be evaluated without dehydrating or removing the water from the sample. The most likely explanation for this could be attributed to the fact that wet samples provide sharp diffraction peaks, which are convenient to study [47].
5. Generally, random mounts are preferred for judging reflections from different planes. On the other hand, oriented mounts are also studied but are not much preferred due to their lower ability to show diffraction effects. The latter employs the preparation of the sample using water, that is, it results in the formation of slurry, which is poured over a glass or ceramic-

based slide constituting the flat surface of the sample holder. The evaporation of water is facilitated until the sample is streaked over the slide.

6. Copper is the most common material used as the XRD target due to its high melting point. A voltage of 40 kV and a current of 40 mA is applied, producing high heat. Further, the wavelength of copper is quite comparable to the interatomic distance.
7. The 2θ values of 10° and 11° for A- and B-type polymorph were chosen by [8] respectively.
8. The C-type polymorph shows few other small peaks at values of 15° and 5.6°. The C-type starch is categorized into C_A and C_B based on their closeness to other polymorphs. Both show similar peaks as typical C-type; however, the former shows an additional peak at 18° (2θ) whereas the latter shows two shoulder peaks at 22° and 24° 2θ.
9. One such study [59] was conducted to highlight starch gelatinization using XRD. The glass tubes were used for the sample preparation of starches. The moisture equilibrium is of utmost importance because high and intermediate moisture samples were prepared using starch and water. The glass capillary tubes were sealed and kept at ambient temperatures for 24 h. Generally, these are kept in a desiccator to avoid extra contact with moisture.
10. It has been noted that the addition of water greatly affects the diffraction peaks obtained. Furthermore, various XRD techniques, including wide-angle X-ray diffraction (WAXD) and small-angle X-ray scattering (SAXS), are being employed for studying molecular changes in the starch. The former technique is utilized for the estimation of long-range ordered structure of starch, whereas the latter is generally conducted to evaluate the repetitive amorphous as well as crystalline lamellae in starch [60].

5 Conclusion and Future Direction

The rheological properties of starch are mainly dependent on its chemical composition. XRD is one of the frequently used techniques for advanced characterization of food materials. Additionally, development in data analysis, artificial intelligence, and microelectronics has explored its effectiveness in characterization/analysis being broadly available and convenient to apply. XRD is widely applicable for the identification of unknown crystalline material like inorganic compound and degree of crystallinity of starch samples. Advancement in hardware part like source of X-ray, detectors, optics, helped in reducing analysis time with increased precision. Latest advancement includes adoption of liquid metal X-ray source that has remarkably decreased heat production that further helps to reach high beam intensity.

XRD with high resolution can offer new opportunities for advanced research regarding material characterization and identification. XRD analysis under nonambient conditions or in progress processes can be of great help in product or process improvement.

References

1. Punia S (2020a) Barley starch: structure, properties and in vitro digestibility – a review. Int J Biol Macromol 155:868–875
2. Punia S (2020b) Barley starch modifications: physical, chemical and enzymatic – a review. Int J Biol Macromol 144:578–585
3. Pérez S, Bertoft E (2010) The molecular structures of starch components and their contribution to the architecture of starch granules: a comprehensive review. Starch-Stärke 62(8): 389–420
4. Blazek J, Gilbert EP (2011) Application of small-angle X-ray and neutron scattering techniques to the characterisation of starch structure: a review. Carbohydr Polym 85(2): 281–293
5. Cheetham NW, Tao L (1998) Variation in crystalline type with amylose content in maize starch granules: an X-ray powder diffraction study. Carbohydr Polym 36(4):277–284
6. Frost K, Kaminski D, Kirwan G, Lascaris E, Shanks R (2009) Crystallinity and structure of starch using wide angle X-ray scattering. Carbohydr Polym 78(3):543–548
7. Brückner S (2000) Estimation of the background in powder diffraction patterns through a robust smoothing procedure. J Appl Crystallogr 33(3):977–979
8. Lopez-Rubio A, Flanagan BM, Gilbert EP, Gidley MJ (2008) A novel approach for calculating starch crystallinity and its correlation with double helix content: a combined XRD and NMR study. Biopolymers 89(9):761–768
9. Karim A, Norziah MH, Seow CC (2000) Methods for the study of starch retrogradation. Food Chem 71(1):9–36
10. Cairns P, Bogracheva TY, Ring SG, Hedley CL, Morris VJ (1997) Determination of the polymorphic composition of smooth pea starch. Carbohydr Polym 32(3–4):275–282
11. Correia P, Cruz-Lopes L, Beirão-da-Costa L (2012) Morphology and structure of chestnut starch isolated by alkali and enzymatic methods. Food Hydrocoll 28(2):313–319
12. Hanashiro I, Ikeda I, Honda O, Kawasaki S, Fujimori K, Takeda Y (2004) Molecular structures and some properties of starches from propagules of mangrove species. J Exp Mar Biol Ecol 309(2):141–154
13. Noda T, Takahata Y, Sato T, Hisamatsu M, Yamada T (1995) Physicochemical properties of starches extracted from sweet potato roots differing in physiological age. J Agric Food Chem 43(12):3016–3020
14. Imberty A, Buléon A, Tran V, Péerez S (1991) Recent advances in knowledge of starch structure. Starch-Stärke 43(10):375–384
15. Hizukuri S, Kaneko T, Takeda Y (1983) Measurement of the chain length of amylopectin and its relevance to the origin of crystalline polymorphism of starch granules. Biochim Biophys Acta 760(1):188–191
16. Hizukuri S (1985) Relationship between the distribution of the chain length of amylopectin and the crystalline structure of starch granules. Carbohydr Res 141(2):295–306
17. BeMiller JN, Whistler RL (eds) (2009) Starch: chemistry and technology. Academic
18. Dome K, Podgorbunskikh E, Bychkov A, Lomovsky O (2020) Changes in the crystallinity degree of starch having different types of crystal structure after mechanical pretreatment. Polymers 12(3):641
19. Purohit SR, Jayachandran LE, Raj AS, Nayak D, Rao PS (2019) X-ray diffraction for food quality evaluation. In: Evaluation technologies for food quality. Woodhead Publishing, pp 579–594
20. Singh V, Ali SZ, Somashekar R, Mukherjee PS (2006) Nature of crystallinity in native and acid modified starches. Int J Food Prop 9(4): 845–854
21. Alcázar-Alay SC, Meireles MAA (2015) Physicochemical properties, modifications and applications of starches from different botanical sources. Food Sci Technol 35:215–236

22. Nara S, Mori A, Komiya T (1978) Study on relative crystallinity of moist potato starch. Starch-Stärke 30(4):111–114
23. Chávez-Salazar A, Bello-Pérez LA, Agama-Acevedo E, Castellanos-Galeano FJ, Álvarez-Barreto CI, Pacheco-Vargas G (2017) Isolation and partial characterization of starch from banana cultivars grown in Colombia. Int J Biol Macromol 98:240–246
24. Derycke V, Vandeputte GE, Vermeylen R, De Man W, Goderis B, Koch MHJ, Delcour JA (2005) Starch gelatinization and amylose–lipid interactions during rice parboiling investigated by temperature resolved wide angle X-ray scattering and differential scanning calorimetry. J Cereal Sci 42(3):334–343
25. Purohit SR, Rao PS (2017) Water absorption and gelatinization kinetics of non-issuable rice and its characterization. J Food Meas Charact 11:2110–2118
26. Zhou Z, Ren X, Wang F, Li J, Si X, Cao R et al (2015) High pressure processing manipulated buckwheat antioxidant activity, anti-adipogenic properties and starch digestibility. J Cereal Sci 66:31–36
27. Nagataki A, Tomita H, Himeda Y, Takemori T, Fukuoka M (2018) A quantification method of retrogradation for cooked rice based on a single isolated peak in X-ray diffraction. J Cereal Sci 79:80–85
28. Jyothi AN, Sajeev MS, Sreekumar JN (2010) Hydrothermal modifications of tropical tuber starches. 1. Effect of heat-moisture treatment on the physicochemical, rheological and gelatinization characteristics. Starch-Stärke 62(1): 28–40
29. Nadiah NI, Uthumporn U, Syahariza ZA (2015) Effect of microwave heating on potato and tapioca starches in water suspension. Int J Adv Sci Eng Inf Technol 5(4):264–271
30. Bashir K, Aggarwal M (2017) Physicochemical, thermal and functional properties of gamma irradiated chickpea starch. Int J Biol Macromol 97:426–433
31. Olu-Owolabi BI, Afolabi TA, Adebowale KO (2011) Pasting, thermal, hydration, and functional properties of annealed and heat-moisture treated starch of sword bean (Canavalia gladiata). Int J Food Prop 14(1):157–174
32. Epp J (2016) X-ray diffraction (XRD) techniques for materials characterization. In: Materials characterization using nondestructive evaluation (NDE) methods. Woodhead Publishing, pp 81–124
33. Sun Y, Ye H, Hu B, Wang W, Lei S, Wang X, Zhou L, Zeng X (2015) Changes in crystal structure of chickpea starch samples during processing treatments: an X-ray diffraction and starch moisture analysis study. Carbohydr Polym 121:169–174
34. DeWitt S, Hahn N, Zavadil K, Thornton K (2015) Computational examination of orientation-dependent morphological evolution during the electrodeposition and electrodissolution of magnesium. J Electrochem Soc 163(3):A513
35. Bunaciu AA, UdriŞTioiu EG, Aboul-Enein HY (2015) X-ray diffraction: instrumentation and applications. Crit Rev Anal Chem 45(4): 289–299
36. Iyengar SS, Buhrke VE, Jenkins R, Smith DK (1997) Sample preparation for clays in preparation of specimens for x-ray fluorescence and X-ray diffraction analysis
37. Kleeberg R (2013) Preparation of powder specimen for quantitative XRD. T U Bergakademie Freiberg, Freiberg
38. Sarko AHCH, Wu HC (1978) The crystal structures of A-, B-and C-polymorphs of amylose and starch. Starch-Stärke 30(3):73–78
39. Bultosa G, Taylor JR (2003) Chemical and physical characterisation of grain Tef [Eragrostis tef (Zucc.) Trotter] starch granule composition. Starch-Stärke 55(7):304–312
40. Noosuk P, Hill SE, Farhat IA, Mitchell JR, Pradipasena P (2005) Relationship between viscoelastic properties and starch structure in rice from Thailand. Starch-Stärke 57(12): 587–598
41. Sun Y, Wu Z, Hu B, Wang W, Ye H, Sun Y, Wang X, Zeng X (2014) A new method for determining the relative crystallinity of chickpea starch by Fourier-transform infrared spectroscopy. Carbohydr Polym 108:153–158
42. Chang F, He X, Huang Q (2013) Effect of lauric acid on the V-amylose complex distribution and properties of swelled normal cornstarch granules. Cereal Sci 58:89–95
43. He W, Wei C (2017) Progress in C-type starches from different plant sources. Food Hydrocoll 73:162–175
44. Pozo C, Rodríguez-Llamazares S, Bouza R, Barral L, Castaño J, Müller N, Restrepo I (2018) Study of the structural order of native starch granules using combined FTIR and XRD analysis. J Polym Res 25:1–8
45. Ai Y, Jane JL (2015) Gelatinization and rheological properties of starch. Starch-Stärke 67: 213–224
46. Yang Z, Gu Q, Hemar Y (2013) In situ study of maize starch gelatinization under ultra-high hydrostatic pressure using X-ray diffraction. Carbohydr Polym 97(1):235–238

47. Taguchi T, Onishi M, Katsuno N, Miwa N, Oomoto C, Sato M, Sekita M, Yamaguchi H, Imaizumi T, Nishizu T (2023) Evaluation of starch retrogradation by X-ray diffraction using a water-addition method. LWT 173:114341
48. Bangar SP, Siroha AK, Nehra M, Trif M, Ganwal V, Kumar S (2021) Structural and film-forming properties of millet starches: a comparative study. Coatings 11(8):954
49. Subroto E, Jeanette G, Meiyanasari Y, Luwinsky I, Baraddiaz S (2020) Review on the analysis methods of starch, amylose, amylopectinin food and agricultural products. Int J Emerg Trends. https://doi.org/10.30534/ijeter/2020/103872020
50. Barroso AG, Garcia RH, Del Mastro NL (2019) X-ray diffraction pattern and relative crystallinity of irradiated arrowroot starch. Braz J Radiat Sci 7(2A)
51. Li W, Bai Y, Mousaa SA, Zhang Q, Shen Q (2012) Effect of high hydrostatic pressure on physicochemical and structural properties of rice starch. Food Bioprocess Technol 5:2233–2241
52. da Costa Pinto C, Sanches EA, Clerici MTPS, Pereira MT, Campelo PH, de Souza SM (2021) X-ray diffraction and Rietveld characterization of radiation-induced physicochemical changes in Ariá (*Goeppertia allouia*) C-type starch. Food Hydrocoll 117:106682
53. Tattiyakul J, Naksriarporn T, Pradipasena P (2012) X-ray diffraction pattern and functional properties of *Dioscorea hispida* Dennst starch hydrothermally modified at different temperatures. Food Bioprocess Technol 5:964–971
54. Li Y, Zhao L, Lin L, Li E, Cao Q, Wei C (2022) Relationships between X-ray diffraction peaks, molecular components, and heat properties of C-type starches from different sweet potato varieties. Molecules 27(11):3385
55. Mutungi C, Passauer L, Onyango C, Jaros D, Rohm H (2012) Debranched cassava starch crystallinity determination by Raman spectroscopy: correlation of features in Raman spectra with X-ray diffraction and 13C CP/MAS NMR spectroscopy. Carbohydr Polym 87(1):598–606
56. Polesi LF, Sarmento SBS, de Moraes J, Franco CML, Canniatti-Brazaca SG (2016) Physicochemical and structural characteristics of rice starch modified by irradiation. Food Chem 191:59–66
57. Rahaman A, Kumari A, Zeng XA, Farooq MA, Siddique R, Khalifa I et al (2021) Ultrasound based modification and structural-functional analysis of corn and cassava starch. Ultrason Sonochem 80:105795
58. Rafiq SI, Jan K, Singh S, Saxena DC (2015) Physicochemical, pasting, rheological, thermal and morphological properties of horse chestnut starch. J Food Sci Technol 52:5651–5660
59. Zobel H, Young S, Rocca L (1988) Starch gelatinization: an X-ray diffraction study. Cereal Chem 65(6):443–446
60. Wang S, Wang J, Zhang W, Li C, Yu J, Wang S (2015) Molecular order and functional properties of starches from three waxy wheat varieties grown in China. Food Chem 181:43–50

Chapter 8

Starch Digestibility Protocols

Priyanka Kajla, Sangeeta Yadav, and Mukesh Gaur

Abstract

Starch is a major energy source in the diet and is an easily absorbed form of carbohydrates. Maltose, maltotriose, and other branched oligosaccharides, glucose, and limit dextrins are produced when digestive enzymes hydrolyze starch in food. These branched oligosaccharides are then converted into glucose by the brush border enzymes. The percentage of starch absorbed in the small intestine is related to starch digestibility. A fraction of starch that can only be digested under specific circumstances is known as digestibility, a crucial metabolic reaction. In vitro and in vivo procedures are typically used to evaluate starch digestibility. Following various sample preparation techniques, different in vitro methodologies have been employed to assess starch digestibility of starchy foods. However, in-vivo approaches use experiments on human subjects to measure glycemic response. The glycemic index (GI), despite being time- and money-consuming, is the widely accepted assessment method for starch digestibility in the food industry. This chapter attempts to organize the starch digestibility methods under the headings of in vitro and in vivo methods. The chapter covers the theory of the method used, materials required, sample preparation, detailed procedure, observations, inferences, and precautions if applicable.

Key words In vitro, In vivo starch digesta, Amylases, Glycemic index, Starch digestibility index, Ileostomy, Nylon bags

1 Introduction

Starch is the major storage polysaccharide present in plants, and starch-rich food is an important energy source in the human diet [1, 2]. Humans primarily obtain their nutritional energy from starch, which in the typical Western diet accounts for 37–70% of all the calories consumed as glucose [3]. Foods made from cereal, grains, tubers, and pulses contain an adequate amount of carbohydrates in the form of starch. Along with varying fat and protein levels, starch is bundled as granules within plant cells confined by cell walls. These granules prevent water and enzymes from being generally accessible, which affects how well starch is digested in the gastrointestinal system [4]. Thus, starch digestibility affects the

Sneh Punia Bangar (ed.), *Standardized Procedures and Protocols for Starch*, Methods and Protocols in Food Science,
https://doi.org/10.1007/978-1-0716-3866-8_8,

metabolic responses variably depending upon food sources, which is a vital consideration for designing healthy diets, viz., low glycemic index foods.

1.1 Overview of Starch Hydrolysis and Digestibility

All of the starch in food is hydrolyzed, first by salivary amylase and then by pancreatic amylase in the small intestine. This hydrolysis results in the production of maltose, maltotriose, and other branched oligosaccharides known as limit dextrins. The brush boundary enzymes such as sucrase-isomaltase and glucoamylase maltase further hydrolyze these oligosaccharides into glucose [5]. According to the pace and duration of the glycemic reaction, which is a general indicator of digestibility, starch and starchy food components can be categorized. The microbiota in the colon ferments starch to varying degrees after it has reached there because a sizeable portion of it is not digested and absorbed in the small intestine. This fraction of starch is termed resistant starch [6]. The amount of starch absorbed in the small intestine is correlated with starch digestibility. One of the most popular methods of classifying starches is based on the kinetics of in vitro digestion. The technique is based on the in vitro digestion of starch, where digestion conditions are simulated, and glucose release is measured at various intervals [7] (Table 1).

The Englyst method of classifying starches is based on physiological processes that occur during the hydrolysis, absorption, and assimilation of starch by digestive enzymes, primarily in the small intestine. The majority of starch hydrolysis is caused by the action of both alpha-amylase and amyloglucosidase enzymes. According to the Englyst categorization of starch fractions, a fraction that is slowly digested results in a sustained release of glucose in the intestinal tract due to the pace and quantity of hydrolysis. The other component is called resistant starch which is fermented in the large intestine and results in short-chain fatty acids that have been related to preventing colon cancer, lowering cholesterol production, preventing fat buildup, and improving mineral absorption. RS has a poor calorie supply and is neither digested nor absorbed in the small intestine [1, 7].

Table 1
Classification of different starch fractions based on digestibility

Classification of starch	Characteristics of digested starch
Rapidly digestible starch (RDS)	Amount of glucose released after 20 min of digestion
Slowly digestible starch (SDS)	Amount of glucose released between 20 and 120 min of in vitro digestion
Resistant starch (RS)	Total starch minus the amount of glucose released within 120 min of in vitro digestion

Fundamental characteristics of starch, such as size and shape of granules, ratio of amylose to amylopectin, chain length of amylose and amylopectin, order of arrangement, and fractionation method, affect its digestibility. These characteristics are crucial when food is consumed as a complex matrix and when processing is carried out utilizing starch ingredients with a low glycaemic index [8]. In some starches (potato starch), the presence of monoesters of phosphate and phospholipids impart a unique gel-forming capacity. Because of the phosphate groups located at the amylopectin branching points, such gels exhibit high peak viscosity and great transparency, thereby affecting the susceptibility to digestive enzymes and hence its starch digestibility. Additionally, the presence of lipids in starches of maize and rice results in the formation of inclusion complexes with amylose during gelation, thereby affecting the digestibility of starch as these complexes limit the accessibility to digestive enzymes. Starch becomes a poor ground/substrate for amylolytic enzymes owing to the complex formation between phosphate groups and amylopectin fraction, which alters the structural and, hence, the digestibility and functional properties of starch [9]. The nutritional value of starch is paramount in both its natural and processed forms of consumption, where the type of the product (food matrix), its processing, and other constituents (such as lipids, proteins, and polyphenols) have a significant impact on the digestibility of starch and the glycaemic response [10]. Nonetheless, in the past decade, there has been growing interest in processing starch and its fractions less liable to starch hydrolyzing enzymes, resulting in a retarted and meager supply of glucose to the bloodstream. People who have diabetes and other health issues associated with obesity are drawn to foods with low digestible starch [8, 11]. The starch in food is gelatinized during processing, and the reorganization of its polysaccharide components begins as soon as the food is cooled. The amylose chains that leak out of the starch granules during the gelatinization process are ordered in a parallel or elongated pattern during retrogradation. This effect is exacerbated by storage and as a result of this configuration, the resistant starch is increased, thereby affecting starch digestibility. Additionally, native starch granules might add to the total RS fraction in the diet. Amylose-lipid complexes that are produced during the thermal digestion of lipid-containing starchy meals can also be found in the RS fraction. The presence of different food matrix components, such as proteins and lipids that can interact with starch, alters the product's sensory qualities as well as how quickly the polysaccharide can be absorbed. However, the usage of lipids results in the formation of amylose-lipid complexes, which have a reduced capacity to digest starch. Starch that is partially gelatinized is frequently seen in food. It contributes to the SDS percentage and preserves the crystalline structure, primarily in the form of A-type grain starches [9, 12]. Another aspect of the food matrix that may slow down starch digestion is its physical structure, which can be influenced by the degree of cooking. One example is pasta, which has a compact and dense structure that causes a surprisingly delayed rate of starch digestion and glycaemic response [13]. The crystalline structure is altered by acid hydrolysis, followed by esterification to create cross-linking between the short amylopectin chains. These modifications in the starch structure also lead to higher RS concentrations [14]. In a typical meal matrix made up of cooked rice, the principal nutrients are protein and starch. According to studies, when rice starch undergoes structural changes during cooking that affect its digestibility, amylopectin promotes a structure that withstands enzymatic hydrolysis compared to rice starch, which has a greater proportion of short chains [15]. Thus, it can be concluded that processing treatments have a profound impact on the digestibility of starch (Fig. 1).

1.2 Starch Structure and Its Influence on Digestibility

The content of slow-digested starch in raw grain starches is substantial. Additionally, it is said that the native grain starches' slow digestion characteristics are due to their semicrystalline structure (A type) with a large concentration of short chains and branches. However, a parabolic connection between amylopectin chains and slowly digestible starch levels has been identified, as amylopectin with long chains consists of significantly more slowly digested

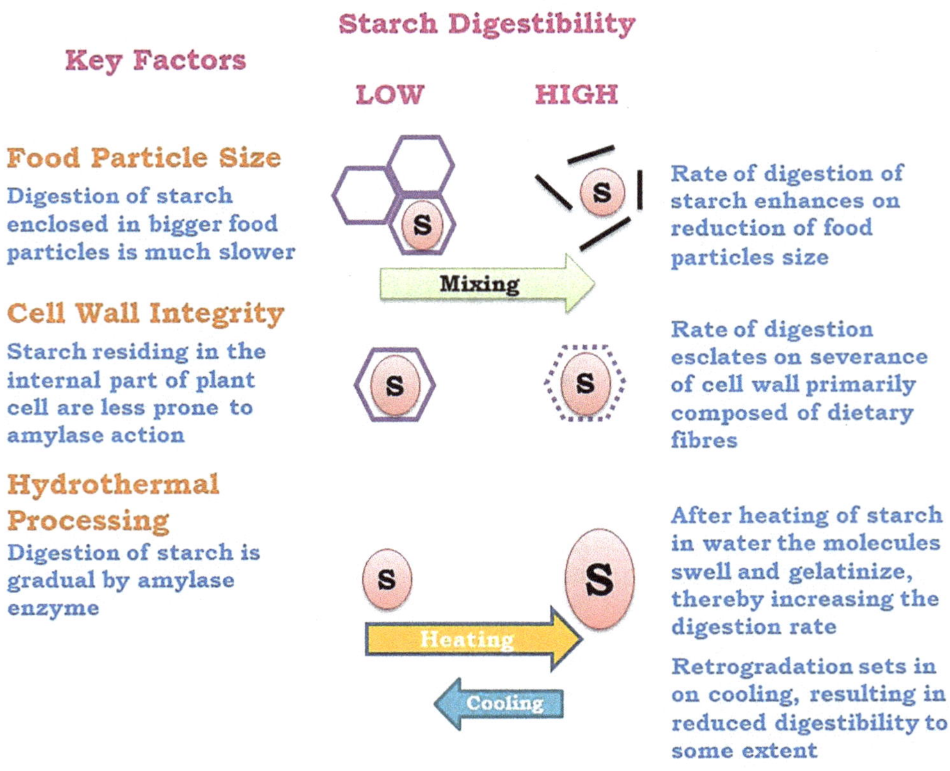

Fig. 1 Key factors affecting digestibility of starch

starch than amylopectin with short chains [16, 17]. Cereal starches have pores and channels that connect the central hilum area to the surface, and these channels are arranged in concentric circles in both their crystalline and amorphous sections. Amylose and amylopectin, the two components of starch, are arranged in concentric rings with crystalline and amorphous regions, and branching patterns affect the digestion properties of starch. When amylolytic enzymes are used to treat starch granules, which travel through channels present inside granules resulting in slow digestion both ways, upended and juxtaposed [17]. Enzymatic digestion typically starts at the surface pores followed by channels deep inside; nevertheless, this simultaneous digestion of crystalline and amorphous regions eventually widens the channel [18]. The structural characteristics of high-amylopectin starch (native waxy maize starch) with multiple short chains and a concurrent high branching density resulted in the synthesis of slowly digested starch, determined by debranching time and enzyme concentration [19]. It is also revealed that raising the enzyme concentration and reducing debranching time also results in an increment of slowly digestible starch SDS, having amorphous and small amounts of partially organized double helical crystalline areas. Thus, it can be concluded that native starches have slow-digestion qualities due to their physical structure, particularly those with a high branching density and an A-type pattern [17, 18]. In order to correlate the structural features of debranched waxy maize starch with in vitro digestibility, pullulanase was used in the investigation. Pullulanase-debranched starches showed a B-type structure in the form of aggregation and reordering of double helices from short linear chains in a crystalline structure, increasing resistant starch content [20]. Six kinds of native waxy rice starches were analyzed for their crystalline and molecular compositions and their effects on digestibility [21]. The cultivar with the average amylopectin chain length and a high percentage of long-branched chains had greater relative crystallinity and RS content than the other cultivars. The degree of polymerization in rice starches was decreased by a combination of acid and heat-moisture treatments, and overall, RS content rose. Shorter chains produced by citric acid hydrolysis are cross-linked by the esterification of citric anhydrides, altering the structure of the starch in ways that digestive enzymes are unable to identify while also raising the RS content [14]. According to a structural–digestibility relationship in rice starches, rice lines with changed starch biosynthesis may have higher RS values because these produce a greater percentage of intermediate and long chains in the amylopectin molecule [15]. According to a study, the amylopectin short chains created weak spots in the crystalline structure of starch, making it more vulnerable to breakdown by digestive enzymes [22]. The digestibility of normal and high amylose maize starches was compared in order to assess the impact of the amylopectin structure on starch digestibility. In comparison to high-amylose starch, normal maize starch had a larger level of short chains, and high proportions of long chains were identified in high-amylose starch, according to the chain-length distribution [20, 23]. The resistance of the starch to hydrolysis by digestive enzymes is related to the existence of long chains in the starch structure. As a result of the explanation above, it can be inferred that amylopectin chain length, distribution, and crystallinity affect how easily starch may be digested.

2 Starch Digestibility and Protocols

In the human digestive system, starch molecules are broken down into smaller oligosaccharides, which are then converted into glucose in the small intestine by the brush border enzymes sucrose-isomaltase and maltase-glucoamylase [24]. To evaluate the nutritional value of foods, the digestibility of starch has been linked to the glycemic index. For starch-based foods, the glycemic index has been employed as a marker of the postprandial glucose response; these items are categorized based on this reaction [18]. Therefore, in vitro and in vivo tests are usually employed to assess starch digestibility.

2.1 Overview and Significance of Estimation of In Vivo Digestibility of Starch

Majority of in vivo studies on people and animals have used digesta collected from the terminal ileum, colon, or feces despite the fact that starch is mostly digested in the small intestine [18, 25]. These digesta merely indicate the structure of resistant starch that is not digested in the small intestine and fail to sufficiently clarify the mechanism of starch digestion therein, especially the differences between rapidly digestible and slowly digestible starch. It takes a long time (ethical approval and volunteer recruitment), costs a lot (paying trained medical staff and disposing of clinical waste), and is expensive to test the digestibility of food commodities in vivo (with human subjects) to measure glycaemic response. Generally, the high intra- and inter-individual variability seen in biological systems, and gastrointestinal starch absorption in particular, requires stringent adherence to testing methodologies and factors in these bioassays. In addition, performing human (or animal) studies has fundamental ethical and technical limitations [26]. A sophisticated, multimodal orchestration of various physicochemical processes is required for the digestion and absorption of starch. Eating habits, a person's general health, their nutritional state, and their reproductive and mental health all have an impact on the gut ability to digest starches efficiently and effectively. To find out how much starch is digestible or, alternatively, how much escapes (resists) digestion in the upper gastrointestinal system, the typical approach is to collect and analyze ileal digesta for starch. The amount found there is then compared to the total amount of starch taken throughout the research period. The majority of investigations usually employ four to eight volunteers kept on a low- or no-starch diet during the experiment. Stomal effluent is then collected over the course of the following 24 h. The contents of the ileostomy bag must be removed often, typically every 2 h or less, and frozen right away to prevent starch from being degraded by human amylases or bacteria that are typically found in intestine contents.

2.2 Overview and Significance of Estimation of In Vitro Digestibility of Starch

Compared to in vivo procedures, in vitro techniques may be able to forecast changes in digestibility resulting from modifications in food processing more rapidly and affordably. in vitro techniques have also been used to categorize foods based on aspects of digestion that mirror in vivo conditions and to pinpoint foods with slow-release carbohydrates. Starchy foods are frequently described using an in vitro method that uses controlled enzymatic hydrolysis to mimic the pace and degree of starch digestion in the human stomach. A schematic representation of starch digestibility using different enzymes in the human gastrointestinal tract is shown in Fig. 2. This technique is useful to anticipate the glycemic response to dietary carbohydrates [27].

Compared to in vivo procedures, in vitro techniques may be able to forecast changes in digestibility resulting from modifications in food processing more rapidly and affordably. In vitro techniques have also been used to categorize foods based on aspects of digestion that mirror in vivo conditions and to pinpoint foods with slow-release carbohydrates. Starchy foods are frequently described using an in vitro method that uses controlled enzymatic hydrolysis to mimic the pace and degree of starch digestion in the human stomach. To anticipate the glycemic response to dietary carbohydrates, this technique is useful. The results of starch digestion in vitro may

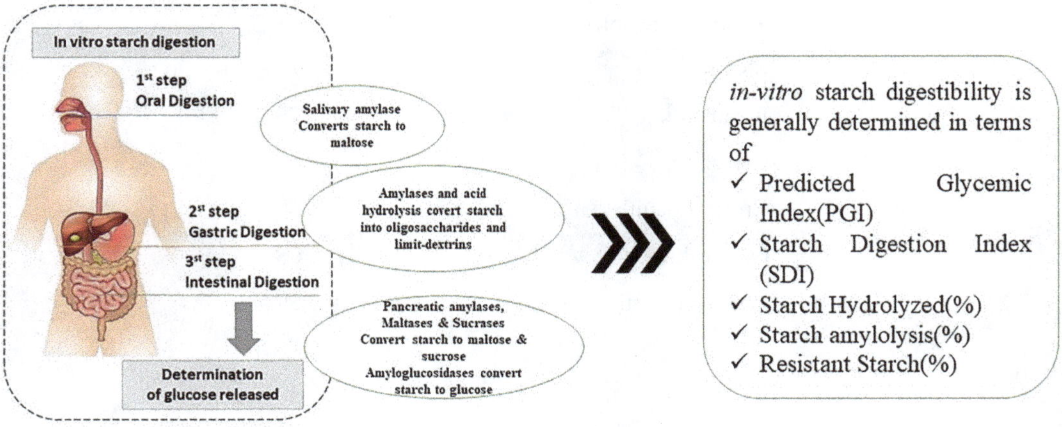

Fig. 2 Starch digestibility using digestive enzymes and expression of results of starch digestibility via in vitro digestibility methods

not be representative of starch digestion in vivo in the small intestine, and differences in the rate of hydrolysis between in vivo and in vitro investigations have been documented [10, 28]. Due to this, a quick, reliable, and inexpensive in vitro digestion assay is required that can not only identify plant breeding materials with a low rate of starch digestibility but also does not necessarily replicate the complex interactions between food and gastrointestinal tract. Numerous in vitro techniques have been employed to assess digestibility of starch from different sources presented in Table 2 [29, 30], which are usually the adopted methods developed in early 1990s [7]; however, the protocols are not reliable and consistent. Different methods of sample preparation involve grinding grains to flour, hydration followed by homogenization, and so on. There is no precise enzyme formula; some assays combine amylases with lipases, proteases, and glucoamylases, while other methods use only amylases. Rapidity and a fair cost per assay are requirements for a high-throughput digestibility method [30].

Table 2
Glycemic index of different starch-rich foods assessed using in vitro digestibility methods

Starchy foods	Starch digestibility (in vitro) as glycemic index (GI)
White bread (WS)	68.2
Chicken sandwich (CS)	45.9
Croissant (CR)	45.5
Doughnut (Donut)	47.9
Samosa vegetables	61.7
Biscuit	71
Bread	75
Cereal	80
Cream biscuit	60
Oatmeal biscuit	53
Pasta spaghetti	59
Potato	75
Pulse	42
Rice	68

Source: [31–34]

3 Materials

3.1 For In Vivo Studies

3.1.1 The Human Ileostomy Model [35]

Subjects

Identify adults who have permanent ileostomies that are conventional, well-established, and functioning, with only a small (10 cm) amount of bowel resection. Recruitment of patients who underwent ileostomy surgery to treat ulcerative colitis is necessary. During the trial, four and eight subjects consumed little or no starch.

3.1.2 Combined CO_2/H_2 Breath Test [36]

Subjects

Each participant in the study must be in good health, have no known gastrointestinal conditions, have abstained from antibiotic use for at least 4 weeks, and have given their informed consent. All breath tests should be conducted the day before testing, following an overnight fast of at least 8 h, and without dietary fiber or products that are naturally high in 13C. The volunteers abstained from smoking, exercising, and other behaviors that are known to affect breath CO_2 and H_2 exhalation (*see* **Note 1**).

3.1.3 In Vivo Glycemic Index Method [37, 38]

Subjects

Men and women aged 21–24 were selected as healthy panelists (*n* varies with test foods). The panelists should have body mass indices (BMIs) between 18.26 and 22.04 kg m^2. BMI is calculated as the product of height (m) squared and weight (kg) (*see* **Note 2**).

Test Foods

Prepare test foods and glucose powder (food grade).

3.1.4 In Situ Mobile Nylon Bag Method [39]

Theory

Invivo small intestine digestibility of mobile bag starch could be predicted using either rumen preincubation or without it. By comparing in vivo data with mobile bag starch disappearances, the mobile bag approach provides a quick way for measuring intestine and whole-tissue starch digestibility.

Samples

The current experiment must conform with the requirements of the Indian Animal Welfare Board.

- Two nursing Holstein cows are equipped with cannulas for post-ruminal digesting in the duodenum (open T-piece implanted 60 cm caudal to the pylorus) and ileum (20 cm cranial to the caecum).

- Two dry Holstein cows fitted with rumen cannulas for rumen preincubation.
- During typical rumen incubation, the cows are fed the recommended dry cow ration (*see* **Note 3**).
- Use nursing cows with a 16 kg dry matter feed intake and an average daily milk output of 27 kg for intestinal incubations (see **Note 4**).
- Examine several meal samples for in vivo research.

3.1.5 Starch Digestibility Using In Vivo Digesta [28]

Raw normal maize starch, MOPS sodium salt (M-9381), D-Glucose (GOPOD) kit (K-GLUC), thermostable R-amylase from *Bacillus licheniformis* (E-BLAAM), amyloglucosidase from *Aspergillus niger* (E-AMGDF), absolute ethanol, 80% ethanol, sodium azide (0.02% w/v), acetate buffer, dimethyl sulfoxide, and pig diet (*see* **Note 5**).

3.1.6 In Vivo Glycemic Response Using Mice [40]

Reagents

High amylose maize starch (HAMS), potato starch (PS), ascorbyl palmitate (AP), and glucose (D-(+)-glucose, $C_6H_{12}O_6$, 99.5%). Total starch assay kit (K-TSTA) and digestible starch and resistant starch assay kit (K-DSTRS). Test strips and a blood glucose meter. Dimethyl sulfoxide (DMSO), ethanol (200 proof), sodium hydroxide (NaOH), sodium maleate ($C_4H_4Na_2O_5$), hydrochloric acid (HCl, 1 N), glacial acetic acid (CH_3COOH), and desiccant.

Sample Preparation

Starch inclusion complexes with guest compound AP are made using two different techniques.

1. DMSO method: Traditional method.
   - In a water bath set to 90 °C, dissolve 500 mg of HAMS in 10 mL of 95% (v/v) DMSO.
   - After that, predissolve the guest compound (AP) (50 mg) in 1 mL of 95% (v/v) DMSO. Mix starch and guest solution for 15 min at 90 °C before adding warm water with vigorous stirring.
   - Keep it overnight, then centrifuge it for 15 min at 3000 rpm to recover the precipitate, and then wash it with aqueous ethanol.
   - Pulverized into fine powders and allowed to dry in a desiccator at room temperature (20 °C) before further examination. The HAMS-AP ICs (high amylose maize starch—ascorbyl palmitate inclusion complexes) are the starch–AP ICs made using the DMSO technique.

2. "Empty" V-type method: Novel method
   - Guest compounds (AP), premade "empty" V6h starch, and 40% (v/v) aqueous ethanol solution are combined in a 1: 2 weight ratio to generate the V-type IC samples. The suspension is then heated to 70 °C for 10 min.
   - Rinse the mixture with ethanol, dry for 48 h in a vacuum desiccator, and crush it into fine powder for further analysis.
   - The starch-AP ICs made using these "empty" V-type techniques are known as V6h-AP ICs.

   Hydrothermal Treatments on Inclusion Complex Samples
   - To starch ICs produced by annealing (ANN) and acid hydrolysis (ACH), also known as ANN-ACH, apply hydrothermal treatments. Starch-AP IC samples are heated in 40% (v/v) aqueous ethanol for 10 min at 75 °C (*see* **Note 6**).
   - Recover the samples that were subjected to ANN treatment, wash with ethanol, and let it dry overnight. Samples that have been dried and annealed are gathered and ground into a fine powder.
   - After being annealed, IC samples should be dispersed in 1 N HCl (2%, w/v) and incubated for 2 days at 40 °C with 2 min of stirring each day.
   - Then wash the recovered sample with ethanol, and let it dry in a desiccator overnight. Then pulverize dried ANN-ACH samples into fine powders for further examination.

3.2 For In Vitro Studies

3.2.1 Using Pepsin Enzyme [34]

Processed Foods

Sample

White bread, croissant, doughnut, chicken sandwich, and veggie samos

Sample Preparation

(a) Boil all samples (50 mg) in sealed tubes in tap water (5 mL) until soft, except white bread, biscuits, and crisp potatoes.

(b) Soxhlet extraction with petroleum ether is used to defat samples containing more than 5% fat (biscuits and crisp potatoes).

(c) Using a homogenizer with a controlled speed, homogenize samples in a liquid media, such as water or buffer in the case of raw samples.

Reagents

Glucose oxidase-peroxidase kit (*see* **Note** 7), Tris-maleate buffer solution (*see* **Note 8**), pepsin, alpha-amylase, glucose, amyloglucosidase.

Whole Grains (Rice and Barley)

Sample Preparation

Whole grains (100 mg); 1:2 whole grains-to-water ratio and incubate at 37 °C.

Reagents

Tris-maleate buffer (0.05 M, pH 6.9) (*see* **Note 8**), HCl–KCl buffer (0.05 M, pH 1.5) (*see* **Note 9**), sodium acetate buffer (0.4 M, pH 4.75) (see **Note 10**), pepsin (4.12 U/mL or 95.9 μg/mL) (*see* **Notes 11** and **12**), aqueous NaOH (0.163 M), α-amylase (0.176 U/mL or 11.7 μg/mL), amyloglucosidase (AMG 2.75 U/mL), GOPOD reagent kit (*see* **Note 13**).

Pasta Products [41]

Sample Preparation

All samples of pasta former foods (pff) should be ground to a thickness of 1 mm in a laboratory mill before being analyzed for dry matter (DM), ash, crude protein (CP), and ether extract (EE) in accordance with AOAC guidelines. An Ankom 220 fiber analyzer is used for estimation of neutral detergent fiber (NDF), acid detergent fiber (ADF), and acid detergent lignin (ADL), as well as heat-stable amylase with an activity of 17.40 Liquefon units/mL for assessment of structural carbohydrates as per AOAC standards.

3.2.2 Using Amylase Enzyme [32]

(a) *Samples* Pulses, tubers, cereals, bread, biscuits and pasta.

Sample Preparation Cooking raw food samples in accordance with the directions on package, followed by grinding the ingredients to fine particles (250–500 μm) followed by sieving. Drain canned products followed by grinding into fine paste. Take such conditions for this study in an effort to protect dietary components intact in a similar manner as these will reach the intestinal lumen. Standardize factors like cooking, chilling conditions, and particle size to promote optimal reproducibility.

Reagent Required Phosphate buffered saline (*see* **Note 14**), pancreatic α-amylase (*see* **Note 15**), 0.3 M Na_2CO_3 (*see* **Note 16**).

(b) *Samples*: Grains [32].

Sample Preparation: Whole grains (100 mg); soak whole grains in water (1:2) and incubate at 37 °C.

Reagent Required α-amylase, pepsin (*see* **Note 17**), NaOH (0.2 M), sodium acetate buffer (0.1 M, pH 6.0), pancreatin (2 mg/mL buffer), AMG (25.4 U/mL or 0.41 mg/mL buffer), GOPOD assay

(c) *Samples:* Maize starch and ingredients of the pig diet [28].

Sample Preparation Porcine pancreatic R-amylase type VI-B pepsin from gastric porcine mucosa (lyophilized powder), pancreatin from porcine pancreas (*see* **Note 18**), MOPS sodium salt and lithium bromide. D-Glucose (GOPOD Format) kit (K-GLUC), thermostable R-amylase from Bacillus licheniformis, amyloglucosidase from Aspergillus niger (*see* **Note 19**), and isoamylase from Pseudomonas sp. dimethyl sulfoxide (GR for analysis ACS).

3.2.3 Using Combination of Enzymes Pepsin: Amylase

Sample

Whole grains [32].

Sample Preparation

Whole grains (100 mg); soak grains in water (1:2) and incubate at 37 °C.

Reagent

0.1 M sodium potassium phosphate buffer (pH 6.9), (*see* **Note 20**) α-amylase, amylolysis, aqueous HCl (pH ~0.92), pepsin, aqueous NaOH (pH ~12.6), pancreatin (28.4 μg/mL), amyloglucosidase (13 U/mL or 0.208 mg/mL), AMG solution (33.3 U/mL in 0.4 M sodium acetate buffer, pH 5.0)

4 In Vivo Starch Digestibility Methods and Protocols

4.1 Human Model

4.1.1 The Human Ileostomy Model [35]

Theory: It is a straightforward approach for determining nutrient balance and a direct way to examine starchy food and meal digestion in the human small intestinal tract.

Procedure

- Participants must consume 100 g of a starchy test food.
- Next, for the following 24 h, collect stomal effluent through an ileostomy bag often, typically every 2 h or less, and freeze it to reduce the possibility of starch breakdown by human amylase enzyme or bacteria that are typically present in contents of the intestinal tract (*see* **Note 21**).
- To establish a baseline for starch excretion, prepare a control sample in which the test foods are not consumed (*see* **Note 22**).
- Starch analysis of ileal digesta (*see* **Note 23**).
- Evaluate it against the overall quantity of starch consumed throughout the research duration.
- Estimate how much food resists digestion in the upper gastrointestinal system or vice versa.

4.1.2 Combined CO_2/H_2 Breath Test [36]

Theory

Using hydrogen breath tests to measure H_2 exhalation, estimations of carbohydrate fermentation in the large intestine can be obtained in a semi-quantitative manner. However, precise measurements cannot be made as the amount of exhaled H_2 varies and can be used through different pathways during methanogenesis or sulfate reduction. A more accurate test is therefore produced by combining the H_2 breath test with another marker, like $13CO_2$. Because the Hatch-Slack pathway fixes CO_2 into a four-carbon intermediate, this test applies to C4 plants and 13C-enriched products like maize and 13C-enriched wheat flour. Breath excretion of $13CO_2$ is a consequence of liver metabolism and absorption, colonic fermentation, and small intestine digestion of starch from the C4 plant. It is feasible to identify the sources of $13CO_2$ excretion by simultaneously checking for H_2 in the breath.

Procedure

- The individuals go through the breath tests randomly, with 1 week elapsed between each sample test.
- Take a sample with 100 mL of water after collecting baseline breath samples (by inhaling via a straw into an Exetainer tube).
- After the meal, take 10 mL breath samples every 30 min for $13CO_2$ analysis, up to 300 min for starch and up to 720 min for resistant starch.
- Collect breath samples (20 mL) for H_2 analysis every 30 min.
- Permit participants to consume a light lunch devoid of any substrates that are naturally enriched with 13C followed by 330 min breath collection point of the resistant starch breath test (*see* **Note 24**).

Analysis of Collected CO_2 & H_2

- Ignite each sample in a bomb calorimeter to measure the 13C abundance of the starch and resistant starch. Then, use isotope ratio mass spectrometry (IRMS) to analyze the CO_2 that results.
- Determine the 13C content of the samples by calculating the 13C/12C ratio in parts per thousand with respect to the international primary standard Pee Dee Belemnite calcium carbonate.
- For the breath tests, utilize gas chromatography (GC) to detect the presence of H_2 and IRMS to analyze the 10 mL breath samples for the presence of $13CO_2$ and 13C. Express the $13CO_2$ findings as the cumulative percentage of administered dosage of 13C recovered over time (%

dose/t), which is computed using the 13C substrate enrichment values determined by combustion, and as the percentage of 13C recovery per hour of the initial dose provided (%dose/h).

- Assume that each person produces 300 mmol of CO_2 per (m^2 body surface area) hour. Use the height-weight formula utilized by Haycock et al. [42] to determine body surface area [SA (m^2) = weight (kg) × 0.5378 × height (cm) × 0.3964].
- Express result in (ppm).

4.1.3 In Vivo Glycemic Index Method [37, 38]

Procedure

- All tests begin at 8:00 am, and the panelists have been fasting for 10 h prior to the exams. Time zero (T0) following a fast is used to measure blood sugar levels.
- On the first day of the trial, instruct the panelists to drink 200 mL of water and 50 g of glucose powder in 10–15 min.
- Take finger-prick capillary blood samples before ingesting any food as well as 15, 30, 45, 60, 90, and 120 min after ingesting the test food. Volunteers should refrain from exercising throughout the test.
- Use a blood glucose meter to examine the blood sample that was transferred onto the test tip.
- Start the next in vivo experiment after 48 h.
- The 50 g of test foods that the panelists each consumed on the third, fifth, seventh, and ninth trial days. Use the same testing approach throughout the entire experiment.

Data Manipulation

- For each volunteer, create curves showing the rise in blood sugar (measured in millimoles per liter) over time (from 0 to 120 min).
- Subtract the blood glucose concentration measured at T0 from the blood glucose concentration values measured at any subsequent time point to determine the rise in blood glucose. All curves will have a blood glucose value of zero at time T0.
- Add the incremental areas under the curve (IAUC) roughly calculated using the trapezoid rule for time periods 0–15, 15–30, 30–45, 45–60, 60–90, and 90–120 min. This will get the total area under each curve, or total glucose response (GR), for each curve. The main and minor foundation will be the values of the blood glucose increases on the y-axis (mol mL1), corresponding to the two time

points, while the height of the rectangular trapezium will be the time difference (in minutes) between the two following time points on the x-axis.

- Calculate the "apparent glycemic index" (GIa), also known as the percentage ratio between the sample's GR and the reference food (GR of the glucose solution), which is used as a standard. Use the following equation to get the glycemic load (GL):

$$GL = (GIa \times \text{grams of carbohydrate in the standard serving size}/100)$$

4.2 Animal Model

4.2.1 In Situ Mobile Nylon Bag Method [39]

Theory

In vivo small intestine digestibility of mobile bag starch could be predicted using either rumen preincubation or without it. By comparing in vivo data with mobile bag starch disappearances, the mobile bag approach provides a quick way for measuring intestine and whole-tissue starch digestibility.

Procedure for Starch Analysis

- For the mobile bag experiments and the starch analyses, grind all feed samples to a consistency that will pass a 1.5 mm screen (Cutter mill).
- 1 g of feedstuff is placed into 20 tared bags, each of which has a pore size of 12 m and a size of 6 cm × 6 cm. The bags are then heat-sealed shut.
- Rumen preincubate (RP) 12 bags of each feedstock in two cows for 4 h before separating into two halves.
- Put mobile bags in bigger bags with 200 mm pore-size stoppers affixed to them.
- Rinse the bags in cold tap water as soon as you remove them from the rumen, then wash them in a washing machine with warm water (around 25 °C).
- Before pepsin-HCl incubation, keep the RP bags for duodenal insertion (eight bags per feedstuff) at −18 °C.
- To simulate abomasal digestion, soak 16 bags (*see* **Note 25**) for 1 h in HCl solution and then for another 2 h in HCl–pepsin.
- Then, until incubation in the duodenum, keep bags at −18 °C.
- After defrosting, insert up to nine bags per cow daily through the duodenal cannula at a rate of two bags every 20 min for 7–9 h.
- To remove the feed sample bags from the ileum, place two iron balls within them.

- To retrieve bags from the ileum, insert a powerful cylindrical magnet with dimensions of 1.0 cm in diameter and 4 cm in length within the ileal cannula.
- After 4.5 h when initial bags are placed in the duodenum, insert the ileal cannula every 30 min for bags that the magnet has detected.
- Note the interval between the entry of the duodenum and ileal recovery. Remove the magnet at the fifth hour. If ileum collection fails, the bag will collect from feces, and the identical bag meant for fecal recovery will be retained at the ileum cannula, resulting in an equal number of bags collected from all the locations.
- Wash bags, dry them at 60 °C, and weigh them to find the amount of DM (Dry Matter) residues after removal from ileum or feces. Then, put the leftovers through a 4 min grinding process in a mixer mill before storing them in airtight glass tubes at room temperature for analysis.
- Drying at 60 °C in a forced air oven will reveal the amount of dry matter (DM) present.
- To calculate the quantity of glucose, use an analyzer to measure the starch content of feedstuffs and residues after rumen and intestinal incubation. Use the formula below to determine the starch's intestinal digestibility after not being rumen-degraded:

$$DS = (UDS - IS)/UDS$$

where DS is the digestibility in the small intestine UDS is the rumen-undegraded starch calculated from rumen preincubation (proportion of feed starch) IS the starch recovered in ileum (proportion of feed starch) (*see* **Note 26**).

4.2.2 Starch Digestibility Using In Vivo Digesta [28]

Procedure

- In vivo digesta should be collected from the small intestines of seven large, white pigs (P-1-7).
- Under anesthesia, divide the small intestine into four parts, SI-1 through SI-4, extending from the stomach to the cecum (*see* **Note 27**).
- Squeeze off the digesta from each piece, then store it at −20 °C until analysis (*see* **Note 28**).

Starch Content

- Analyze the starch content of in vivo digesta in duplicate using the AACC Method.
- Five times dilute the digesta from SI-1 and SI-2 before analysis; analyze from SI-3 directly.

- To eliminate the soluble sugars from the digesta, carefully weigh it (0.2 mL) and mix it into 3.0 mL of 100% ethanol. For 10 min, centrifuge at 4000 g. In a fume hood at room temperature, permit the precipitate to dry overnight. It is then wetted with 80% ethanol (0.2 mL), followed by the dissolution of dimethyl sulfoxide (2 mL) for 5 min while stirring in a hot water bath.
- Place the dissolved digesta in a boiling water bath and incubate it for 12 min while stirring with 0.1 mL of thermostable R-amylase in 2.9 mL of MOPS buffer (50 mM, pH 7.0), which also contains calcium chloride (5 mM) and sodium azide (0.02% w/v).
- Combine the digesta with the acetate buffer (4.0 mL, 200 mM, pH 4.5) containing sodium azide (0.02% w/v) and mix for 30 min while amyloglucosidase (0.1 mL) is incubated in a water bath at 50 °C.
- Adjust the resultant volume to 10.0 mL. Using the megazyme D-glucose (GOPOD Format) kit and the manufacturer's instructions, calculate the total amount of glucose generated from the starch in the digesta.
- Multiply the glucose by 0.9 to get any hydroglucose unit.
- Use the weight difference between the digesta before and after it has dried overnight at 110 °C to calculate its moisture content. The following formula should be used to determine the starch content (wb and db):

$$\%\text{starch content (wet basis)} = \frac{\text{total weight of glucose} \times 0.9}{\text{weight of digesta}} \times 100\%$$

$$\%\text{starch content (dry basis)} = \frac{\text{total weight of glucose} \times 0.9}{\text{weight of digesta}} \times (1 - \%\text{moisture content}) \times 100\%$$

4.2.3 In Vivo Glycemic Response Using Mice [40]

In-Vivo Glycemic Response

The in vivo GI and glycemic response of the treated IC samples are evaluated to demonstrate their effects on reducing mice blood glucose levels and potential as novel RS5.

Study Design

- Purchase male C57BL/6 J mice that are 7 weeks old. Individually house the mice in solid floor cages with a typical 12 h light/dark cycle in a climate-controlled setting. Handle the animals in accordance with the National Research Council's Guide for the Care and Use of Laboratory Animals.

- Give mice unrestricted access to food and water, and feed them for a week to acclimatize to their regular diets. Keep track of the mice's body weights (BWs) following the 1-week acclimatization period.
- The following six groups are then randomly selected from the mouse population:
  - Group 1 ($n = 3$; control group for glucose): This group of mice received a single dose of pure 260 glucose solution as a standard to the mice.
  - Group 2 ($n = 6$; RS2 group): As in the RS2 comparison group, mice in this group received a single dosage of noncomplexed HAMS.
  - Group 3 ($n = 6$; cooked HAMS group): This group of mice received a single dosage of 264 cooked HAMS (cooked for 5 min).
  - Group 4 ($n = 6$, cooked PS group): This group of mice received a single dosage of prepared PS (cooked for 5 min).
  - Group 5 ($n = 6$; DMSO IC group): This group of mice received a single dosage of ANN-ACH HAMS-AP IC (made by the DMSO technique).
  - Group 6 ($n = 6$; "empty" V-type IC group): Mice in this group were given a single dosage of 270 ANN-ACH V6h-AP IC (made using the "empty" V-type procedure).
- Due to their superior efficiency, only the two treated ICs are employed in the treatment groups, as the Resistant Starch levels of raw IC samples are substantially lower than those of treated ICs. Before the trial day, each mouse should undergo an overnight fast (9 p.m. to 9 a.m.). To ensure that each mouse received the same quantity of carbs, all groups received the same dose of 2 g total carbohydrate (TC)/kg BW.

Group 1 mice receive a dosage of 2 g glucose/kg body weight, whereas doses for groups 2 to 6 mice are calculated depending on the total starch content and IC samples. Measure each mouse's body weight the night before the experiment. Dosages should be administered orally. All suspensions should be made in water. After dose injection, provide mice with unrestricted access to water; however, during the experiment, food is not allowed. Blood glucose levels in these six groups were monitored using a hand-held glucometer before oral administration and 15, 30, 60, 90, 120, 180, and 240 min later.

Treatment Preparation

Boil raw HAMS and PS samples in 100 C water (10%, w/v) 296 for 5 min to prepare cooked HAMS and PS. Prepare the ANN-ACH HAMS-AP and ANN-ACH V6h-AP IC samples according to the instructions in starch inclusion complexes and Hydrothermal treatment of inclusion complex samples. And make all of the sample treatments into suspensions with DI water at a 10% (w/v) concentration.

Blood Glucose Measurement

- Measure blood glucose levels in mice at time points of 0 (baseline level after fasting and prior to oral medication), 15, 30, 60, 90, 120, 180, and 240 min after dosage administration using a hand-held blood glucometer.
- Gently cut the tail of the mouse with a scalpel to obtain blood from the mouse tail vein.
- Collect blood droplets by lightly rubbing or squeezing the tail. Press the glucometer device firmly on the test strip's tail tip and the blood drop to the front edge to gather the results.

Conduct three independent measurements for each mouse at each time point.

Calculation of In-Vivo GI

Plot the glycemic response curves for each treatment group after determining the mean blood glucose level at each time point. Using the fit curve feature in the Origin Pro software, calculate the AUC blood glucose level for each mouse group. The following equation is used to compute the in vivo GI values of sample treatments:

$$\text{in-vivo GI} = \text{AUC Sample}/\text{AUC reference(Glucose)}$$

Statistical Analysis

Using SPSS, analyze the data from all in vivo studies using one-way analysis of variance (one-way ANOVA), followed by the Tukey multiple comparison test. Compare the patterns after ingestion of the different starches using the paired *t*-test, with a significance at $P \geq 0.05$. Using Microsoft Office Excel 2007, calculate the starch content of in vivo digesta.

5 In Vitro Starch Digestibility Methods and Protocols

5.1 Using Pepsin Enzyme [34]

5.1.1 Processed Foods

Resistant Starch (RS) and Digestible Starch (DS)

Protein from samples is removed using pepsin. Then incubate samples with α-amylase for 16 h to hydrolyze digestible starch followed by alkali (2 M KOH) treatment of the residue to solubilize resistant starch; Further incubate it with amyloglucosidase. Glucose thus produced is measured via glucose oxidase test (GOD-PAP reagent). RS is determined by multiplying glucose (mg) by 0.9. To calculate DS, divide the difference between TS and RS (*see* **Note 29**).

Total Starch (TS)

Shake 50 mg of the sample vigorously in 6 mL of 2 M KOH for 30 min at room temperature. Add 3 mL of 0.4 M sodium acetate buffer (pH 4.75) and 60 μL of amyloglucosidase to a controlled shaking water bath, and then incubate for 45 min at 60 °C. Use the Glucose GOD-PAP kit to convert starch to glucose (the conversion factor is 0.9 from glucose to starch).

Procedure

- Boil samples (50 g) for 60 min in a solution containing 20 mg pepsin.
- Fill 25 mL container with a Tris-maleate buffer solution.
- In a shaking water bath, incubate the aforementioned sample at 37 °C with 5 mL of Tris-maleate solution containing 3.3 IU amylase.
- Every 30 min from 0 to 180 min, take a 1 mL of aliquot sample from each tube and place it in a 100 °C water bath for 5 min to deactivate the enzyme. Place the sample in the refrigerator until further examination.
- After 15 min of centrifugation at 4500 g, add 60 L of amyloglucosidase and incubate at 60 °C for 45 min to hydrolyze the digested starch into glucose.
- To determine the amount of glucose in the supernatant, use a glucose oxidase-peroxidase kit. The quantity of glucose generated per 100 g of sample hydrolyzed at various intervals of 0, 30, 60, 90, 120, and 180 min is a common way to represent the rate of starch digestion. The following formula can be used to determine the predicted glycemic index (PGI) and starch digestion index (SDI):

$$\text{Predicted Glycemic Index (PGI)} = \{39.21 + 0.803^{*}(\text{H90})\}$$

$$\text{Starch Digestion Index (SDI)} = \frac{\text{RDS} \times 100}{\text{TS}}$$

where

RDS = Rapidly digestible starch at 30 min

TS = Total starch

5.1.2 Whole Grains (Rice and Barley) [33]

Procedure

- Boiling water is added to a sample (100 mg), which is then cooked for 23 min.
- Let the cooked sample cool for 5 min before equilibrating water bath for an additional 5 min at 37 °C.
- Add 0.5 mL HCl–KCl buffer and mince the sample with 20 downward strokes with stainless-steel spatula. Then, again add HCl–KCl buffer (9.7 mL) containing 40 U of pepsin, and mix continuously for 1 h at 90 rpm using a magnetic stirrer. Add 2 mL of aqueous NaOH solution to bring the pH to 6.9, then add 14.8 mL of Tris-maleate buffer containing 2.6 U of amylase.
- Keep the water bath at 37 °C and 90 rpm for 180 min to allow starch to be hydrolyzed.
- Collect 0.1 mL aliquot at regular intervals of time viz., 0, 30, 60, 90, 120, and 180 min, and immerse it in boiling water for 5 min for enzyme deactivation.
- Centrifuge the aliquot at 13,500 rpm for 10 min to convert the starch into glucose. In sodium acetate buffer, mix 10 mL of above supernatant with 30 mL of amyloglucosidase, vortex it, and then incubate at 50 °C for 20 min. Followed by incubation, add 60 μL of Milli-Q water, vortex it, and then 10 μL sub-aliquot is transferred to microfuge tube (0.6 mL) for centrifugation.
- Then use the GOPOD reagent kit to calculate glucose.
- In order to eliminate interferences caused by the digesting media, incubate a blank reagent under the same parameters as those mentioned above and measure the absorbance.
- Determining starch hydrolysis using glucose levels and estimated GI (pGI).

Glucose is determined as follows:

$$\text{Absorbance of sample}/\text{Absorbance of glucose standard} \times 100 \times 10/1000$$

(where fraction 10/1000 is used to convert mcg glucose/0.1 mL aliquot to mg/mL)

Starch hydrolyzed (% SH) can be calculated using equation:

$$\text{SH} = (\text{glucose} \times \text{DF} \times \text{V} \times 162/180)/(\text{weight of sample} \times (\%\text{TS})/100) \times 100$$

where glucose: in mg/mL DF: dilution factor used on the sample aliquot *V*: volume of the digestion medium at the time point 162/180: conversion factor from glucose to starch TS: total starch content of the sample.

5.1.3 For Pasta Products [41]

Determine Starch Content Using Polarimetry

$$\text{Nonstructural carbohydrates (NSC)} = 100 - (\%\text{NDF} + \%\text{CP} + \%\text{EE} + \%\text{Ash})$$

Using the formulas suggested by Noblet and Perez and NRC, digestible energy (DE) and metabolizable energy (ME) can be estimated as:

$$\text{DE (kcal/kg)} = 4.168\ (91 \times \text{Ash}) + (19 \times \text{CP}) + (39 \times \text{EE}) - (36 \times \text{NDF})$$

$$\text{ME (kcal/kg)} = \text{DE} - (68 \times \text{CP})$$

Procedure

- Free glucose (FS, 0 min), quickly digestible starch (RDS, within 20 min of incubation), and slowly digestible starch (SDS, between 20 and 120 min) are all not further hydrolyzed. This includes resistant starch (RS $>$ 240 min).
- Incubate ground samples for 30 min at 37 °C with continual stirring in acidified (HCl 0.05 mol/L) solution containing pepsin and guar gum.
- Further incubate samples with an enzyme solution containing 30 g of pancreatin, amyloglucosidase, and invertase (1 U hydrolyzes sucrose to 1 mg of fructose in 5 min at 20 °C, pH 4.5).
- According to the instructions on the glucose oxidase kit package, terminate starch digestion by adding 100% ethanol to 1 mL of the solution and measuring the glucose released by calorimetric method at 540 nm.
- Using the values of glucose released after 20 min (G20), 120 min (G120), FG (free glucose), and TG (total glucose), the amount of total starch, rapidly digestible starch, slowly digestible starch, and resistant starch can be estimated.

$$\text{TS} = (\text{TG} - \text{FG}) \times 0.9$$

$$\text{RDS} = (\text{G20} - \text{FG}) \times 0.9$$

$$\text{SDS} = (\text{G120} - \text{G20}) \times 0.9$$

$$\text{RS} = (\text{TG} - \text{G120}) \times 0.9$$

where

TS = Total starch

FG = Free glucose

RDS = Rapidly digestible starch

SDS = Slowly digestible starch

RS = Resistant starch

5.2 Using Amylase [32]

Procedure

- Weigh samples in such a way that each 15 mL corning tube contain 100 mg of starch content and then add 10 mL phosphate-buffered saline.
- To equilibrate, maintain sample tubes at 37 °C for 20 min on a rotating mixer (20 rpm, 30° angle). Before the assay begins, take a 200 μL blank aliquot of all the samples and place it in a microcentrifuge tube (1.5 mL) having 200 μL of the stop solution.
- Add porcine pancreatic amylase and prepare in phosphate buffer saline to provide the digesting mixture an activity of 4 U/mL (10 mg/mL starch) (*see* **Note 30**). After adding the amylase, quickly place the tubes back in the incubator mixer at 37 °C. After 5, 10, 15, 20, 25, 30, 45, 60, and 90 min, transfer aliquots (200 μL) of the digesting mixture into an equivalent amount of the stop solution.
- Centrifuge the collected aliquots at 15,000 g for 5 min to remove any residual starch, and then keep the resulting supernatants at 20 °C for estimation of the extent of starch amylolysis.

$$(\text{Starch amylolysis}\%)_{\text{t}} = \frac{[\text{maltose}]_{\text{t}} \times 100}{[\text{maltose}]_{\text{substrate}}}$$

where

$[\text{maltose}]_t$ = maltose equivalent concentration (after baseline correction) measure in the liquid phase of the reaction mixture at a particular time t; and

$[\text{maltose}]_{\text{substrate}}$ = theoretical maltose equivalent concentration that would be present at the beginning of reaction, assuming that all starch in the given sample is converted into maltose.

Procedure

- Use 50 mL tube with a foil top for soaking milled grains (500 mg) at room temperature for 10 min, while soaking brown rice for 20 min, in 6 mL distilled water.
- This is followed by boiling samples for 20 min.
- After removing excess water, cool the sample for at least 5 min.
- Fill the sample tubes with 0.5 mL of alpha-amylase, and place all in a water bath at 37 °C. To simulate the buccal phase, mash the grains with a spatula for 20 s. Then, add 5 mL of pepsin.
- Set the speed of the magnetic stirrer at 200 rpm, add spin bars, and mix.

- After 30 min, stir at a speed of 700 rpm while adding 5 mL of NaOH to balance the pH, 20 mL of sodium acetate buffer followed by addition of 5 mL pancreatin-amyloglucosidase solution.
- After 180 min of letting the reaction run, take 0.2 mL of aliquots at each of the following ten time intervals: 0, 5, 10, 20, 30, 45, 60, 90, 120, and 180 min. Place these aliquots in a 0.6 mL microfuge tube. Before centrifuging the tubes at 13,000 rpm (4 °C) for 10 min, put the tubes in an ice bath to cool the reaction.
- If necessary, dilute samples with 0.1 M sodium acetate buffer. Add aliquots of 50 μL and 5 μL of amyloglucosidase, and then incubate for 20 min at 50 °C.
- Use the GOPOD test kit at 510 nm to calculate the amount of hydrolyzed starch.
- Calculate the estimated GI (pGI) value.

$$\text{Glucose (in form of GI)} : \text{Absorbance of sample}/\text{Absorbance of glucose standard} \times 100 \times 10/1000$$

(where the fraction 10/1000 = convert mcg glucose/0.1 mL aliquot to mg/mL)

Calculate %starch hydrolyzed (SH):

$$\text{SH} : (\text{glucose} \times \text{DF} \times \text{V} \times 162/180)/(\text{sample weight} \times (\%\text{TS})/100) \times 100$$

where glucose is provided in mg/mL DF denotes the dilution factor used on the sample aliquot *V* denotes the volume of the digestion medium at the time point 162/180 denotes the conversion factor from glucose to starch TS denotes the total starch content of the sample.

Samples

Maize starch and ingredients of the pig diet [28].

Procedure

- For 15–20 s at room temperature, preincubate a pig diet containing 10 g (db) starch with an artificial saliva solution (5 mL) containing porcine pancreatic R-amylase (1000 U/mL), potassium chloride (21.1 mM), calcium chloride (1.59 mM), and magnesium chloride (0.2 mM) in carbonate buffer (14.4 mM, pH 7).
- Preincubate the resultant at 37 °C for 30 min with pepsin (4 mg/mL) dissolved in a 25 mL hydrochloric acid solution (0.02 M). Then, titrate it with NaOH solution (25 mL, 0.02 M), and dilute it with acetate buffer solution (125 mL, 0.2 M, pH 6), containing calcium chloride (200 mM) and magnesium chloride (0.49 mM).

- Use pancreatin (8 mg/mL) and amyloglucosidase (112 U/mL) to digest the results in the same 25 mL acetate buffer solution at 37 °C.
- Remove 10 mL of aliquots at regular time intervals of 0, 1, 2, 5, and 25 h.
- Centrifuge each aliquot at 4000 g for 10 min after mixing it with 100% ethanol (30 mL). In a fume hood at room temperature, let the precipitate (in vitro digesta) dry overnight before storing it in a dry location for future analysis.
- Use the megazyme D-glucose (GOPOD Format) kit to measure the amount of glucose present in the supernatant.
- Glucose converts to anyhydroglucose, the starch monomer unit, with a molar mass conversion of 0.9.
- Calculate the degree of hydrolysis using the following equation:

$$\%\text{Hydrolysis} = \frac{\text{total weight of glucose in supernatant} \times 0.9 \times 100\%}{\text{Dry weight of starch in diet}}$$

$$\%\text{starch content(wb)} = \frac{\text{total weight of glucose} \times 0.9 \times 100\%}{\text{weight of digesta}}$$

$$\%\text{starch content(db)} = \frac{\text{total weight of glucose} \times 0.9 \times 100\%}{\text{weight of digesta} \times (1 - \%\text{moisture content})}$$

5.3 Using Combination of Enzymes Pepsin–Amylase

Procedure

- Pour boiling water into a 50 mL tube with 300 mg of milled rice and keep it for 23 min. The sample should then be left at room temperature for 5 min.
- Add 1 mL of sodium potassium phosphate buffer (0.1 M) and mince with a stainless-steel spatula in 35 downward strokes for 10 s. Allow the mixture to equilibrate for 5 min in a 37 °C water bath.
- Then add 2 mL of α-amylase, and allow amylolysis for 75 s in a water bath with constant stirring using magnetic spin bar at 37 °C at 60 rpm.
- Make the digesting media acidic by adding 3 mL of HCl to halt the reaction.
- Then add 3 mL of pepsin and place the mixture in a 37 °C water bath at 60 rpm for 30 min.
- Use 15 mL of aqueous NaOH(aq) to bring the pH to 6.9 to stop the enzymatic reaction.
- Next, add 15 mL of enzyme solution containing amyloglucosidase and pancreatin in 0.1 M sodium potassium

phosphate buffer, and let the mixture incubate for an additional 180 min (at 37 °C and 120 rpm).

- Collect sample aliquots (0.70 mL) and place them in ice-bath for 0, 30, 60, 90, 120, 150, and 180 min (immediately before adding pancreatin-AMG). Then, centrifuge the sample at 13,500 rpm for 10 min at 4 °C.
- Transfer a 1 μL volume of the supernatant to a 0.6 mL microcentrifuge tube. Add 9 mL of the AMG solution (33.3 U/mL in a 0.4 M sodium acetate buffer, pH 5.0), vortex the mixture, and incubate for 20 min at 50 °C.
- Add GOPOD reagent to the mixture and reagent blank to quantify the amount of glucose.

$$\text{Glucose} = \text{Absorbance of sample}/\text{Absorbance of glucose standard} \times 100 \times 10/1000$$

Calculate the percentage of starch hydrolyzed (SH) using the following equation:

$$\text{SH} = (\text{glucose} \times \text{DF} \times \text{V} \times 162/180)/(\text{sample weight} \times (\%\text{TS})/100) \times 100$$

where

Glucose: in mg/mL.

DF: dilution factor used on the sample aliquot.

V: volume of the digestion medium at the time point.

162/180: conversion factor from glucose to starch.

TS: total starch content of the sample.

6 Conclusion

The in vitro starch digestibility screening approach exhibits potential for quick GI value prediction and is advised for mechanistic investigations and early stages of food product development. So it may be possible to predict relative differences in the GI of different foods using in vitro techniques that mimic digestive conditions. Although it is becoming more common to use digestion models that mimic the biochemical conditions (digestive enzymes, pH, mixing, rate, etc.) of oral, gastric, and intestinal digestion, their methodological complexity and high reagent requirements prevent their use for high-throughput screening. Therefore, it is critical that in vitro tests are evaluated based on how well these operate across a wide range of food products with various structural and digestive susceptibilities. Due to superior reproducibility and practical reasons (in vivo investigations on animals or humans are far more expensive and include significant ethical issues), in vitro digesting techniques are frequently utilized to gather this information instead of in vivo experiments.

7 Notes

1. The Women's and Children's Hospital's Research Ethics Committee authorized the procedure, which adhered to the Helsinki Declaration.
2. The primary requirements for participation in the study (namely, being healthy, being free of dietary allergies and intolerances, not using drugs, and not being in a specific physiological situation, such as breastfeeding) should be explained to volunteers. Obtain the written consent from every study subject. Each volunteer is free to leave the experiment at any moment and without giving a reason.
3. Dry cow normal feed (kg/day): 2.0 kg spring barley straw, 4.0 kg artificially dried grass hay, 2.8 kg concentrate (including g/kg: 400 barley, 400 oats, 100 soybean meal, 30 rapeseed meal, 30 beetroot molasses, and 40 mineral combinations), and 0.150 kg vitamin mix every week. The chemical makeup of the ration (g/kg dry matter [DM]) was 137 crude protein, 465 nondigestible fiber, and 137 starch.
4. A typical diet for lactating cows consists of 385 maize silages, 154 grass-clover silages, 63 lucerne silages, 162 rolled barley, 86 dried sugar beetroot pulp, 82 rapeseed meal, 54 soya bean meal, and 15 mineral-vitamin blend (g/kg DM). The chemical makeup of the ration was 239 starch, 375 NDF, and 189 CP (g/kg DM).
5. Pig was given a diet that included cooked red meat (23.6%, wet basis), sunflower seed oil (8.9%, wet basis), sucrose (7.0%, wet basis), tallow (5.0%, wet basis), wheat bran (4.0%, wet basis), sodium chloride (0.3%, wet basis), and vitamin and mineral mix (0.2%, wet basis) on the night before sacrifice as well as at 2 and 4 h prior to sacrifice.
6. The temperatures and ethanol concentrations have been demonstrated to improve starch crystallinity the most while maintaining the V-type structure of starch ICs.
7. By multiplying mg of glucose by 0.9, one can determine how much starch is digested.
8. The solution will be diluted with 800 mL of distilled water and 121.14 g of Tris base. Using HCl, bring the solution's pH down to where you want it to be (normally 7.0). Adjust it to 1 L in volume.
9. About 7.45 g of KCl and 0.772 g of HCl should be dissolved in 800 mL of distilled water. HCl and NaOH are used to further alter pH. Fill the container to 1 L with distilled water.

10. Buffer solution with sodium acetate pH 4.5. Add 90 mL of acetic acid to dissolve 63 g of anhydrous sodium acetate in water, set the pH to 4.5, and then dilute with 1000 mL of water.
11. Pepsin: 800–2500 U/mg protein, from the stomach mucosa of pigs. 1 U will result in an absorbance change of 0.001/min at 280 nm at a pH of 2.0 when evaluated as soluble products using hemoglobin as a substrate, sodium acetate, sugars, sucrose >99.5%, arabinose, and glucose.
12. Pepsin solution: 0.05 M hydrochloric acid containing 5 mg/mL pepsin and 5 mg/mL guar gum is the first enzyme solution. To avoid lumps, put 1.2 g of guar gum in a beaker and moisten it with 2 mL of ethanol. A quantity of 1.2 g of pepsin should be added gradually while 240 mL of 0.05 M hydrochloric acid is being continuously mixed with a magnetic stirrer. To maintain the guar gum suspended, this solution needs to be prepared and continually stirred within 20 min of use.
13. Glucose oxidase/peroxidase must be kept. It should be reacted at 2 to 8 °C. Each capsule includes buffer salts, 100 purpurogallin units of horseradish peroxidase, and 500 units of *Aspergillus niger* glucose oxidase. The capsule's contents should be dissolved in 39.2 mL of deionized water and placed in an amber container. The solution can be left unfrozen for at least 6 months at −20 °C and remains stable for up to 1 month at 2–8 °C. If turbidity appears, throw it away.
14. Saline with a phosphate buffer pH 7.4: Put 8.0 g of sodium chloride, 2.38 g of sodium hydrogen phosphate, and 0.19 g of potassium dihydrogen phosphate in water to dissolve them. Use the same solvent to dilute to 1000 mL. If necessary, change the pH (2–2.3).
15. A vortex mixer is used to suspend six 50 mL centrifuge tubes containing 3 g of USP pancreatin in 20 mL of 0.1 M calcium chloride. A magnetic stirring bar is then added, and the mixture is mixed for 10 min. Take 17 mL of the supernatant from each tube, centrifuge at 900 g for 10 min, and then mix. Add 4 mL invertase and 8 mL glucoamylase. Mix thoroughly; this solution must be ready within an hour of usage and must reach equilibrium at 37 °C just before use.
16. 0.3 M Na_2CO_3, pH 9: The 0.3 M Na_2CO_3, often known as the "stop solution," is used to quickly halt the amylolysis at each time point by inactivating the amylase activity in the aliquots.
17. Use pepsin of 19.5 U/mL or 0.45 mg/mL of 0.1 M sodium potassium phosphate buffer, pH 1.5 at 37 °C.

18. Pancreatin 8× USP from porcine pancreas (pancreatic amylase: 1 U releases the equivalent of 1 mg glucose/min from soluble starch at 40 °C measured as reducing groups by the method of Dahlqvist).
19. Glucoamylase of 10,000 U/mL, from *Aspergillus niger* (1 U releases 1 μmol glucose/min from starch at 55 °C and pH 4.5).
20. Use 55.5 U/mL of α-amylase or 3.7 mg/mL of 0.05 M sodium potassium phosphate buffer, pH 6.9 at 37 °C.
21. The first signs of ingested food arrive in the colon between 3 and 4 h after consumption, and the final traces of a meal reach the colon in the next 5 h. After 8 h after consumption, remove any uneaten test food from the terminal ileum. The pace of food passing through the upper part of the gut varies across and among individuals, and this variation must be taken into account. Therefore, sample collection is generally preferred at interval of at least 14 h but ideal interval is 24 h.
22. It is imperative that the quantification method chosen is sufficient to ensure complete solubilization and hydrolysis of all starch present in ileal digesta collection because starch leaving the small intestine is resistant to amylolysis, particularly of retrograded and intact granular starches. The approach must not only be quantitative for starch but also be able to identify glucose and other byproducts of complete or partial digestion of starch. However, starch may not always be the source of the glucose in ileal digesta. By subtracting baseline "starch" excretion data when sample/test foods are consumed, this issue is indirectly resolved.
23. Orocecal transit time, along with the quantity of ingested starch, is a key factor in determining the amount of resistant starch that is produced in the upper gut. The amount of starch that is resistant to digestion remained largely constant.
24. Throughout the breath test period and other breath tests, no further food or drink is allowed.
25. Eight bags had intact feed without preincubation in the rumen and eight bags had already been incubated in the rumen.
26. Categorize the feed into four groups: grains, wheat that has been treated with NaOH, maize, and legumes. Utilize the mixed static in vitro digestibility assay (SAS) technique to examine the impact of the recovery site and rumen preincubation within the feed group. Feed, cow, site, rumen preincubation, and other potential interactions are all included in the statistical model. A compound symmetry covariance structure will be used to treat the bags at the feed-cow site as repeated measurements.

27. SI-1 is 1 m away from the stomach end. SI-2 spans the length of the small intestine from the end of SI-1 to its halfway. The remaining portion of the small intestine, known as SI-3, includes everything past the cecum end. From the cecum end, SI-4 is the final meter.
28. Due to the small quantity of the digesta and the extremely low starch content in the digesta, the digesta from SI-4 section is not examined in this study.
29. To eliminate protein from the samples, this procedure is completed to calculate the digestible starch (DS).
30. The amount of amylase required to release 1.0 mg of maltose from starch in 3 min at pH 6.9 and 20 °C is referred to as one unit.

References

1. Butterworth PJ, Bajka BH, Edwards CH, Warren FJ, Ellis PR (2022) Enzyme kinetic approach for mechanistic insight and predictions of in vivo starch digestibility and the glycaemic index of foods. Trends Food Sci Technol 120:254–264
2. Grundy MM-L, Edwards CH, Mackie AR, Gidley MJ, Butterworth PJ, Ellis PR (2016) Re-evaluation of the mechanisms of dietary fibre and implications for macronutrient bioaccessibility, digestion and postprandial metabolism. Br J Nutr 116(5):816–833
3. Copeland L (2016) Food carbohydrates from plants. In: Wild harvest: plants in the hominin and pre-agrarian human worlds, pp 19–30
4. Edwards CH, Ryden P, Mandalari G, Butterworth PJ, Ellis PR (2021) Structure–function studies of chickpea and durum wheat uncover mechanisms by which cell wall properties influence starch bioaccessibility. Nature Food 2(2): 118–126
5. Lee BH, Bello-Pérez LA, Lin AHM, Kim CY, Hamaker BR (2013) Importance of location of digestion and colonic fermentation of starch related to its quality. Cereal Chem 90(4): 335–343
6. Lovegrove A, Edwards C, De Noni I, Patel H, El S, Grassby T, Zielke C, Ulmius M, Nilsson L, Butterworth P (2017) Role of polysaccharides in food, digestion, and health. Crit Rev Food Sci Nutr 57(2):237–253
7. Englyst HN, Kingman S, Cummings J (1992) Classification and measurement of nutritionally important starch fractions. Eur J Clin Nutr 46: S33–S50
8. Emami S, Perera A, Meda V, Tyler RT (2012) Effect of microwave treatment on starch digestibility and physico-chemical properties of three barley types. Food Bioprocess Technol 5:2266–2274
9. Bello-Perez LA, Flores-Silva PC, Agama-Acevedo E, Tovar J (2020) Starch digestibility: past, present, and future. J Sci Food Agric 100(14):5009–5016
10. Singh J, Dartois A, Kaur L (2010) Starch digestibility in food matrix: a review. Trends Food Sci Technol 21(4):168–180
11. Hu P, Zhao H, Duan Z, Linlin Z, Wu D (2004) Starch digestibility and the estimated glycemic score of different types of rice differing in amylose contents. J Cereal Sci 40(3):231–237
12. Zhang G, Ao Z, Hamaker BR (2006) Slow digestion property of native cereal starches. Biomacromolecules 7(11):3252–3258
13. Marti A, Abbasi Parizad P, Marengo M, Erba D, Pagani MA, Casiraghi MC (2017) In vitro starch digestibility of commercial gluten-free pasta: the role of ingredients and origin. J Food Sci 82(4):1012–1019
14. Van Hung P, Chau HT, Phi NTL (2016) In vitro digestibility and in vivo glucose response of native and physically modified rice starches varying amylose contents. Food Chem 191: 74–80
15. Tsuiki K, Fujisawa H, Itoh A, Sato M, Fujita N (2016) Alterations of starch structure lead to increased resistant starch of steamed rice: identification of high resistant starch rice lines. J Cereal Sci 68:88–92

16. Zhang G, Ao Z, Hamaker BR (2008) Nutritional property of endosperm starches from maize mutants: a parabolic relationship between slowly digestible starch and amylopectin fine structure. J Agric Food Chem 56(12): 4686–4694
17. Zhang G, Hamaker BR (2009) Slowly digestible starch: concept, mechanism, and proposed extended glycemic index. Crit Rev Food Sci Nutr 49(10):852–867
18. Miao M, Jiang B, Cui SW, Zhang T, Jin Z (2015) Slowly digestible starch—a review. Crit Rev Food Sci Nutr 55(12):1642–1657
19. Miao M, Jiang B, Zhang T (2009) Effect of pullulanase debranching and recrystallization on structure and digestibility of waxy maize starch. Carbohydr Polym 76(2):214–221
20. Liu W, Hong Y, Gu Z, Cheng L, Li Z, Li C (2017) In structure and in-vitro digestibility of waxy corn starch debranched by pullulanase. Food Hydrocoll 67:104–110
21. You S-Y, Lim S-T, Lee JH, Chung H-J (2014) Impact of molecular and crystalline structures on in vitro digestibility of waxy rice starches. Carbohydr Polym 112:729–735
22. Jane J-l, Wong K-s, McPherson AE (1997) Branch-structure difference in starches of A-and B-type X-ray patterns revealed by their Naegeli dextrins. Carbohydr Res 300(3): 219–227
23. Lin L, Guo D, Huang J, Zhang X, Zhang L, Wei C (2016) Molecular structure and enzymatic hydrolysis properties of starches from high-amylose maize inbred lines and their hybrids. Food Hydrocoll 58:246–254
24. Quezada-Calvillo R, Robayo-Torres CC, Ao Z, Hamaker BR, Quaroni A, Brayer GD, Sterchi EE, Baker SS, Nichols BL (2007) Luminal substrate "brake" on mucosal maltase-glucoamylase activity regulates total rate of starch digestion to glucose. J Pediatr Gastroenterol Nutr 45(1):32–43
25. McCleary BV, Monaghan DA (2002) Measurement of resistant starch. J AOAC Int 85(3): 665–675
26. Magallanes-Cruz PA, Flores-Silva PC, Bello-Perez LA (2017) Starch structure influences its digestibility: a review. J Food Sci 82(9): 2016–2023
27. Aarathi A, Urooj A, Puttaraj S (2003) In vitro starch digestibility and nutritionally important starch fractions in cereals and their mixtures. Starch-Stärke 55(2):94–99
28. Hasjim J, Lavau GC, Gidley MJ, Gilbert RG (2010) In vivo and in vitro starch digestion: are current in vitro techniques adequate? Biomacromolecules 11(12):3600–3608
29. Sumczynski D, Bubelová Z, Fišera M (2015) Determination of chemical, insoluble dietary fibre, neutral-detergent fibre and in vitro digestibility in rice types commercialized in Czech markets. J Food Compos Anal 40:8–13
30. Toutounji MR, Butardo VM Jr, Zou W, Farahnaky A, Pallas L, Oli P, Blanchard CL (2019) A high-throughput in vitro assay for screening rice starch digestibility. Food Secur 8(12):601. s
31. Englyst K, Goux A, Meynier A, Quigley M, Englyst H, Brack O, Vinoy S (2018) Inter-laboratory validation of the starch digestibility method for determination of rapidly digestible and slowly digestible starch. Food Chem 245: 1183–1189
32. Edwards CH, Cochetel N, Setterfield L, Perez-Moral N, Warren FJ (2019) A single-enzyme system for starch digestibility screening and its relevance to understanding and predicting the glycaemic index of food products. Food Funct 10(8):4751–4760
33. Pautong PA, Añonuevo JJ, de Guzman MK, Sumayao R Jr, Henry CJ, Sreenivasulu N (2022) Evaluation of in vitro digestion methods and starch structure components as determinants for predicting the glycemic index of rice. LWT 168:113929
34. Ali A, Waly MI, Bhatt N (2022) Correlation between the in-vitro starch digestibility, predicted glycemic index and in-vivo glycemic index of selected Omani snack foods.International. J Chem Biochem Sci 21:112–119
35. Bird AR, Lopez-Rubio A, Shrestha AK, Gidley MJ (2009) Resistant starch in vitro and in vivo: factors determining yield, structure, and physiological relevance. In: Modern biopolymer science. Elsevier, pp 449–510
36. Symonds EL, Kritas S, Omari TI, Butler RN (2004) A combined 13CO2/H2 breath test can be used to assess starch digestion and fermentation in humans. J Nutr 134(5): 1193–1196
37. Yu Z, Fan Y, Wang X, Xia M, Cai Y (2020) In vitro and in vivo digestibility of pea and chickpea powder prepared by cooking and drying treatment. Int J Food Prop 23(1):1187–1199
38. Fois S, Piu PP, Sanna M, Roggio T, Catzeddu P (2021) In vivo and in vitro starch digestibility of fresh pasta produced using semolina-based or wholemeal semolina-based liquid sourdough. Food Secur 10(10):2507
39. Ghoorchi T, Lund P, Larsen M, Hvelplund T, Hansen-Møller J, Weisbjerg M (2013) Assessment of the mobile bag method for estimation of in vivo starch digestibility. Animal 7(2): 265–271

40. Guo J, Ellis A, Zhang Y, Kong L, Tan L (2023) Starch-ascorbyl palmitate inclusion complex, a novel type 5 resistant starch, reduced in vitro starch digestibility and improved in vivo glycemic response in mice. Carbohydr Polym 321: 121289
41. Vastolo A, Cutrignelli MI, Serrapica F, donné Kiatti D, Di Francia A, Masucci F, Calabro S (2022) Evaluation of in vitro starch digestibility and chemical composition in pasta former foods. Front Vet Sci 9
42. Haycock GB, Schwartz GJ, Wisotsky DH (1978) Geometric method for measuring body surface area: a height-weight formula validated in infants, children, and adults. J Pediatr 93(1):62–66

Chapter 9

Physically Modified Starch

Palak Mahajan and Kalyani Sharma

Abstract

The demand for starch is never-ending and is increasing at an alarming rate worldwide for its food and nonfood applications. However, native starch has some limitations, such as low gel strength, viscosity, water binding capacity, and thickening properties. This leads the researchers to find various ways of modifying the starch to improve its beneficial qualities and correct the drawbacks of natural starches. The starch modification industry is changing continuously, providing various physical, chemical, and enzymatic methods and sometimes their hybrids. Among these, the physical modification methods possess higher safety, environmental friendliness, and cost-effectiveness without leading to the production of chemical wastes. The physical modification methods can be divided into two categories: thermal and nonthermal. The former category includes heat-moisture treatments, annealing, microwave heating, and osmotic pressure treatment, while the latter includes ultrasonication, ultrahigh-pressure treatments, and pulsed electric fields. This chapter covers recent advancements in thermal and nonthermal physical modification of starch, their principles, procedures, and their effect on starch functionalities.

Key words Physical modification, Ultrasound, Annealing, Heat moisture treatment, Modified starch

1 Introduction

Starch is an abundant organic compound found in nature after cellulose. It is mainly present in seeds, roots, tubers, leaves, fruits, and vegetables. Starch granules are made up of amylose and amylopectin fractions [1]. Both these components differ significantly in their properties and, thereby, in functionalities [2]. The fraction of these components in starch also determines its nutritional properties, health benefits, and, thus, its end-use. High amylose starch is known to have more resistant starch (RS) and lesser rapid (RDS) and slow digestibility starch (SDS), resulting in a food with a low glycemic index (GI) [3].

However, the industrial exploitation of starch is restricted owing to the lack of properties such as low water solubility, lower swelling power, weak pasting behaviors, and syneresis, leading to weak and unstable gel. This leads to the need to modify the starch

Sneh Punia Bangar (ed.), *Standardized Procedures and Protocols for Starch*, Methods and Protocols in Food Science, https://doi.org/10.1007/978-1-0716-3866-8_9,

to achieve the required characteristics [1, 4]. Modification plays an important role in enhancing its positive attributes by eliminating its shortcomings and unlocking new paths for starch applications (Table 1). It leads to decreased retrogradation, lower syneresis, and improved paste clarity, gel texture, and strength, making it suitable for various applications [4, 5].

Table 1
Effect of starch modification on various properties, related advantages, and applications

Effect	Advantages	Applications	References
Functional properties	Enhanced solubility, gelatinization behavior, texture, and viscosity	Provide food formulations with the desired functions, such as stabilizing, thickening, or gelling.	[6]
Processability	Enhanced processability, including better flowability, dispersibility, and handling simplicity	Suitable for a variety of manufacturing procedures, including coating, extrusion, and spray drying	[7]
Stability	Reduce retrogradation, avoid syneresis, and increase resistance to heat, shear, and other processing conditions to increase the stability of starch-based products	Structural alterations and degradation that improve product quality and shelf life	[8]
Cost optimization	By transforming low-value or underutilized starch sources into higher-value products, modification approaches can assist in maximizing the utilization of starch and lower costs	The commercial value of native starches can be raised by converting them into modified starches with certain functions	[2]
Tailored applications	Allows starch characteristics to be manipulated to satisfy the unique needs of various applications	In food items like sauces, dressings, baked goods, or frozen sweets, where distinct textural qualities or stability are essential, modified starches can be created for use	[2, 3]
Bioavailability and digestibility	Enhance amylose content, decrease enzyme accessibility, increase bioavailability, and improve starch digestion	Essential in the creation of starch-based components for the food and medicinal sectors	[5]
Environmental considerations	Contribute to sustainable practices	Modified starches can be used to make biodegradable packaging, coatings, and films, which can reduce the use of nonbiodegradable materials and encourage the use of greener alternatives	[1]

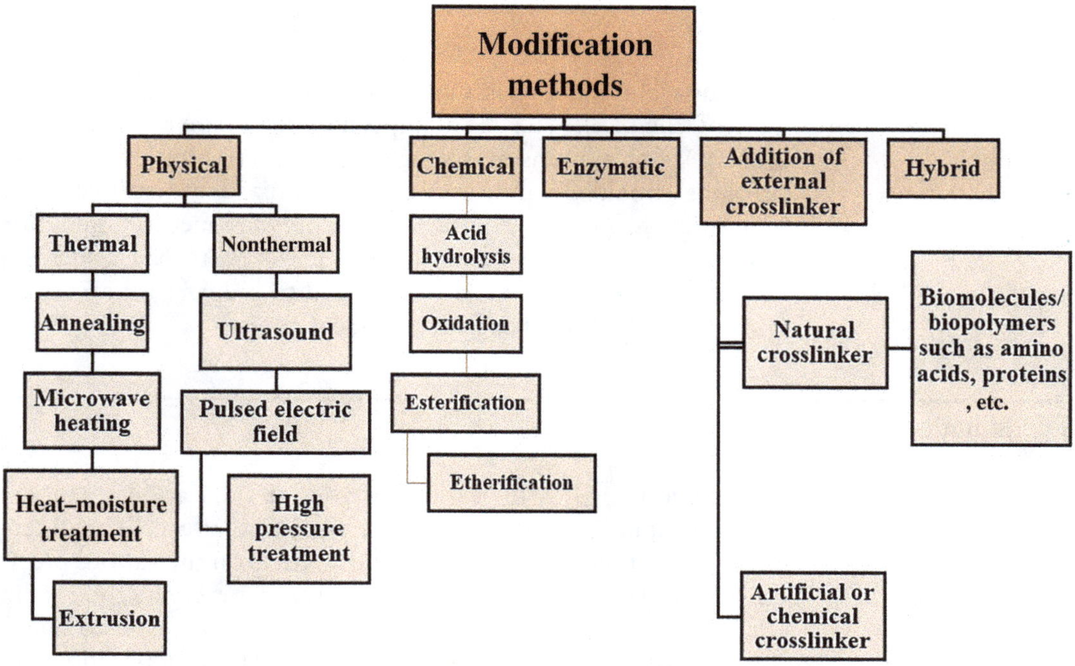

Fig. 1 Methods for the starch modification

The various ways to modify the starch can be categorized into physical, chemical, enzymatic, addition of natural and synthetic crosslinkers, and sometimes their combinations (Fig. 1). Among these, the physical method is simple, inexpensive, and nontoxic. The industry of starch modification is always changing and offers many opportunities to produce unique starches with new functionality and value-added qualities [5]. Therefore, feasible solutions for large-scale starch modification without causing negative impacts on the environment are required. Thus, these requirements lead to the acceptance of physical modification methods. In physical modification, native starch granules are treated using pressure, shear, and irradiation, as well as various temperature and moisture combinations. Since they do not entail any chemical treatment that could be dangerous for human usage, physical modification techniques are typically preferred [9–11].

Depending on the use of heat treatment, physical modification techniques can be divided into two categories, namely, thermal and nonthermal [12, 13]. Thermal physical modification methods, such as annealing (ANN), heat moisture treatment (HMT), extrusion, osmotic heating, and microwave heating (MWH), employ the use of heat to modify the starches. Nonthermal physical modification methods, such as ultrasounds, high hydrostatic pressure (HHP), and pulse electric field (PEF), use high temperatures [14]. This chapter discusses the frequently used modification techniques for starch, particularly focusing on the physical methods, their principles, procedures, and their effect on starch properties.

2 Materials

The physical modification methods do not employ the use of any kind of chemical. The most important requirement is water to prepare starch suspension or for starch conditioning. All the suspensions should be prepared using deionized water, free from impurities (*see* **Note 1**). In the osmotic pressure method, the solution typically contains solutes such as sugars or salts. NaCl, sucrose, and sodium sulfate are most widely used to prepare osmotic solutions and should be of analytical grade.

3 Methods

This section focuses on various physical modification methods, their principles, procedures, and the associated effects. The different treatment methods have different impacts on the various properties of the starches (Table 2). The first step for every method is the selection of starch source. Choose the appropriate type of starch based on the desired modification and the specific application (*see* **Note 16**).

3.1 Heat–Moisture Treatment (HMT)

This process involves subjecting starch to controlled heat and moisture conditions. The method includes agitating the low-moisture starch granules at temperatures of about 80 to 140 °C [21]. The process can be divided into three stages: moistening, heat treatment, and equilibration-cooling.

3.1.1 Principle

The basic principle of the HMT method involves subjecting starch to controlled heating in the presence of moisture. The moisture content is typically maintained between 10% and 40%, and the processing temperature varies from 84 to 140 °C [16]. The key principles of this method include moisture absorption, gelatinization, and molecular rearrangement. Starch granules have a natural ability to absorb moisture from their surroundings. Structural variation of native starch upon heat moisture treatment is shown in Fig. 2. During gelatinization, heat and moisture cause molecular rearrangement within the starch granules. This rearrangement involves the breakdown and reformation of starch molecular bonds, leading to changes in the organization and structure of amylose and amylopectin molecules, and therefore alters the functional, pasting, textural, and structural properties and the functionalities of the modified starch [22].

3.1.2 Procedure

1. Condition the starch by adjusting its moisture content to a specific level. For this, the dry starch is sprayed with a fine mist of water to achieve the desired moisture content. Usually,

Table 2
Effect of various physical modification treatments on starch properties and structure

Treatment	Effect on starch properties	Effect on starch structure	References
Gentle heating	Shows little impact on the physico-mechanical characteristics of starch	Morphology of starch granules and amylose:amylopectin ratio affected significantly	[15]
High-pressure treatment	Enhances starch's ability to gelatinize and make a paste	Upon cooling, produces cream-textured spreadable gel	[11]
High-pressure treatment	Improves the solubility, thermal stability, gelatinization, pasting temperature, pasting time, and susceptibility of starch to chemical and enzymatic attack	Causes the starch to vary in morphology and break down molecularly and intensifies polarization	[16]
Extrusion heating	Enhances water solubility and starch digestibility while reducing viscosity and swelling power	By randomly rupturing the chain, it leads to the breakdown of the starch polymer's amylose and amylopectin; it also results in a total loss of crystallinity and a high level of granule disruption	[7]
High-pressure treatment	Rheological characteristics of starch alter as a result of the pressure treatment; makes the starch gels harder and chewier and increases their freeze-thaw stability	Influences the microstructure of starch under pressure and with a time-dependent effect; crystals of amylopectin melt as a result, and birefringence is lost	[17]
Osmotic pressure treatment	Increases the gelatinization temperature	Transforms crystalline starch from B-type to A-type and distorts the morphologies of starch granules	[18]
Ultrasonication	Decreases the starch's digestion and pasting ability while increasing swelling capacity; it depresses the solubility while raising the temperature and enthalpy of gelatinization	The granules of starch are distorted	[19]
Annealing	Improves thermal stability and susceptibility to amylase while decreasing granule swelling and amylose leaching	Improves molecular order, permits future recrystallization, restructuring, of both starch chains, and increases uniformity among crystallites	[16]
Pulse electric field	Enthalpy, gelatinization temperature, and crystallinity of the granules are all decreased; the field strength increases along with the granule sizes		[20]

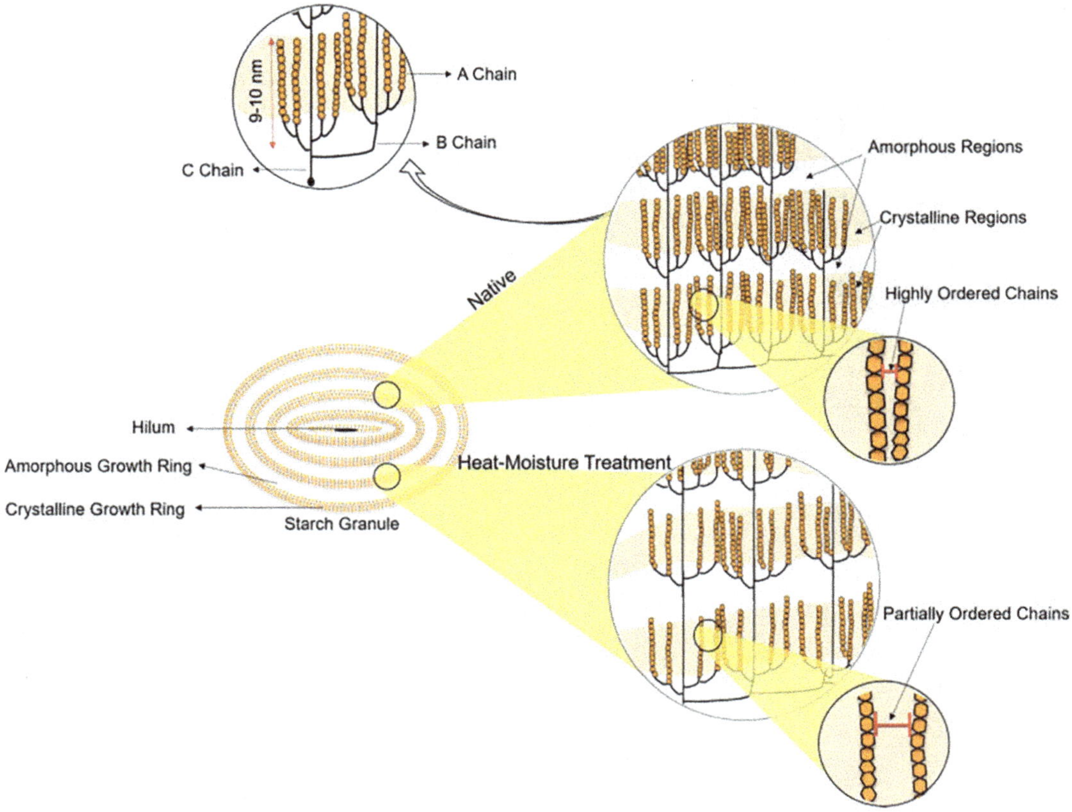

Fig. 2 Structure of native and heat-moisture–treated starch chains. (Reprinted with permission from Elsevier [23])

the moisture content is adjusted within 10–40% depending upon the objective of the study [16]. The moisture content is critical for gelatinization and allows for the penetration of heat during subsequent processing steps (*see* **Notes 1, 2, 3,** and **4**).

2. The conditioned starch is then heated to a specific temperature. The temperature can range from 80 to 150 °C depending upon desired modification [22].
3. Cool the gelatinized and moisture-adjusted starch to stop the gelatinization process. Cooling can be done by air-drying or using a cooling system.
4. After cooling, dry the modified starch to reduce its moisture content to a stable level, usually below 10%. Drying can be accomplished through various methods, such as hot air drying at 40 °C or freeze-drying (*see* **Notes 5, 6**, and **7**).

3.1.3 Effect on Starch Properties

HMT can lead to the modification of starch properties and, thus, their related applications. The specific properties associated with HMT depend on various factors, including starch type, processing

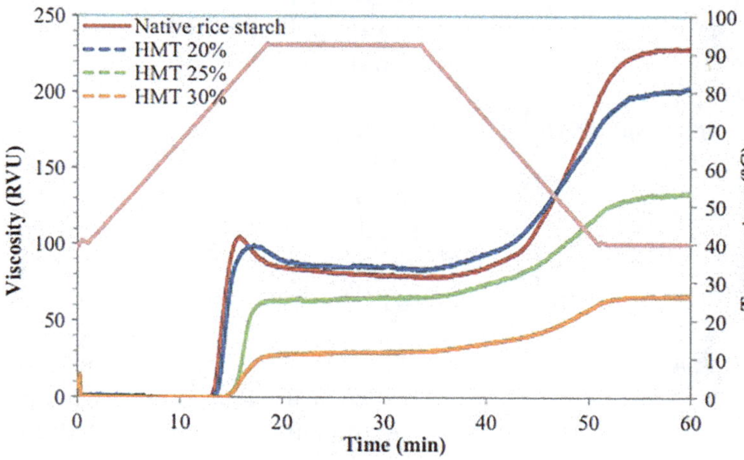

Fig. 3 Effect of HMT on the pasting properties of rice starch [24]

conditions, and intended application. The effects of HMT on some of the starch properties are as follows:

1. Pasting properties: The treatment has been reported to show a strong effect on the pasting properties of the rice starch maintained at 20, 25, and 30% moisture content and heated at 100 °C [24] shown in Fig. 3. Peak viscosity decreased and paste temperature rose as the moisture content rose. The starch needs greater heat for structural disintegration as the forces among the intra-granule linkages are strengthened. Thus, a high paste temperature suggests that the starch granules contain more forces and cross-links [22].
2. Improved stability: Modified starches obtained through HMT tend to exhibit improved stability, especially against factors like high temperature, shear, and pH variations. This enhanced stability makes them suitable for applications requiring heat resistance or prolonged shelf life.
3. Enhanced solubility: Modified starches obtained through HMT often exhibit improved solubility compared to native starch. This improved solubility enables their use in applications that require quick dispersion, such as instant food products, beverages, or powdered mixes.
4. Swelling and absorption: HMT can influence the swelling and water absorption properties of starch. The impact of HMT on the swelling power of potato, cassava, rice, sorghum, and maize starches has been investigated by researchers [22, 25–28]. The swelling power of HMT starches was reduced in each of these studies, according to the authors. An increase in the interactions between amylose and amylopectin molecules

strengthened the intramolecular bonds and decreased hydration and swelling power leading to the changes in the configuration of the crystalline regions of starch [29].

3.2 Annealing (ANN)

The annealing process employs heating and cooling of starch under controlled conditions of moisture content and treatment temperature to induce molecular reorganization [30]. The molecular chains of the starch fractions can be effectively reorganized using the ANN process. The ANN process involves the following steps: dispersion, heating, gelatinization, holding, cooling, and drying.

3.2.1 Principle

The basic principle of the ANN method involves subjecting starch to controlled heating generally at a temperature between 25 °C and 80 °C and cooling in the presence of moisture, usually at 40–90% [22, 31]. The heating phase allows for the complete gelatinization of starch granules, while the cooling phase controls retrogradation and stabilizes the modified starch. The heating and cooling cycles during ANN induce structural changes in starch granules, disrupting the semicrystalline structure, rearrangement of amylose-amylopectin molecules, and modifications in molecular organization [22].

3.2.2 Procedure

1. Prepare a starch suspension: Measure the required amount of starch powder based on the desired moisture content. Place the starch in a clean and dry container. Gradually add the required amount of water to the starch, usually at 40–90% moisture content (*see* **Notes 1**, **2**, **3**, and **4**). Continue stirring or mixing the starch suspension until it becomes homogenous, with a consistent texture and appearance.
2. Heat the starch dispersion gradually while stirring to ensure uniform heating. The time duration can vary from a few hours to days [31]. For instance, banana starch was annealed at 65% for 24 h [32], corn starch at 45 and 50 °C for 72 h [33], oat starch at 45 °C for 3 and 24 h [34], and wheat starch at 30, 40, and 50 °C for 24 h [35].
3. After the required heating duration, cool the starch dispersion to stop gelatinization. This can be achieved by reducing the heat source and/or using a cooling system. The cooling rate should be controlled to minimize retrogradation.
4. Once the ANN process is complete, dry the modified starch to remove excess moisture (*see* **Notes 5**, **6**, and **7**).

3.2.3 Effect on Starch Properties

The ANN treatment can lead to various properties and characteristics in the modified starch. The specific properties associated with the ANN treatment of starch modification depend on factors such as temperature, duration, and starch type. The effects of ANN on common properties are as follows:

1. Thermal properties: ANN treatment can result in modified starches with an altered gelatinization temperature compared to native starch. The onset gelatinization temperature of wheat starch dramatically increased when the annealing temperature rose from 30 to 50 °C. Peak (T_p) and conclusion (T_c) temperatures, on the other hand, increased considerably after annealing at 50 °C as opposed to remaining virtually unchanged at 30 and 40 °C. The enthalpy change of wheat starch was not considerably altered by annealing at 30 and 40 °C, but it was significantly reduced by 50 °C, going from 10.3 to 5.6 J/g. These findings showed that wheat starch's ordered structures were only slightly impacted by annealing at 30 and 40 °C but significantly lowered by annealing at 50 °C [35].
2. Enhanced paste clarity: ANN treatment can lead to modified starches with improved paste clarity. The process helps to reduce the retrogradation of starch, resulting in a more translucent or clear paste.
3. Enhanced viscosity stability: Annealed starches often exhibit improved viscosity stability during prolonged heating or processing. The modified starches maintain their thickening properties and viscosity over an extended period, making them suitable for applications where prolonged cooking or processing is required.
4. Altered textural properties: ANN treatment can modify the textural properties of starch. ANN results in a rearrangement of starch molecules, reducing swelling power and solubility. This decrease in gel volume encourages a rise in gel hardness [36]. Rice starch that had been exposed to ANN showed an increase in gel hardness [27].
5. Morphological properties: The effect of annealing temperatures (30, 40, and 50 °C) was observed on the wheat starch [35]. Grooves and depressions were observed on the surface of the native starch. No significant difference was observed for the starch treated at 30 and 40 °C; however, starch annealed at 50 °C showed evidence of destruction and fusion of starch granules (Fig. 4).

3.3 Osmotic Pressure Treatment (OPT)

In OPT, starch modification is done by subjecting starch to an osmotic solution, typically containing solutes (sugars or salts). It can induce structural and functional changes in starch by altering the osmotic environment surrounding the starch granules [37].

3.3.1 Principle

The key principle of the OPT for starch modification is osmosis. In the context of starch modification, starch granules act as the semipermeable membrane, and solutes in the osmotic solution create an osmotic pressure that affects the water movement in and out of the

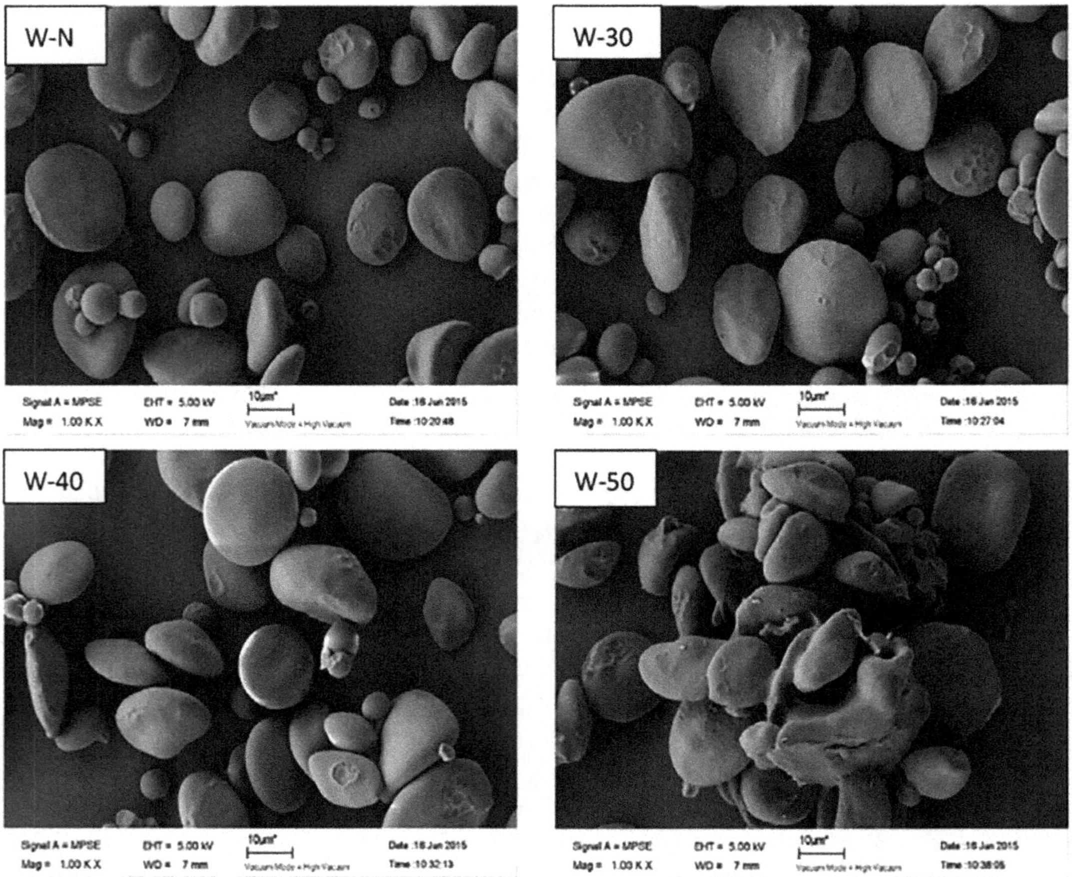

Fig. 4 Effect of annealing temperatures 30 °C (W-30), 40 °C (W-40), and 50 °C (W-50) on the morphological properties of wheat starch. (Reprinted with permission from Elsevier [35])

starch granules. This leads to swelling or shrinkage of the granules, changes in granule morphology, and alterations in the molecular arrangement of starch molecules [37].

3.3.2 Procedure

1. Weigh the exact amount of starch to be treated (*see* **Notes 1 and 4**).
2. Prepare an osmotic solution by dissolving specific solutes. Most commonly, sodium sulfate (Na_2SO_4) is used. Fasuan and Akanbi [38] prepared a Na_2SO_4 solution by dissolving 120 g Na_2SO_4 in 200 mL distilled water (*see* **Note 8**).
3. Immerse the starch into the osmotic solution (*see* **Note 9**). The starch:osmotic solution ratio can be varied to observe their effects. For example, in a study, 100 g starch was suspended in 200 mL saturated sodium sulfate solution [37], while in another, amaranth starch varying from 50 to 150 g was dispersed in 150 to 250 mL saturated Na_2SO_4 solution [38].

4. Allow the starch to remain in the osmotic solution for a specific duration, mainly for 10–20 min. The treatment time can vary depending on the desired modification and the specific starch being used (*see* **Note 10**). The time–temperature combinations can be varied, and usually, the temperature varying from 105 to 120 °C corresponding to the calculated osmotic pressures of 328 and 341 atm, respectively, has been reported [37].
5. After the desired treatment duration, remove the starch from the osmotic solution. Rinse the starch with water or a suitable solvent to remove excess osmotic solution.
6. After OPT, filter, wash with water, and dry the modified starch to remove excess water or solvents (*see* **Notes 5**, **6**, and **7**).

3.3.3 Effect on Starch Properties

The osmotic pressure causes changes in the starch structure, which in turn causes a significant effect on the starch functionality. The effects of osmotic pressure-assisted techniques for starch modification are as follows:

1. Moisture content control: OPT can be used to control the moisture content of starch by subjecting it to a hypertonic or hypotonic solution. This process can result in modified starches with controlled water content, impacting their functionality and storage stability.
2. Improved texture and structure: OPT can influence the texture and structure of starch [38]. For example, osmotic dehydration can result in modified starches with reduced water content, leading to enhanced crispness or texture in certain applications, such as snacks or baked goods.
3. Modified chemical composition: Depending on the osmotic solution used and treatment conditions, osmotic pressure-assisted techniques may result in the exchange or uptake of solutes by the starch. This can lead to modified starches with altered chemical compositions, potentially impacting their functional properties and interactions in food systems [37, 38].

3.4 Microwave Heating (MWH)

This method is used to modify starch, where starch is exposed to microwave radiation to induce heating and bring about structural and functional changes. It provides benefits such as homogenous operation throughout the entire sample volume, rapid processing, deep penetration, and improved product quality [39].

3.4.1 Principle

The basic principle of the MWH method for the physical modification of starch involves subjecting starch to electromagnetic wave within the frequency range of 300 GHz and 300 MHz [40]. The treatment has been reported for the modification of corn, potato, chestnut, Bambara groundnut starches, and many more [39–42]. Starch, being a polar molecule, is capable of absorbing

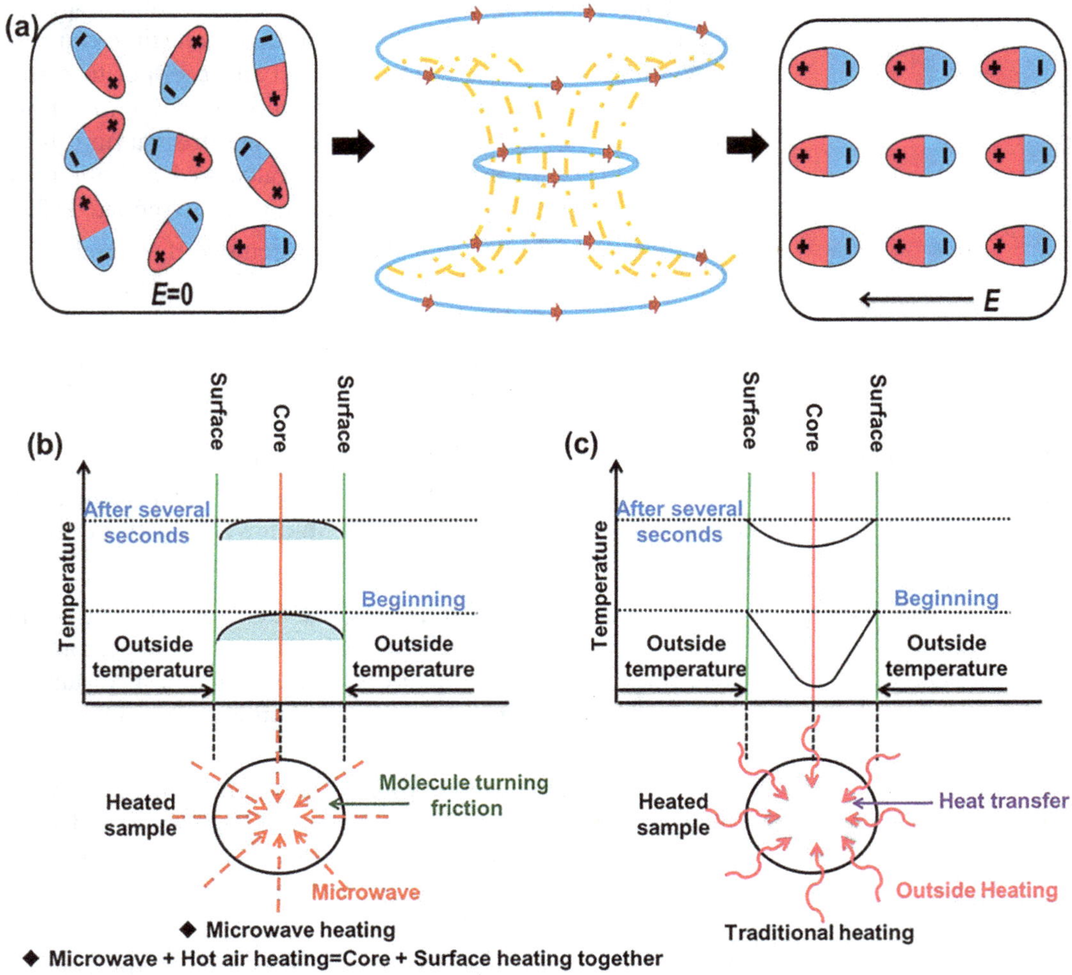

Fig. 5 Mechanism of microwave heat treatment (**a** and **b**) and traditional heating (**c**) [43]

microwave energy and interacting with the alternating electromagnetic field generated by microwaves. Exposure of the sample to the microwave causes the alignment of the polarized or charged atoms with the electric field, causing the rapid change in the direction of the microwaves (Fig. 5). The mechanism of starch modification by this method can be classified into four stages [46]: (1) dielectric relaxation phenomenon of water molecule causing initial starch heating, (2) rapid surge in temperature which is accompanied by loss of moisture from the starch granule interior, (3) creation of high pressure inside the granules causes expansion which occurs from the center, and (4) starch granule degradation. This rapid and selective heating leads to structural and functional changes in starch, providing opportunities for tailored modifications and improved functionality.

3.4.2 Procedure

1. Prepare a starch suspension or dispersion by dispersing the starch in the water to maintain a moisture content of 20–30% (*see* **Notes 1, 2, 3,** and **4**) [44, 45].
2. Equilibrate the moistened starch at least for 2 h, before heating [40].
3. Transfer the starch suspension into a suitable container that is microwave-safe and allows for efficient microwave energy absorption. Oyeyinka et al. [45] reported the use of perforated polyethylene foil specially designed for microwave ovens.
4. Ensure that the starch suspension is spread evenly in the container to facilitate uniform heating (*see* **Note 11**).
5. Place the starch suspension container in a microwave oven and initiate the heating process (*see* **Notes 10** and **12**). Oyeyinka et al. [45] treated the Bambara groundnut starch at 700 W output and a frequency of 2450 MHz. The time duration for the treatment varied from 10 to 60 s.
6. Intermittently pause the heating process to stir or agitate the starch suspension to ensure even heating and avoid localized hotspots.
7. After the desired heating time, remove the starch suspension from the microwave oven and allow it to cool.
8. After cooling, the modified starch suspension may require further processing steps. These can include steps such as filtration, drying, or other treatments to remove excess water or solvents and obtain the modified starch in the desired form (*see* **Notes 5**, **6**, and 7).

3.4.3 Effect on Starch Properties

The MWH treatment involves subjecting starch to microwave radiation, which generates heat and leads to various changes in the starch structure and properties such as:

1. Enhanced gelatinization: MWH can induce efficient and uniform gelatinization of starch. The heat generated by microwaves rapidly raises the temperature of the starch, leading to its swelling.
2. Gelatinization temperature: The gelatinization temperatures of Bambara groundnut starch were dramatically raised by microwaving. With longer microwaving times, the gelatinization temperatures rose while the enthalpy dropped. An increase in gelatinization temperature following microwaving may be attributed to reorganization in the amorphous and crystalline areas, which may have reinforced contacts along the starch chains between amylose–amylose and amylose–amylopectin [45].

3. Improved digestibility: MWH has been shown to enhance the digestibility of starch by modifying its structure. It can lead to starch with increased susceptibility to enzymatic hydrolysis which can be beneficial for applications requiring increased starch digestibility.
4. Altered granule morphology: MWH can cause changes in the morphology and structure of starch granules. These changes may include swelling, partial melting, or even fragmentation, leading to modified starches with unique morphological properties [46].

3.5 Extrusion

Amorphous or semi-amorphous starch particles can be produced fast from a starch suspension using extrusion technology. It involves the application of heat, shear, and pressure to starch as it passes through a narrow barrel under controlled conditions. The combination of these forces causes gelatinization, molecular rearrangement, and textural [47].

3.5.1 Principle

The principle behind the extrusion method of starch modification is the application of mechanical and thermal forces, which lead to physical and chemical changes in the starch structure [13]. The process typically involves high temperatures ranging from 100 to 200 °C. These elevated temperatures cause the starch granules to swell and undergo gelatinization. During extrusion, starch is forced through a narrow orifice under high pressure. The shear forces involved promote molecular alignment, breakdown of starch molecules, and disruption of the starch granule structure.

3.5.2 Procedure

1. Precondition the starch by adding water or steam to adjust the moisture content (*see* **Notes 1, 2, 3**, and **4**). This step helps to facilitate the gelatinization process during extrusion. Gonzalez and Perez [48] conditioned the lentil starch to a moisture content of 25%.
2. Blend the preconditioned starch with other ingredients, such as additives, flavors, or colorants, if desired. This step ensures uniform distribution of ingredients throughout the starch matrix.
3. Feed the starch mixture into the extruder using a hopper or feeder. The extruder consists of a barrel with a screw that conveys the material along its length.
4. Subject the starch mixture to heat and mechanical shear in the barrel. For the modification of lentil starch, the temperature was maintained at 150 °C, and the screw speed was 90 rpm [48].
5. After the cooking zone, the modified starch is rapidly cooled to stop gelatinization.

6. The starch can be shaped into its final form using dye or shaping equipment attached to the extruder.
7. Once the modified starch has been shaped, cut it into desired lengths, such as pellets or flakes.
8. Dry the extrudate at a temperature of 45 °C to reduce the moisture content and stabilize the modified starch (*see* **Notes 5**, **6**, and **7**).

3.5.3 Effect on Starch Properties

The extrusion treatment can affect the properties and characteristics of the modified starch as discussed below:

1. Enhanced gelatinization: The high temperature and shear forces in the extrusion process promote the gelatinization of starch. This leads to modified starches with improved solubility, viscosity, and thickening properties.
2. Improved stability: The extrusion treatment can enhance the stability of starch by modifying its structure. The process can result in modified starches with increased resistance to retrogradation, which is the tendency of starch gels to undergo syneresis or become firm upon cooling.
3. Increased digestibility: The extrusion process can partially hydrolyze starch molecules, resulting in modified starches with increased digestibility. The treatment can break down the starch into smaller fragments, making it more susceptible to enzymatic digestion.
4. Altered granule morphology: The extrusion treatment can cause changes in the morphology and structure of starch granules. These changes may include partial disruption, fragmentation, or reshaping of the granules.

3.6 Ultrasonication

The use of ultrasounds, known as ultrasonication, is a green technology and can be used to modify starch by exposing the starch to different frequencies, intensities, water, temperature, and time combinations [14]. In food processing, ultrasounds are used in the frequency range of 20 kHz to 10 MHz. They can be created using piezoelectric or magnetostrictive transducers, which are projected or transported directly or indirectly over the fluid [49].

3.6.1 Principle

As sinusoidal ultrasonic vibrations move through the aqueous medium, it undergoes a continuous wave-like motion as shown in Fig. 6. During the compression cycle, bubbles contract, but not all of them entirely disappear into the liquid. The bubbles expand over numerous cycles until the oscillation of the bubble wall approaches the applied frequency of the sound waves, at which point the bubbles explode during one compression cycle. Due to the compression and rarefaction process, the system suffers cavitation, causing the bubbles to collapse [50]. The bubble rupture alters the

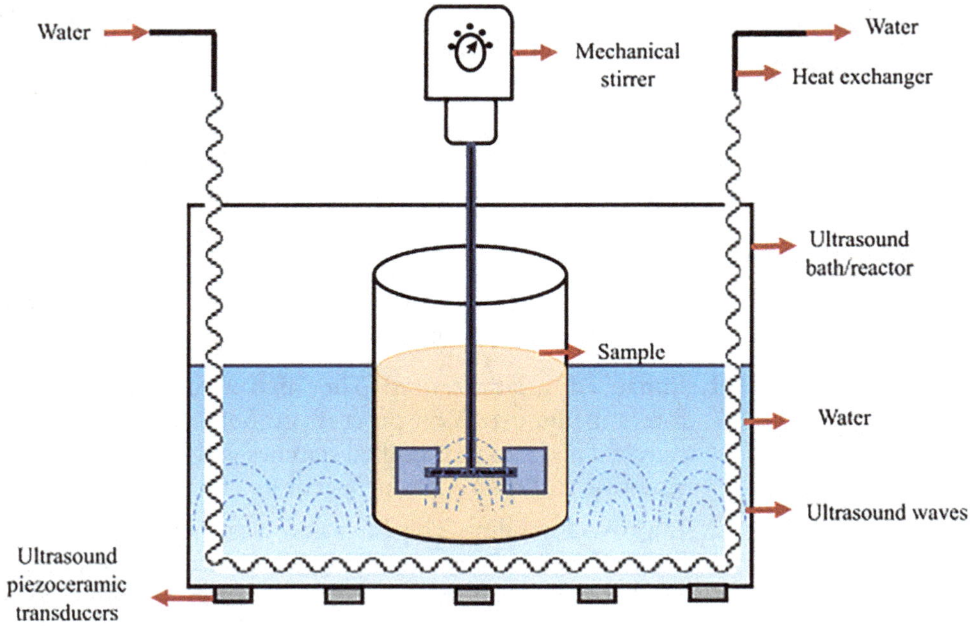

Fig. 6 Ultrasound system. (Reprinted with permission from Elsevier [52])

physical and chemical properties and also produces powerful shearing forces, huge energetic waves, and locally substantially increased turbulence and temperature [51]. The mechanical forces generated by cavitation can disrupt the starch granule structure, causing the release of starch molecules.

3.6.2 Procedure

1. Prepare a starch suspension or dispersion by dispersing the starch in water or a suitable solvent (*see* **Notes 1, 2, 3**, and **4**). Zhu et al. [53] prepared 10% potato starch suspension, Hu et al. [54] reported 5% corn starch suspension, while Sujka [55] and Sujka and Jamroz [56] prepared 30% starch suspension of rice, corn, potato, and wheat starches.
2. Set up an ultrasonic bath or reactor, which consists of a container filled with the starch suspension. The container is equipped with ultrasonic transducers that generate high-frequency sound waves as shown in Fig. 6.
3. Place the starch suspension into the ultrasonic bath or reactor and activate the ultrasonic transducers. Zhu et al. [53] treated the potato starch suspension for 20 min with different power levels of 60, 105, and 155 W and at a frequency of 20 kHz.
4. After the ultrasound treatment, the modified starch suspension may require further processing steps. These can include steps such as centrifugation, filtration, and drying to remove excess water or solvents and obtain the modified starch in the desired form (*see* **Notes 5**, **6**, and 7).

3.6.3 Effect on Starch Properties

While the use of ultrasound for starch modification is still an area of active research, it has shown potential in altering the properties of starch, such as:

1. Enhanced enzymatic digestibility: Ultrasound treatment can disrupt the starch granule structure and increase the surface area available for enzymatic attack. This can lead to modified starches with enhanced enzymatic digestibility.
2. Increased solubility and dispersion: Ultrasound treatment can improve the solubility and dispersibility of starch in water or other solvents. The mechanical forces generated by ultrasound waves can break down the starch granule structure, facilitating the release of starch molecules into the surrounding medium. This results in modified starches with improved solubility and dispersibility.
3. Altered rheological properties: When starch pastes are subjected to ultrasound treatment, the molecules of the starch polysaccharide are depolymerized, and the paste's viscosity is significantly and quickly reduced [16]. Chain cleavages have been attributed to mechanical forces as well as the formation of OH radicals. The reductions in paste viscosities and consistency coefficients are unquestionably the result of the same processes, particularly the production of OH radicals, which take place in starch granules.
4. Altered granule morphology: Ultrasound treatment can cause changes in the morphology and structure of starch granules. The mechanical forces generated by ultrasound waves can induce granule swelling, partial disruption, or reshaping [16]. These changes can result in modified starches with unique morphological properties, which may impact their functional properties and applications.

3.7 High Hydrostatic Pressure (HHP)

In food-processing industries, HHP at room temperature is used to acquire modification and sterilization by exposing the food product to a pressure of 100–1000 MPa. The process is known to destroy the noncovalent bonds, resulting in structural changes and starch gelatinization [57]. The schematic view of the high hydrostatic pressure treatment is shown in Fig. 7.

3.7.1 Principle

The basic principle of HHP on starch modification involves the disruption of the starch granule structure. The hydration, as well as pressure, distorts the crystalline region of starch. During this, the secondary and tertiary structures are broken, whereas the covalent bonds remain intact. The manner in which the starch polymer remains in the granule structure decides the behavior of starch under the treatment [58]. When starch is exposed to HHP, the pressure forces water molecules into the granules, increasing the

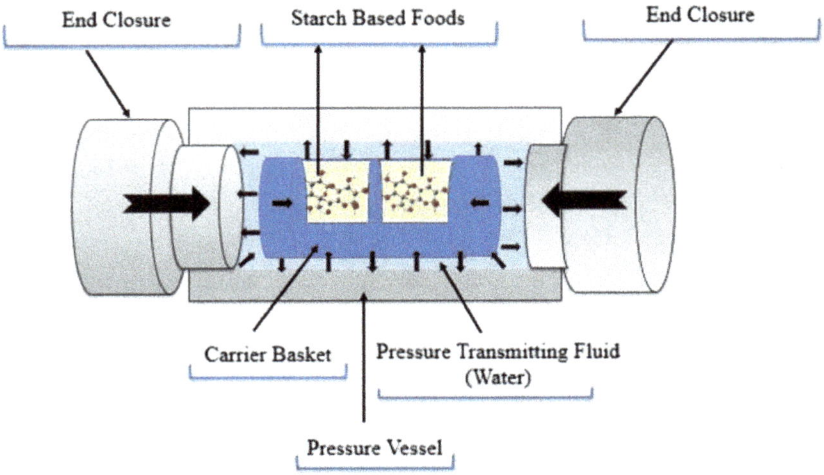

Fig. 7 Representation of high hydrostatic pressure processing [59]

internal pressure. This internal pressure can cause swelling of the granules, leading to the disruption of the crystalline regions within the starch.

3.7.2 Procedure

1. Prepare the starch suspension by adding the required amount of water (*see* **Notes 1, 2, 3,** and **4**). For example, the effect of HHP on 20% (w/w) mung bean starch suspension [60] and the effect on starch suspension (w/v) prepared in the ratio of 1: 3 were studied [57].
2. Transfer the starch suspension into a suitable container that can withstand high pressure, such as flexible pouches or metal cylinders (*see* **Note 13**).
3. Place the packaged starch samples into a high-pressure vessel, which is specialized equipment designed to withstand high pressures (*see* **Notes 14** and **15**). Li et al. [60] carried out HHP treatment of mung bean starch at varying pressures of 120–600 MPa at an interval of 120 MPa for 30 min at room temperature. A variation of pressure from 100 to 600 MPa and maintained for 30 min at a controlled temperature of 30 °C was reported [57].
4. Control the pressure level and treatment duration based on the desired modification and the specific starch being used.
5. After the desired treatment duration, slowly release the pressure from the vessel to atmospheric pressure (*see* **Note 16**).
6. After the pressure release, the modified starch samples may require further processing steps such as filtration, drying, or other treatments to remove excess water or solvents and obtain the modified starch in the desired form (*see* **Notes 5, 6,** and **7**).

3.7.3 Effect on Starch Properties

The high hydrostatic pressure (HHP) treatment involves subjecting starch to elevated pressures. Some potential properties associated with HHP treatment for starch modification are:

1. Enhanced gelatinization: HHP treatment can promote the gelatinization of starch. The applied pressure disrupts the starch granule structure, leading to increased water absorption, swelling, and release of starch molecules. This can result in modified starches with improved solubility, viscosity, and thickening properties.
2. Pasting properties: In comparison to native mung bean starch, the Rapid Visco-Analyzer (RVA) viscograms for the HHP-treated starch showed a substantial increase in the peak viscosity, trough viscosity, final viscosity, and setback and a decrease in the breakdown viscosity and pasting temperature [60]. The sample subjected to a pressure of 600 MPa, however, shows the greatest values for peaking time and pasting temperature and the lowest value for peak viscosity. These variations in granular structures throughout the transformation of the crystalline structure led to changes in viscosity characteristics, pasting temperature, and peak time.
3. Altered rheological properties: HHP treatment can influence the rheological properties of starch, such as viscosity, shear-thinning behavior, and gelation. The treatment can induce changes in the interactions between starch molecules, resulting in modified starches with different thickening, gelling, or textural properties.
4. Altered granule morphology: HHP treatment can cause changes in the morphology and structure of starch granules. The native mung bean granules are reported to possess kidney and ellipse shapes with smooth surfaces without pores. However, exposing them to high pressure of 600 MPa for 30 min resulted in the collapse of starch shapes. Further, the morphologies of starch granules changed from being regular to having deeper grooves as the pressure level rose from 120 to 600 MPa. This meant that the greater pressure caused a more severe loss of granule integrity [60].

3.8 Pulse Electric Field (PEF)

Pulse electric field technology is another nonthermal technique that has been widely employed for food preservation. The technique kills and inactivates the pathogenic microorganisms and enzymes, resulting in minimal destruction of the nutrients.

3.8.1 Principle

The basic principle of the PEF method involves subjecting starch to short, intense electric pulses [14]. The general system of the treatment is shown in Fig. 8. This process can induce structural and functional changes in starch by applying high-voltage pulses to

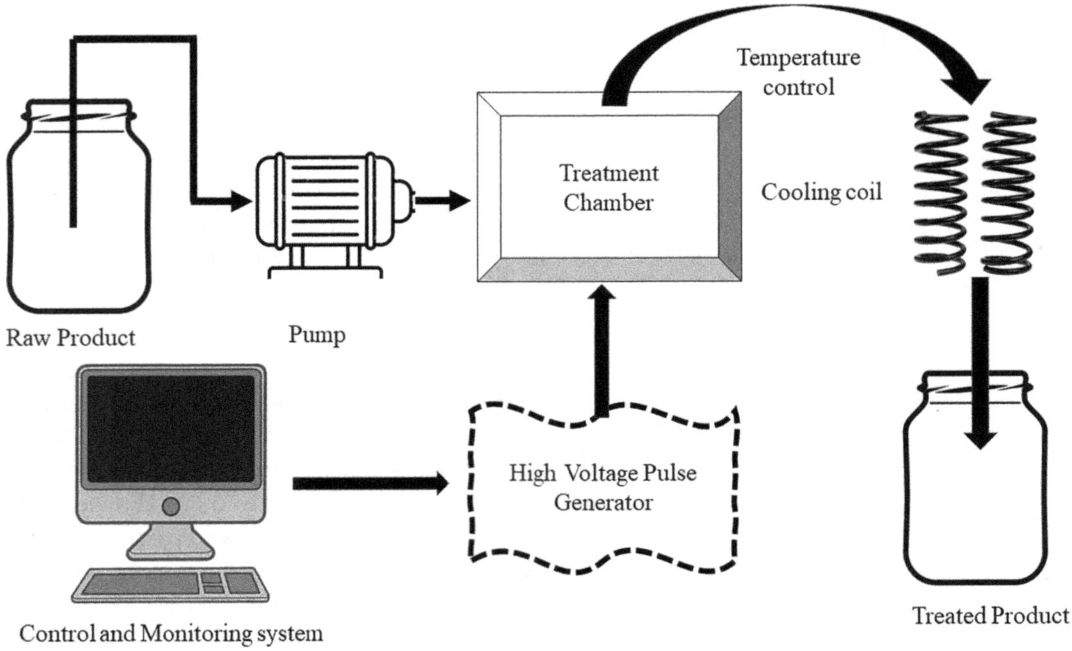

Fig. 8 Process of pulse electric field system

disrupt the starch granules and modify their properties [14]. When starch is exposed to intense electric pulses, the electric field causes the formation of temporary pores or openings in the cell membrane of the starch granules. This phenomenon is known as electroporation. The electroporation of starch granules during PEF treatment results in the rearrangement of molecular structures within the granules. The disruption of the starch granule matrix can alter the organization of amylose and amylopectin molecules, potentially leading to changes in the crystallinity and molecular arrangement of the starch [61].

3.8.2 Procedure

1. Prepare a starch suspension or dispersion by dispersing the starch in water (*see* **Notes 1, 2, 3**, and **4**). The water-to-starch ratio may vary. For instance, 8% (w/w), 10% (w/w), and 40% (w/w) for corn starch, waxy rice starch, japonica rice, wheat, potato, and pea starch have been reported in the literature [61–64].
2. Transfer the starch suspension into a suitable treatment chamber or container that allows for the application of electric pulses (*see* **Note 17**).
3. Place two electrodes in the treatment chamber, ensuring proper positioning and alignment with respect to the starch suspension.

4. Apply short, intense electric pulses to the starch suspension using a high-voltage pulse generator.
   (a) Control the pulse duration, intensity, and frequency based on the desired modification and the specific starch being used. The effect of the electric field of intensities 30, 40, and 50 kV/cm on the tapioca and corn starch has been reported [20, 61].
   (b) Monitor and control the temperature of the starch suspension during the PEF treatment. The temperature is usually kept below 50 °C to avoid starch swelling and gelatinization.
5. After the PEF treatment, subject the modified starch suspension to filtration and drying (*see* **Notes 5**, **6**, and 7).

3.8.3 Effect on Starch Properties

While PEF treatment is not as extensively studied for starch modification compared to other methods, it has shown the potential to alter the properties of starch as follows:

1. Enzymatic digestibility: PEF treatment can disrupt the starch granule structure, leading to increased surface area and affecting the accessibility of enzymes to starch molecules. PEF treatment on esterified potato starch was used to assess the impact on its digestibility [65]. Granules with greater surface roughness and improved glycemic digestibility were produced by esterified starch treated with PEF.
2. Morphological properties: Native tapioca starch granules had a smooth, round, and atypically formed surface morphology. The surface of the tapioca starch granules exhibited some roughness or damage following PEF treatment at 30 kV/cm. As a result of the treatment at 40 kV/cm, some pits and smaller starch particles grew larger, demonstrating that PEF had changed the native starch's granule structure. The tapioca starch appears to lose its granular shape after PEF treatment at 50 kV/cm and is replaced by some grouped pieces with gel-like properties [20].
3. Increased solubility: PEF treatment can improve the solubility of starch in water or other solvents. The electric pulses can disrupt the starch granule structure and facilitate the release of starch molecules into the surrounding medium, leading to modified starches with improved solubility and dispersibility.
4. Improved swelling and hydration: PEF treatment can enhance the swelling and hydration properties of starch. The electric pulses can disrupt the starch granule structure, allowing for increased absorption of water and improved swelling capacity.

4 Notes

1. Use only deionized water for the starch modification, wherever required. The presence of impurities in the water may lead to a change in the starch structure and bonding, which affects the functional, pasting, textural, and morphological properties.
2. The equipment should be properly cleaned before use. Wipe using acetone. There should be no chance of contamination.
3. The glassware used in the starch isolation and modification must also be washed and sanitized properly.
4. Weighing of the starch and water for preparing the starch suspension and conditioning purposes should be carried out accurately. Slight differences in the weight can lead to significant changes in the starch properties.
5. Modified starch should be packaged in suitable containers or bags to protect it from moisture and other environmental factors.
6. Proper storage conditions, such as cool and dry environments, are essential to maintain the quality and stability of the modified starch.
7. Dry the sample at a temperature of 45 °C or lower; otherwise, it would lead to gelatinization.
8. The fresh solution, for example, the osmotic solution, must be used each time to avoid discrepancies in the result.
9. Ensure that the starch is fully submerged and that the osmotic solution covers the starch particles uniformly.
10. The heating process should be carried out under controlled conditions to avoid overheating or scorching. Ensure that the samples are properly arranged and spaced within the HHP.
11. It is important to avoid overcrowding the container, as it can affect the efficiency of MWH.
12. Control the microwave power level and the heating time based on the desired modification and the specific starch being used.
13. The container should be sealed to prevent leakage during the HHP treatment and are mostly vacuum packed.
14. Ensure that the samples are properly arranged and spaced within the HHP vessel to allow for efficient pressure transmission.
15. The specific parameters, such as temperature, moisture content, and processing time, can vary depending on the type of starch, the desired modifications, and the end-use applications of the modified starch.

16. The release of pressure should be done gradually to prevent sudden decompression, which can affect the integrity of the starch samples.
17. The PEF chamber should be designed to withstand high voltage and ensure proper alignment of the electrodes.

5 Conclusion

Physical methods of starch modification offer a versatile and efficient approach to tailoring the properties of starch for various food and industrial applications. These methods involve minimal use of chemical additives or enzymes, preserving the natural composition and clean-label characteristics of starch. The procedures for physical starch modification vary depending on the specific method employed. The step-by-step procedures for the various physical modification methods are discussed in the chapter, along with the mechanism of the method and their effect on the starch properties. The important precautions to be followed by modifying the starch are also given in the "Note" section. The effect of physical modification on starch properties is significant and diverse. As the demand for modified starches with specific functionalities continues to grow, physical methods are expected to play an increasingly important role in the development of innovative starch-based ingredients for diverse applications across the food and industrial sectors. By understanding the effects of physical modification on starch properties, food scientists and manufacturers can develop innovative starch-based ingredients and products with enhanced quality, stability, and functionality. These modified starches play a critical role in meeting the diverse needs and preferences of consumers, contributing to the continuous advancement of the food and industrial sectors.

References

1. Mahajan P, Bera MB, Panesar PS, Chauhan A (2021a) Millet starch: a review. Int J Biol Macromol 180:61–79
2. Mahajan P, Bera MB, Panesar PS (2022a) Structural, functional, textural characterization and in vitro digestibility of underutilized Kutki millet (*Panicum sumatrense*) starch. LWT 154: 112831
3. Mahajan P, Bera MB, Panesar PS, Dixit H (2022b) Structural, functional, textural characterization and in vitro starch digestibility of underutilized Kutki millet (*Panicum sumatrense*) flour. J Food Meas Charact 16:4800–4812
4. Mahajan P, Bera MB (2023) Application of free volume concept and manipulation of network structure parameters for optimum loading of gallic acid in the modified kutki (*Panicum sumatrense*) millet starch hydrogel matrix. Food Hydrocoll 135:108218
5. Mahajan P, Bera MB, Panesar PS (2021b) Modification of Kutki millet (*Panicum sumatrense*) starch properties by the addition of amino acids for the preparation of hydrogels and its characterization. Int J Biol Macromol 191:9–18
6. Akpa JG, Dagde KK (2012) Modification of cassava starch for industrial uses. Int J Eng Technol 2:913–919

7. Lawal MV (2019) Modified starches as direct compression excipients–effect of physical and chemical modifications on tablet properties: a review. Starch-Stärke 71:1800040
8. Shrestha AK, Halley PJ (2014) Starch modification to develop novel starch-biopolymer blends: state of art and perspectives. Starch Polym:105–143
9. Wang Q, Li L, Zheng X (2020) A review of milling damaged starch: generation, measurement, functionality and its effect on starch-based food systems. Food Chem 315:126267
10. Karapantsios TD (2006) Conductive drying kinetics of pregelatinized starch thin films. J Food Eng 76:477–489
11. Majzoobi M, Radi M, Farahnaky A, Jamalian J, Tongtang T, Mesbahi G (2011) Physicochemical properties of pre-gelatinized wheat starch produced by a twin drum drier. J Agric Sci Technol 13:193–202
12. Yan H, Zhengbiao GU (2010) Morphology of modified starches prepared by different methods. Food Res Int 43:767–772
13. Akintayo AAO (2014) Recent trend in the physical and chemical modification of starches from different botanical sources: a review. Starch-Stärke 66:41–57
14. Zia-ud-Din XH, Fei P (2017) Physical and chemical modification of starches: a review. Crit Rev Food Sci Nutr 57:2691–2705
15. Jekle M, Mühlberger K, Becker T (2016) Starch–gluten interactions during gelatinization and its functionality in dough like model systems. Food Hydrocoll 54:196–201
16. BeMiller JN, Huber KC (2015) Physical modification of food starch functionalities. Annu Rev Food Sci Technol 6:19–69
17. Liu W, Guo Z, Zeng S, Zheng B (2013) The influence of ultra-high pressure treatment on the physicochemical properties of areca taro starch. Zhongguo Liangyou Xuebao 28:80
18. Neelam K, Vijay S, Lalit S (2012) Various techniques for the modification of starch and the applications of its derivatives. Int Res J Pharm 3:25–31
19. Nadir AS, Helmy IMF, Nahed MA, Wafaa MMA, Ramadan MT (2015) Modification of potato starch by some different physical methods and utilization in cookies production. Int J Curr Microbiol Appl Sci 4:556–569
20. Han Z, Zeng XA, Fu N, Yu SJ, Chen XD, Kennedy JF (2012) Effects of pulsed electric field treatments on some properties of tapioca starch. Carbohydr Polym 89:1012–1017
21. Ismailoglu SO, Basman A (2015) Effects of infrared heat-moisture treatment on physicochemical properties of corn starch. Starch-Stärke 67:528–539
22. da Rosa ZE, Dias ARG (2011) Impact of heat-moisture treatment and annealing in starches: a review. Carbohydr Polym 83:317–328
23. Schafranski K, Ito VC, Lacerda LG (2021) Impacts and potential applications: a review of the modification of starches by heat-moisture treatment (HMT). Food Hydrocoll 117: 106690
24. Puncha-arnon S, Uttapap D (2013) Rice starch vs. rice flour: differences in their properties when modified by heat–moisture treatment. Carbohydr Polym 91:85–91
25. Olayinka OO, Adebowale KO, Olu-Owolabi BI (2008) Effect of heat-moisture treatment on physicochemical properties of white sorghum starch. Food Hydrocoll 22:225–230
26. Gunaratne A, Hoover R (2002) Effect of heat–moisture treatment on the structure and physicochemical properties of tuber and root starches. Carbohydr Polym 49:425–437
27. Hormdok R, Noomhorm A (2007) Hydrothermal treatments of rice starch for improvement of rice noodle quality. LWT-Food Sci Technol 40:1723–1731
28. Chung HJ, Liu Q, Hoover R (2009) Impact of annealing and heat-moisture treatment on rapidly digestible, slowly digestible and resistant starch levels in native and gelatinized corn, pea and lentil starches. Carbohydr Polym 75: 436–447
29. Waduge RN, Hoover R, Vasanthan T, Gao J, Li J (2006) Effect of annealing on the structure and physicochemical properties of barley starches of varying amylose content. Food Res Int 39:59–77
30. Xu M, Saleh AS, Gong B, Li B, Jing L, Gou M, Jiang H, Li W (2018) The effect of repeated versus continuous annealing on structural, physicochemical, and digestive properties of potato starch. Food Res Int 111:324–333
31. Yao T, Sui Z, Janaswamy S (2018) Annealing. In: Physical modifications of starch. Springer, Singapore, pp 73–89
32. De la Rosa-Millan J, Agama-Acevedo E, Osorio-Díaz P, Bello-Pérez LA (2014) Effect of cooking, annealing and storage on starch digestibility and physicochemical characteristics of unripe banana flour. Rev Mex Ing Quim 13:151–163
33. Yu K, Wang Y, Wang Y, Guo L, Du X (2016) Effects of annealing and additives on the gelatinization, structure, and textural characteristics of corn starch. Int J Food Prop 19:1272–1281

34. Vamadevan V, Bertoft E, Soldatov DV, Seetharaman K (2013) Impact on molecular organization of amylopectin in starch granules upon annealing. Carbohydr Polym 98:1045–1055
35. Wang S, Wang J, Wang S, Wang S (2017) Annealing improves paste viscosity and stability of starch. Food Hydrocoll 62:203–211
36. Chung KM, Moon TW, Chun JK (2000) Influence of annealing on gel properties of mung bean starch. Cereal Chem 77:567–571
37. Pukkahuta C, Shobsngob S, Varavinit S (2007) Effect of osmotic pressure on starch: new method of physical modification of starch. Starch-Stärke 59:78–90
38. Fasuan TO, Akanbi CT (2018) Application of osmotic pressure in modification of Amaranthus viridis starch. LWT 96:182–192
39. Wang M, Sun M, Zhang Y, Chen Y, Wu Y, Ouyang J (2019) Effect of microwave irradiation-retrogradation treatment on the digestive and physicochemical properties of starches with different crystallinity. Food Chem 298:125015
40. Oyeyinka SA, Akintayo OA, Adebo OA, Kayitesi E, Njobeh PB (2021) A review on the physicochemical properties of starches modified by microwave alone and in combination with other methods. Int J Biol Macromol 176:87–95
41. Li YD, Xu TC, Xiao JX, Zong AZ, Qiu B, Jia M, Liu LN, Liu W (2018) Efficacy of potato resistant starch prepared by microwave–toughening treatment. Carbohydr Polym 192:299–307
42. Mutlu S, Kahraman K, Öztürk S (2017) Optimization of resistant starch formation from high amylose corn starch by microwave irradiation treatments and characterization of starch preparations. Int J Biol Macromol 95:635–642
43. Zhang F, Zhou T, Liu Y, Leng J (2015) Microwave synthesis and actuation of shape memory polycaprolactone foams with high speed. Sci Rep 5:11152
44. Zailani MA, Kamilah H, Husaini A, Seruji AZRA, Sarbini SR (2022) Functional and digestibility properties of sago (*Metroxylon sagu*) starch modified by microwave heat treatment. Food Hydrocoll 122:107042
45. Oyeyinka SA, Umaru E, Olatunde SJ, Joseph JK (2019) Effect of short microwave heating time on physicochemical and functional properties of Bambara groundnut starch. Food Biosci 28:36–41
46. Braşoveanu M, Nemţanu MR (2014) Behaviour of starch exposed to microwave radiation treatment. Starch-Stärke 66:3–14
47. Garcia-Valle DE, Agama-Acevedo E, del Carmen Nuñez-Santiago M, Alvarez-Ramirez J, Bello-Pérez LA (2021) Extrusion pregelatinization improves texture, viscoelasticity and in vitro starch digestibility of mango and amaranth flours. J Funct Foods 80:104441
48. Gonzalez Z, Perez E (2002) Evaluation of lentil starches modified by microwave irradiation and extrusion cooking. Food Res Int 35:415–420
49. Kardos N, Luche JL (2001) Sonochemistry of carbohydrate compounds. Carbohydr Res 332: 115–131
50. Patist A, Bates D (2008) Ultrasonic innovations in the food industry: from the laboratory to commercial production. Innov Food Sci Emerg Technol 9:147–154
51. Zuo YYJ, Hébraud P, Hemar Y, Ashokkumar M (2012) Quantification of high-power ultrasound induced damage on potato starch granules using light microscopy. Ultrason Sonochem 19:421–426
52. Raghunathan R, Pandiselvam R, Kothakota A, Khaneghah AM (2021) The application of emerging non-thermal technologies for the modification of cereal starches. LWT 138: 110795
53. Zhu J, Li L, Chen L, Li X (2012) Study on supramolecular structural changes of ultrasonic treated potato starch granules. Food Hydrocoll 29:116–122
54. Hu A, Jiao S, Zheng J, Li L, Fan Y, Chen L, Zhang Z (2015) Ultrasonic frequency effect on corn starch and its cavitation. LWT-Food Sc Technol 60:941–947
55. Sujka M (2017) Ultrasonic modification of starch–impact on granules porosity. Ultrason Sonochem 37:424–429
56. Sujka M, Jamroz J (2013) Ultrasound-treated starch: SEM and TEM imaging, and functional behaviour. Food Hydrocoll 31:413–419
57. Hu X, Xu X, Jin Z, Tian Y, Bai Y, Xie Z (2011) Retrogradation properties of rice starch gelatinized by heat and high hydrostatic pressure (HHP). J Food Eng 106:262–266
58. Pei-Ling L, Xiao-Song H, Qun S (2010) Effect of high hydrostatic pressure on starches: a review. Starch-Stärke 62:615–628
59. Suri S, Singh A (2023) Modification of starch by novel and traditional ways: influence on the structure and functional properties. Sustain Food Technol 1:348–336
60. Li W, Zhang F, Liu P, Bai Y, Gao L, Shen Q (2011) Effect of high hydrostatic pressure on physicochemical, thermal and morphological properties of mung bean (Vigna radiata L.) starch. J Food Eng 103:388–393

61. Han Z, Zeng XA, Zhang BS, Yu SJ (2009) Effects of pulsed electric fields (PEF) treatment on the properties of corn starch. J Food Eng 93:318–323
62. Zeng F, Gao QY, Han Z, Zeng XA, Yu SJ (2016) Structural properties and digestibility of pulsed electric field treated waxy rice starch. Food Chem 194:1313–1319
63. Wu C, Wu QY, Wu M, Jiang W, Qian JY, Rao SQ, Zhang L, Li Q, Zhang C (2019) Effect of pulsed electric field on properties and multi-scale structure of japonica rice starch. LWT 116:108515
64. Li Q, Wu QY, Jiang W, Qian JY, Zhang L, Wu M, Rao SQ, Wu CS (2019) Effect of pulsed electric field on structural properties and digestibility of starches with different crystalline type in solid state. Carbohydr Polym 207: 362–370
65. Hong J, Zeng XA, Han Z, Brennan CS (2018) Effect of pulsed electric fields treatment on the nanostructure of esterified potato starch and their potential glycemic digestibility. Innov Food Sci Emerg Technol 45:438–446

Chapter 10

Chemically Modified Starch

Ranjan Kaushik, Ankit Kumar, Rekha Phogat, Rakesh Gehlot, and Neha Rani

Abstract

Starch is a homopolymer of glucose found naturally in abundance, having unique physicochemical properties, is biodegradable, low cost, can be obtained from many sources, and is widely accepted in the pharmaceutical and food industries. Native starches are unsuitable for most applications due to the presence of many hydroxyl groups and glycosidic bonds; these bonds, when subjected to hydrolysis in many treatments, result in loss of viscosity and thickening power, leading to retrogradation and other problems related to solubility. The presence of certain groups of primary importance like the carboxylic, ester, ether, and amino groups can improve hydrophobicity and related functional properties of starches. Chemical modifications of starch have the advantage of introducing new hydrophobic groups *via* esterification and etherification, and creating strong bonds within molecules *via* crosslinking, condensing reaction, etc.; thus, the shortcomings of native starches related to its functional properties can be eliminated. In recent years, various procedures have been formulated by to extract the starch with chemical means. Thus, to provide a complete overlay of chemical modification (hydrolysis, cross-linking, esterification, etherification, oxidation, and grafting, etc.,) of starch, the chapter has been designed which also provides a complete overview of improved functional properties of starches after modification.

Key words Chemical modification, Hydrophobicity, Retrogradation, Viscosity

1 Introduction

Starch $(C_6H_{10}O_5)_n$ is a complex carbohydrate in which glucose units are linked together in long chains *via* glycosidic bonds. Amylose (amorphous), linear-chained glucose joined by 1,4-glycosidic bonds, and amylopectin (crystalline), a branched chain of glucose unit joined by 1,6-glycosidic bonds, make up the two primary components of starch. Together, they create an insoluble, quasi-crystalline structure [1]. Native starch has many applications in food and other industries other than being a part of our daily diet due to its unique functional properties. But the native starch has one or the other boundaries, like lower swelling and solubility, lower thermal resistance, lower shear resistance, high

Sneh Punia Bangar (ed.), *Standardized Procedures and Protocols for Starch*, Methods and Protocols in Food Science, https://doi.org/10.1007/978-1-0716-3866-8_10,

retrogradation, and syneresis [2]. The qualities of natural starches have been enhanced via starch modification, which is frequently accomplished through chemical, physical, enzymatic, or a mix of these methods [3]. However, physical and enzymatic methods have certain drawbacks particularly with production cost, robustness, and maximum achievable degree of substitution and the heat evolved in the processes [4].

Chemical modification is achieved by coupling a functional group to the native starch molecule. The hydroxyl (-OH) groups at various positions on the glucose units provide multiple reactive sites for chemical modification [5]. Derivatization includes etherification, esterification, cross-linking, oxidation, hydrolysis, and dual modification by following organic reactions such as additions, eliminations, substitutions, and rearrangements. These modifications provide specific functionality and nutritional properties to the native starch granule [6].

1.1 Etherification of Starch

It is a condensation reaction which involves the reaction of starch's hydroxyl groups with alkyl, aryl halides, or other suitable electrophiles [7]. The general reaction for etherification can be represented as follows:

$$R-OH+R'-X\rightarrow R-O-R'+HX$$

where R-OH is starch molecule, R′-X is alkyl halide, R-O-R′ is ether product, and HX is acid produced as a byproduct. Due to the presence of bulky alkyl groups leading to increased steric hindrance and altered molecular interactions, etherification of starch results in increased water resistance, film-forming capabilities, and thermoplasticity, reduced retrogradation, improved stability against heat and shear, enhanced emulsification and binding properties, and potential for controlled release, making it suitable for various applications in coatings, packaging, adhesives, and food industries [8] (Table 1).

1.2 Esterification of Starch

It is also a condensation reaction which involves the reaction between the hydroxyl groups (-OH) of starch and carboxylic acid (RCOOH) or its derivative, usually in the presence of a catalyst [42]. The general reaction for esterification can be represented as follows:

$$RCOOH+R'OH\rightarrow RCOOR'+H_2O$$

where RCOOH is the organic compound having carboxyl group (-COOH) in the molecule, R′OH is the starch molecule, RCOOR′ represents starch ester, and H_2O is the byproduct of the reaction.

Esterification weakens the intermolecular bonding that holds the starch granules together, resulting in alterations functional properties of native starch. It introduces lipophilic groups into the starch molecule. These enhance properties like improved water

Table 1
Different types of starch modifications with methods to identify modified starch and its effect on functional properties

Type of modification (starch derivative)	Method used to identify modification	Degree of substitution (molar substitution values)	Functional property enhanced from native starch	References
Etherification				
Butyl etherification of maize starch	Wettability test, proton nuclear magnetic resonance spectroscopy, and FTIR	Waxy starch-2.0 high amylose starch-2.1	Improved compatibility of waxy and high amylose starches with PLA; enables the fabrication of composite films with excellent tensile strength and an elongation at break	[9]
Hydroxypropyl cassava starch	Microscopy	0.27	High viscosity upon cooling with low amylose content, high water retention capacity, swelling power, and freeze–thaw resistance	[10]
Sago starch solvent-based etherification	FTIR, scanning electron microscopy (SEM), differential scanning calorimetry (DSC), and TGA–DTA	0.35	Larger sago granules, high thermal stability, and better flow properties	[11]
High amylose corn starch etherification	Thermogravimetric analysis, infra-red, ^{1}H-NMR spectroscopy, DSC	2.7	Higher decomposition temperature, clear glass transition temperature, and flexible film with good tensile strength	[12]
1-Allyloxy-2-hydroxy-propyl-starches from amylose-enriched maize starch	^{1}H-NMR, SEM, X-ray diffraction, DSC	0.11–0.13	Increased solubility index, swelling power, decreased gelatinization temperatures, and enhanced pasting properties	[13]

(continued)

Table 1
(continued)

Type of modification (starch derivative)	Method used to identify modification	Degree of substitution (molar substitution values)	Functional property enhanced from native starch	References
Methyl methacrylate etherification of maize starch	Mathematical calculations	0.27	High swelling power	[14]
Different sequences in 1-butyl-3-methylimidazolium chloride [BMIM]Cl ionic liquid media of *Zea mays* starch	FT-IR, XRD, SEM, ^{13}C NMR, ^{31}P-NMR	0.79	Swelling capacity, water solubility indices change to the positive side, reduction in amylose content, crystallinity, and the morphology is completely destroyed	[15]
Esterification				
Waxy maize starch produced by esterifying saturated fatty acid chlorides	FT-IR	Octanoyl, myristoyl, and stearoyl chlorides (0.141, 0.189, and 0.168, respectively)	Better emulsifying properties, crystalline structure of amylopectin destroy, hydrophobicity increment	[16]
1. Acetylated distarch adipate 2. Starch sodium octenyl succinate 3. Starch acetate	DSC, thermogravimetric analysis (TGA)	1. 0.03–0.04 2. 0.0182–0.0202 3. 0.2–1.5	Improved dough quality through lower retrogradation and gelatinization temperatures, decreased gel-forming propensity, and higher paste clarity	[17]
Citric acid esterification of cassava starch	DSC	0.129–0.178	Better freeze–thaw stability; reduced swelling power and solubility	[18]
Octenyl succinic esters of waxy and sweet soluble maize starches	FT-IR, Zeta potential, and particle size analysis	0.0192	Zeta-potential and emulsifying activity increases, oil-in-water emulsion activity develops, reduction of rapidly	[19]

(continued)

Table 1
(continued)

Type of modification (starch derivative)	Method used to identify modification	Degree of substitution (molar substitution values)	Functional property enhanced from native starch	References
			digestible starch, and increase in slowly digestible starch and resistant starch content	
Acetylation of breadfruit starch	Proximate analysis, rapid visco-analyzer	0.23	Improves the solubility and swelling capacity, increases the retention of alkaline water in the mixtures, and decreases the pasting temperature, hot paste viscosity, cold paste viscosity, and gelation	[20]
Maleic anhydride esterified corn starch	SEM, FTIR, X-ray photoelectron spectroscopy (XPS), and laser particle size analysis	0.3386	An enhanced thermoplasticity reduces gelatinization temperature and enhanced esterification thermogravimetric rate	[21]
Citrate-esterified debranched waxy maize starch	Particle size measurements and SEM, FTIR TGA), wide-angle X-ray diffraction, in vitro digestion tests	0.7	B-type crystalline pattern achieved, higher resistance to amylolysis	[22]
Cross-linking				
Starch cross-linked phosphate	X-ray diffraction, DSC, paste properties	0.20	Lowers paste clarity, increased swelling power, with slightly higher water-binding capacity	[23]

(continued)

Table 1
(continued)

Type of modification (starch derivative)	Method used to identify modification	Degree of substitution (molar substitution values)	Functional property enhanced from native starch	References
Banana starch cross-linked with sodium trimetaphosphate/ sodium tripolyphosphate, d-epichlorohydrin, and phosphoryl chloride	X-ray diffraction, DSC, SEM	0.214%	Improves swelling and solubility values, slightly lower enthalpy of gelatinization	[24]
Cross-linked sago starch using $POCl_3$	TGA, SEM, X-ray powder diffractometer, and titrimetric and FT-IR spectroscopy	0.55	Higher thermal stability, high swelling at low temperatures	[25]
Cross-linked wheat starch using STMP and STPP	DSC, α-amylase digestibility test	–	Reduced swelling capacities, elevated temperatures, lower enthalpies of gelatinization, 13–22% slowly digested starch (SDS), and 36–66% resistant starch (RS) content	[26]
Cassava starch-based film produced by cross-linking using casting technique	SEM, contact angle and wettability, FT-IR, X-ray diffraction	–	More hydrophobic, superior mechanical and tensile characteristics, and lower elongation values	[27]
Corn starch cross-linking with STMP/STPP	X-ray diffraction, SEM	10–12%	Decreased swelling factor, solubility, and paste clarity of corn starch	[28]
Oxidation				
Ozone treatments of cassava starch	Rapid visco analyzer (RVA)	40 ppm/90 min and 118 ppm/60 min	Higher viscosities and lower cooking stability, decreased starch microbial contamination	[29]

(continued)

Table 1
(continued)

Type of modification (starch derivative)	Method used to identify modification	Degree of substitution (molar substitution values)	Functional property enhanced from native starch	References
RVA	RVA	–	Swelling ability of oxidized sago and tapioca starches decreased; increase in oxidized corn starch leads to decrease in solubility of tapioca starch; intrinsic viscosity of all oxidized starches also decreased	[30]
Oxidation with sodium hypochlorite of fermented cassava starch	RVA, texture analysis	–	Increased expansion of starches and reduced biscuit hardness, with improved pasting and textural properties	[31]
Hypochlorite oxidation under alkaline conditions	Viscosity by rheometer, DSC	–	Viscosity reduction; lowering of gelatinization temperature and the enthalpy of gelatinization of oxidized starches; oxidized starch effectively retards starch retrogradation	[32]
Hydrolysis				
Starch nanocrystals (SNC) by acid hydrolysis	SEM, Zeta potential, rheometer	–	Increased apparent viscosity, storage modulus, heat stability, and ionic strength	[33]

(continued)

Table 1
(continued)

Type of modification (starch derivative)	Method used to identify modification	Degree of substitution (molar substitution values)	Functional property enhanced from native starch	References
Acid hydrolysis of corn starch	Brabender Viscoamylograph; compression and dynamic oscillatory tests	–	Higher degree of crystallinity and lower amylose content, decrease in storage modulus, and decrease in Loss modulus, gelling temperature, and strength of the acid-thinned starches	[34]
Wax maize, normal maize, high AM maize, potato, and mung bean starches acid hydrolysis	XRD, DSC, SEM	–	A-type starches are more resistant to acid hydrolysis compared to B-type starches; hydrolysis increases the proportion of short chains	[35]
Hydrolysis of sweet potato with water under hydrothermal conditions	HPLC, gel permeation chromatography (GPC)	–	Formation of glucose, maltose, fructose, 5-hydroxy methylfurfural, and furfural	[36]
Hydrolysis of rice starch by alkali method	Structural analysis	–	Contains less amylose, has higher start and peak gelatinization temperatures; the modified starch hydrolyzes more quickly with HCl and amylolytic enzymes	[37, 38]
Dual modification				
Acid-hydrolysis and hydroxypropylation of sago starch	DSC, rheology tests, GPC	0.007–0.15	Decreases molecular weight of starch especially amylopectin, high solubility, and low tendency for retrogradation	[39]

(continued)

Table 1 (continued)

Type of modification (starch derivative)	Method used to identify modification	Degree of substitution (molar substitution values)	Functional property enhanced from native starch	References
Microwave- and ultrasound irradiation-assisted esterification of carboxymethyl cold-water-soluble potato starch with octenyl succinic anhydride	Rheology and emulsion tests	–	Shorten the esterification time, excellent emulsifying and surfactant performance properties	[40]
Acetic acid and ultrasound	Rheological properties		Improved water-binding capacity, swelling power, viscosity, paste clarity, gelatinization temperature, and gelatinization enthalpy	[41]

resistance, film-forming ability, stability under different conditions, and altered rheological behavior [43, 44].

1.3 Cross-Linking of Starch

The modification of starch via cross-linking involves the formation of covalent bonds between starch molecules or between starch molecules and other cross-linking agents [45]. Cross-linking reagents mainly include phosphorus oxychloride ($POCl_3$), epichlorohydrin (EPI), sodium tripolyphosphate (STPP), and sodium trimetaphosphate (STMP) for cross linking of starches [46]. The general reaction for the cross-linking of starch can be represented as follows:

$$[\text{Starch}] + [\text{Cross} - \text{linking agent}] \rightarrow [\text{Cross} - \text{linked starch}]$$

Due to the formation of strong covalent bonds between starch molecules, a three-dimensional network structure is produced that improves the functional properties of the native starch, such as increased resistance to shear, mechanical stress, high and low temperature, and pH, increased water resistance, enhanced thermal stability, reduced retrogradation, improved viscosity stability, and potential for controlled release [47].

1.4 Oxidation of Starch

During the process, three hydroxyl groups on the starch molecule are available which are targeted for oxidation. The oxidizing agent attacks the glycosidic bonds, hydrolyzing them to alcohol (-OH) functions, and the C-C bonds of glucose unit, oxidizing them to carbonyl functions of aldehydes, ketones, and carboxylates. Higher pH favors the formation of carboxylate groups over aldehydes and ketones. The modified starch is commonly known as oxidized starch or bleached starch [46, 48]. The general reaction for the oxidation of starch can be represented as follows:

$$[\text{Starch}] + [\text{Oxidizing agent}] \rightarrow [\text{Oxidized starch/bleached starch}]$$

Oxidation of starch can improve properties such as increased water solubility, increased retrogradation, enhanced viscosity, improved film-forming capabilities, increased adhesiveness, and better stability under certain conditions [8]. Oxidation introduces carbonyl and carboxyl groups and causes depolymerization. Oxidation of native starch in food and other industries is mainly done by using NaOCl and different investigators use it in different concentrations and on starches from different plant sources following different procedures. The variations in concentration and source of starch lead to modified starches with varied functional properties.

1.5 Hydrolysis of Starch

The hydrolysis modification of starch is an addition reaction which involves the cleavage of the glycosidic bonds between the glucose units in the starch molecule by adding H^+/OH^- ions, resulting in the breakdown of starch into smaller, more water-soluble fragments [49]. Hydrolysis of starch with acid is called roasting which results in formation of acid-modified starch, while hydrolysis with sodium or potassium hydroxide results in formation of alkaline-modified starch [50]. The general reaction for the hydrolysis modification of starch can be represented as follows:

$$[\text{Starch}] + [\text{Acid/Alkali}] \rightarrow [\text{Hydrolyzed starch}]$$

Hydrolysis of starch can improve properties such as reduced molecular weight, increased solubility, enhanced digestibility, improved thickening and gelling capabilities, and increased sweetness due to the formation of simpler sugars like maltose and glucose. These modified properties make hydrolyzed starches suitable for applications in food, beverages, pharmaceuticals, and various industrial processes [51]. Hydrolysis of various starch is mostly performed using either acid or alkali methods. Hydrolysis by acid is accomplished mainly by using hydrochloric acid (HCl) and hydrolysis with alkali utilizes sodium hydroxide (NaOH).

1.6 Dual Chemical Modification of Starch

The dual modification of starch refers to a process where two or more of the above modification reactions or the combination of physical and chemical modification methods occur simultaneously or successively, leading to the introduction of new linkages in the starch molecule [52].

Modified starches find applications in food, paper, textiles, and adhesives industries, where these modified properties are desirable. The behavior of modified starch depends on the source from which it is extracted, concentration of the reactants, pH, reaction time, catalysts used, and the degree of substitution (DS), [53, 54]. Using microscopy, differential scanning calorimetry (DSC), and the rapid viscosity analyzer (RVA)/dynamic rheometer, the users may measure the physicochemical properties of modified starches [55].

2 Materials

2.1 Butyl Etherification [9]

1. High amylose maize starch (70% amylose) and waxy maize starch (100% amylopectin).
2. Sodium hydroxide (molar ratio to starch is 1:2): For 100 g, add 40 g of NaOH to 60 g of starch.
3. Sodium sulfate (molar ratio to starch is 1:2): For 100 g, add 71 g of Na_2SO_4 to 29 g of starch.
4. Butylene oxide (1,2-epoxybutane) (molar ratio to starch is 1: 7): Mix 25.8 g of butylene oxide (1,2-epoxybutane) with 74.2 g of maize starch.
5. Acetone.
6. Parr stirred reactor.
7. Freeze dryer.

2.2 Hydroxy-propylation of Corn Starch and Cassava Starch [5]

1. Native cassava starch and corn starch.
2. 1-Chloro-2,3-epoxypropane (epichlorohydrin).
3. Sulfuric acid.
4. Chloropropilene glycol (CPG): Prepare by refluxing 1-chloro-2,3-epoxypropane, sulfuric acid, and water.
5. NaOH (50% (w/v)): Dissolve 50 g of NaOH in 100 mL of distilled water.
6. Static reactor.
7. Flash dryer.

2.3 Double Etherification with CHPTMA and CHPS-Na [56]

1. Food-grade corn starch (11.7% moisture and 53 mPa·s viscosity).
2. Aqueous solution of N-(3-chloro-2-hydroxypropyl) trimethylammonium chloride (CHPTMA) (60%): Mix 60 g of CHPTMA in 100 mL of distilled water.

3. 3-chloro-2-hydroxy-1-propanesulfonic acid sodium salt (CHPS-Na).
4. Sodium hydroxide (3% (w/w)): Dissolve 3 g of NaOH in 100 mL of distilled water.
5. Anhydrous sodium sulfate solution: 32.4 g.
6. Hydrochloric acid (2 M): Dissolve 16.7 mL of 37% HCl and make the final volume pf 100 mL with distilled water.
7. Methanol (80% (v/v)): Mix 80 mL of methanol with 20 mL of distilled water to make a final volume of 100 mL.
8. Ethanol (90% (v/v)): Mix 90 mL of ethanol with 10 mL of distilled water to make a final volume of 100 mL.
9. Four-necked round-bottom flask.
10. Water bath.
11. Oven.

2.4 Benzyl Chloride in Ethanol (Solvent-Based Etherification) [11]

1. Sago starch powder.
2. Sodium sulfate (0.5 moles): Dissolve 71.02 g of sodium sulfate in water and adjust the volume to 600 mL.
3. Sodium hydroxide solution.
4. Benzyl chloride.
5. Methanol (80% (v/v)): Mix 80 mL of methanol with 20 mL of distilled water to make a final volume of 100 mL.
6. Ethanol (90% (v/v)): Mix 90 mL of ethanol with 10 mL of distilled water to make a final volume of 100 mL.
7. Ethanol solution (70% (v/v)): Mix 70 mL of ethanol with 20 mL of distilled water to make a final volume of 100 mL.
8. A reactor with a condenser and internal thermometer.
9. Vacuum suction filtration system.

2.5 Etherification with Dimethyl Sulfoxide (DMSO) [12]

1. High amylose corn starch (70% amylose and 30% amylopectin): 50 g.
2. Sodium hydroxide (fine powder).
3. 1-Bromopropane.
4. HCl (6 N): Add 122.66 mL of 37.5% HCl and make the final volume of 250 mL with distilled water.
5. Diethyl ether.
6. Saturated aqueous sodium chloride.
7. Toluene.
8. Dimethyl sulfoxide (DMSO): 1500 mL.
9. A round, four-neck flask with a condenser and thermometer.

2.6 1-Allyloxy-2-Hydroxy-Propyl-Starches [13]

1. Amylose rich maize starch (70% amylose).
2. Maize starch (27% amylose).
3. Waxy maize starch (0.9% amylose).
4. Allyl glycidyl ether (0.1625 g/g dry starch):
   (a) Add 0.1625 g allyl glycidyl in 1 g of dry starch.
5. NaOH (1.5% or 4% mol/mol dry starch):
   (a) For 1.5%, dissolve approximately 0.6 g of NaOH in 100 mL distilled water.
   (b) For 4%, dissolve approximately 1.6 g of NaOH in 100 mL of distilled water.
6. Na_2SO_4 (5% or 35% g/g dry starch):
   (a) For 5%, dissolve 5 g of Na_2SO_4 in 100 g of distilled water.
   (b) For 35%, dissolve 35 g of Na_2SO_4 in 100 g of distilled water.

2.7 Etherification with Methyl Methacrylate [14]

1. Maize starch.
2. Methyl methacrylate: Prepare freshly distilled methyl methacrylate with 0.01% hydroquinone (under vacuum).
3. Sodium hydroxide.
4. Hydrochloric acid (0.1 N): Add about 0.82 mL of 37% pure HCl and make the final volume of 100 mL with distilled water.
5. Isopropyl alcohol.
6. Acetic acid.
7. Potassium hydroxide.
8. Ethyl alcohol.
9. Oven.
10. Stoppered bottle.

2.8 Hydroxypropyl Banana Starch [57]

1. Green and unripe bananas that are 90 d after flowering are ready for harvest.
2. Sodium bisulfite (0.25 g/L): Dissolve 0.25 g of sodium bisulfite in sufficient water to make a total volume of 1000 mL.
3. NaOH (0.1%): Dissolve 0.1 g of NaOH in 100 mL of distilled water.
4. Na_2SO_4: 15 g.
5. Propylene oxide: 12 mL.
6. HCL (2 M): Dissolve 16.7 mL of 37% hydrochloric acid in distilled water and make final volume to 100 mL.

7. Centrifuge.
8. Refrigerator.
9. Vacuum oven.

2.9 Hydroxypropylation of Maize Starch [15]

1. Starch solution 5% (w/w): Prepare by mixing 1-butyl-3-methylimidazolium chloride [AMIM]Cl with maize starch.
2. Sodium hydroxide (the molar ratio of NaOH to AGU is 1.2): 48 g of NaOH per 61.03 g of AGU of starch.
3. Sodium monochloroacetate (the molar ratio of SMCA to AGU-1.00): 87.46 g of SMCA per 61.03 g of AGU of starch.
4. Ethanol (95 weight%, 100 mL): Mix 95 mL of pure ethanol with 5 mL of distilled water.
5. Three-necked flask with a condenser and a magnetic stirrer.
6. Vacuum oven.
7. Centrifuge.

2.10 Esterification with Citric Acid [18]

1. Cassava starch: 20 g.
2. Citric acid (10 to 40% of starch dry weight): For 10%, mix 2 g citric acid in 20 g of starch, and for 40%, mix 8 g citric acid in 20 g of starch.
3. NaOH (10 M): Dissolve 4 g of NaOH pallets in distilled water and then adjust the volume to 100 mL.
4. Distilled water.
5. Forced-air oven.
6. Centrifuge.
7. Grinder.

2.11 Citrate Esterification [22]

1. Debranched waxy maize starch: 5 g.
2. Pullulanase enzyme.
3. Citric acid.
4. NaOH solution (10 M): Dissolve 40 g of NaOH in 100 mL of distilled water.
5. Distilled water.
6. Absolute ethanol.
7. Autoclave.
8. Centrifuge.
9. Oven.

2.12 Esterification with Octenyl Succinic Anhydride [58]

1. Potato starch (amylopectin content 75.5 ± 2.5% and amylose content 19.5 ± 0.2%).
2. Octenyl succinic anhydride.

3. NaOH solution (3%): Dissolve 3 g NaOH in 100 mL of distilled water.
4. HCl solution (3%): Add 3 mL of hydrochloric acid to 100 mL of distilled water.
5. Aqueous alcohol (70%): Add 70 mL of alcohol mainly pure ethanol with 30 mL distilled water.
6. Oven dryer.
7. Nylon sieve: 120-mesh size.

2.13 Esterification with Saturated Fatty Acid Chlorides [16]

1. Waxy maize starch (WMS-approximately 98% amylopectin).
2. Octanoyl, myristoyl, and stearoyl chlorides (97%): 97 parts is pure substance while containing impurities in other three parts.
3. NaOH (0.25 M): Dissolve 0.15 g of NaOH in 100 mL of distilled water.
4. Ethanol: 150 mL.
5. Methanol: 50 mL.
6. Freeze dryer.
7. Centrifuge.
8. Magnetic stirrer.
9. Phosphorus pentoxide (P_2O_5): To place in desiccator.

2.14 Octenyl Succinic Anhydride Modification [19]

1. Sugary maize soluble starch (SMSS) and commercial waxy maize starch (WMS).
2. 2-Octen-1-ylsuccinic anhydride.
3. Isopropanol.
4. NaOH solution (0.3%): Dissolve 0.3 g of NaOH in 100 mL of distilled water.
5. HCl (2.5 M): Add 20.9 mL of concentrated HCl (37–38% w/w) to 79.1 mL of distilled water.
6. Centrifuge.
7. Oven dryer.

2.15 Maleic Anhydride Esterified Starch by the Dry Method [59]

1. Corn starch: 20 g.
2. Maleic anhydride: 5 g.
3. Acetone.
4. Conical flask: 250 mL.
5. Suction filter.
6. Oven dryer.

2.16 Esterification with Ferulic Acid Chloride [57]

1. Potato starch.
2. Ferulic acid and Thionyl chloride.
3. Catalyst: Pyridine in dimethyl sulfoxide.
4. Chloroform.
5. Alcohol (95%): Add 95 mL of pure ethanol to 5 mL of distilled water.
6. Alcohol (70%): Add 70 mL of pure ethanol to 30 mL of distilled water.
7. Vacuum dryer.

2.17 Esterification with Glutaric Acid [60]

1. Adlay starch: 10 g.
2. Glutaric acid: 10 mL.
3. NaOH (10 M): Add 4 g of NaOH to 100 mL of distilled water.
4. Distilled water.
5. Forced air oven.

2.18 Starch Cross-Linked Phosphate [23]

1. Green, unripen, and ready-to-harvest banana fruit (*Musa valery*).
2. Sodium bisulfite solution (0.25 g/L): Add 0.25 g sodium bisulfite per liter of distilled water. The solution is used in isolation of starch where cut slices of banana are dipped in 2:1 (v/w) proportion.
3. Aqueous hydrochloric acid or sodium hydroxide (10%): Add 10 mL of HCl to 90 mL of distilled water; 10 g of NaOH to 100 mL of distilled water.
4. Sodium sulfate: 10 g.
5. Trimetaphosphate: 4 g.
6. Centrifuge.
7. Vacuum oven.

2.19 Cross-Linked Oat Starch with $POCl_3$ [61]

1. Oat starch: 15 g (db).
2. $POCl_3$: 0.5 and 1 g/kg.
3. Na_2SO_4: 0.3 g.
4. Sodium hydroxide (0.5 M): Dissolve 2 g NaOH in 100 mL of distilled water.
5. Phosphoryl chloride (0.5 and 1 g/kg of starch): 0.5 g of phosphoryl chloride in 1 kg of starch; 1 g of phosphoryl chloride in 1 kg of starch.
6. Dilute HCl (0.1 M): Add 8.35 mL of 37% HCl in 1000 mL distilled water.
7. Oven dryer.

2.20 Cross-Linked Starch with Sodium Trimetaphosphate (STMP) [24]

1. Banana starch: 50 g.
2. Sodium trimetaphosphate: 5.96 g.
3. Sodium tripolyphosphate: 0.6 g.
4. Distilled water: 70 mL.
5. Sodium sulfate: 5.0 g.
6. Sodium hydroxide (1.0 M; 2.0% on starch basis): Dissolve 4 g of NaOH in 100 mL of distilled water.
7. Hydrochloric acid (1.0 M): Add 8.35 mL of 37% HCl to distilled water and make the final volume to 100 mL.
8. Oven dryer.
9. Round bottom flask.

2.21 Cross-Linked Starch with Epichlorohydrin [24]

1. Banana starch: 100 g.
2. Potassium hydroxide solution: Add 6 g of potassium hydroxide in 40 mL of water.
3. Sodium chloride: 1.5 g.
4. Epichlorohydrin: 7.5 mL.
5. Distilled water: 200 mL.
6. Oven dryer.
7. Filter paper: Whatman No. 41.

2.22 Cross-Linking with STMP/STPP of Corn Starch [28]

1. Corn starch: 50 g.
2. STMP/STPP (99:1): Solution consists of 99 parts by weight of STMP and 1 part by weight of STPP.
3. Distilled water: 70 mL.
4. NaOH (0.1 N): Dissolve 0.4 g of NaOH in 100 mL distilled water.
5. HCl (0.1 N): Add 0.835 mL of 37% HCl to distilled water and make the final volume to 100 mL.
6. Shaking water bath.
7. Oven dryer.

2.23 Cross-Linked Sago Starch with $POCl_3$ [25]

1. Sago starch: 50 g (db).
2. POCl3 (0.5-2.5% (w/w)): For 0.5% (w/w), add 0.5 g of $POCl_3$ in 100 g of the starch.
3. Sodium hydroxide (0.5 N): Dissolve 2 g of NaOH in 100 mL of distilled water.
4. HCl (1 N): Add 8.35 mL of 37% HCl to distilled water and make the final volume to 100 mL.
5. Vacuum oven.
6. Stirrer.

2.24 Cross-Linked Tapioca Starch with STMP/STPP [62]

1. Tapioca starch: 200 g (db).
2. Mixture of sodium tripolyphosphate (STPP) and sodium tri-metaphosphate (STMP).
3. NaOH (0.1 N): Dissolve 0.4 g of NaOH in 100 mL of distilled water.
4. HCl (0.1 N): Add 0.835 mL of 37% HCl to distilled water and make the final volume of 100 mL.
5. Oven dryer.

2.25 Cross-Linking with Phosphoryl Chloride ($POCl_3$) of Taro Starch [63]

1. Starch: 30 g (db).
2. Na_2SO_4: 0.6 g.
3. Sodium hydroxide (0.5 M): Dissolve 2 g of NaOH in 100 mL of water.
4. Hydrochloric acid (0.1 M): Add 0.835 mL of 37% HCl to distilled water and make a final volume of 100 mL.
5. Phosphoryl chloride ($POCl_3$).
6. Micropipette.
7. Oven dryer.

2.26 Oxidation of Hybrid Maize Starch [64]

1. Hybrid maize seeds starch: 100 g.
2. Sodium hydroxide (2 M): Dissolve 8 g of NaOH in 100 mL of distilled water.
3. Sodium hypochlorite: 10 mL.
4. Hydrochloric acid (0.5 M): Add 4.175 mL of HCl to distilled water and make a final volume of 100 mL.
5. Sulfuric acid (1 M): Dissolve 5.43 mL in distilled water and make the final volume of 100 mL.
6. Oven dryer.

2.27 Oxidation of Sweet Potato Starch [65]

1. Sweet potato (white cultivar) starch: 100 g.
2. NaOH (2.0 M): Dissolve 8 g of NaOH in 100 mL of distilled water.
3. NaOCl: 10 g.
4. H_2SO_4 (1 M): Dissolve 5.43 mL in distilled water and make the final volume of 100 mL.

2.28 Oxidation of Mucuna Seed Beans Starch [66]

1. Mucuna seed beans.
2. NaOH (0.03 M): Dissolve 0.12 g of NaOH in 100 mL of distilled water.
3. NaOCl: 10 g.
4. Crushed ice and NaCl.

5. HCl (0.3 M): Add 2.505 mL of concentrated hydrochloric acid (HCl, 37% (w/w)) to 100 mL of distilled water.
6. Whatman No. 4 filter paper.

2.29 Oxidation of Cassava Starch [32]

1. Native cassava starch.
2. Sodium hypochlorite containing 10% active chlorine (w/w).
3. NaOH solution.
4. Sodium bisulfite.
5. HCl.
6. Oven dryer.

2.30 Oxidation of Breadfruit Starch [66]

1. Breadfruit starch.
2. NaOH (0.3 M): Dissolve 0.12 g of NaOH in 100 mL of distilled water.
3. HCl (0.3 M): Add 2.505 mL of concentrated hydrochloric acid (HCl, 37% (w/w)) to distilled water and make the final volume to 100 mL.
4. NaOCl (containing 6% active chlorine, i.e., 0.6 g Cl/100 g starch).
5. Crushed ice and NaCl.
6. Whatman No. 4 filter paper.

2.31 Oxidation of Fermented Cassava Starch [31]

1. Cassava starch.
2. HCl (0.1 N): Add 0.835 mL of 37% HCl to distilled water and make the final volume of 100 mL.
3. NaOH (0.1 N): Dissolve 0.4 g of NaOH in 100 mL distilled water.
4. Forced-air oven.

2.32 Oxidation by Ozone [30]

1. Corn, tapioca, and sago starch.
2. Ozone (O_3).
3. Corona discharge reactor.
4. Potassium iodide (KI) traps: 2.

2.33 Acid Hydrolysis of Corn Starch [34]

1. Corn starch: 400 g (db).
2. Hydrochloric acid (HCl) solution: different concentrations (0.06 N, 0.14 N, and 1.0 N HCl).
3. NaOH (1 N): Dissolve 4 g of NaOH in 100 mL distilled water.
4. Deionized water.
5. Convection oven.

2.34 Starch Nanocrystals with Acid Hydrolysis [39]

1. Sago starch (13% moisture): 400 g (db).
2. HCl (0.14 N): Add 1.96 mL of concentrated hydrochloric acid (37%) to 100 mL of distilled water.
3. NaOH (1%): Add 1 g of NaOH to 100 mL of distilled water.
4. Whatman No. 4 filter paper.
5. Oven dryer.

2.35 Acid Hydrolysis of Sweet Potato Starch [65]

1. White cultivar of sweet potato.
2. HCl (0.15 M): add approximately 1.480 mL of 37% HCl solution to 100 mL of water.
3. Magnetic stirrer.
4. Oven dryer.

2.36 Alkali Hydrolysis of Rice Starch [37, 67]

1. High amylose rice starch.
2. 0.1% or 0.4% (w/v) sodium hydroxide solution that contains 0.1% sodium azide (w/v).
3. Sodium azide solution (0.1%): Add 0.1 g of sodium azide to 100 mL of distilled water.
4. Oven dryer.

2.37 Alkali Hydrolysis of Pea Seed Starch [68]

1. Pea seeds.
2. Sodium hydroxide (0.1 M): Add 0.835 mL of 37% HCl to distilled water and make the final volume to 100 mL; also add 0.1% sodium azide to the solution.
3. Ethanol (95%): Add 95 mL pure ethanol to 5 mL of distilled water.
4. Whatman No. 1 filter paper.

2.38 Dual Modification Using Cross-Linking/Acetylation/Hydroxypropylation Methods of Rice Starch [69]

1. Rice starch: 500 g (db).
2. Acetic anhydride (5% solid).
3. NaOH (4%): Add 4 g of NaOH to 100 mL of distilled water.
4. Propylene oxide: 5% solid.
5. Phosphorous oxychloride (0.02%): Add 0.02 g of $POCl_3$ to 98.98 g of starch.
6. HCl (1 N): Dilute approximately 8.75 mL of concentrated HCl (37–38% concentration) and make the final volume to 100 mL with distilled water.

2.39 Dual Modification Using Cross-Linking/ Hydroxypropylation (HPT) Methods of Sweet Potato Starch [70]

1. Sweet potato starch: 1 kg.
2. Trimetaphosphate (0.002 M): Approximately 0.612 g for 100 mL solution.
3. Na_2SO_4 (0.70 M): Approximately 135.58 g for 100 mL solution.
4. NaOH (4% (w/w)): 4 g of NaOH is dissolved in every 100 g of the solution.
5. Sodium trimetaphosphate (STMP) (0.002 M): Dissolve 0.612 g of STMP in 100 mL solution.
6. H_2SO_4 (10 N): Add approximately 18.4 mL of concentrated sulfuric acid (95–98% concentration) and make the final volume of 100 mL.
7. Propylene oxide (0.2 M): approximately 10.8 g for 100 mL solution.

2.40 Dual Modification of Wheat Starch Oxidation/ Cross-Linking [71]

1. Wheat starch: 30 g.
2. Sodium phytate (2%): Add 2 g per 100 mL solution.
3. HCl (0.1 M): 3.65 g of HCl per liter of distilled water.
4. Deionized water.
5. $AgNO_3$.
6. NaOH.
7. Ferrous sulfate ($FeSO_4$).
8. H_2O_2.

2.41 Dual Modification of Tapioca Starch Using Succinylation/Cross-Linking Methods [72]

1. Tapioca starch.
2. Hydrochloric acid (0.1 M, 1.08 N).
3. Sodium hydroxide (3%): Add 3 g of NaOH to 100 mL of distilled water.
4. Sodium sulfate (15%): Add 15 g of sodium sulfate per 100 mL of solution.
5. Propylene oxide.

2.42 Dual Modification of Waxy Cassava Starch Using Acid Hydrolysis/ Succinylation Method [73]

1. Waxy cassava starch: 100 g (db).
2. Hydrochloric acid.
3. Sodium hydroxide.
4. 2-octen-1-yl succinic anhydride (3% (db)).
5. Phenolphthalein indicator.

2.43 Acetylated Distarch Adipate [74]

1. Potato starch.
2. Acetic acid anhydride.

3. Adipic acid.
4. Distilled water.
5. Air dryer.

2.44 Acetic Acid and Ultrasound [41]

1. Isolated starch.
2. Distilled water.
3. Probe-type ultrasonic processor: Tip diameter of the probe is 13 mm.
4. Centrifuge.
5. Oven.
6. Acetic acid.

2.45 Microwave and Other Chemical Modifications [75]

1. Talipot starch.
2. DH_2O.
3. Microwave.
4. Hydrogen peroxide (H_2O_2).
5. Octenyl succinic anhydride.
6. STMP/STPP.

2.46 Succinylation and Annealing [10]

1. Corn starch (12.39% moisture content): 50 g, 62.5 g.
2. Distilled water.
3. Succinic anhydride: 1 g.
4. Sodium carbonate: 1 g.
5. Ethanol: 95%.
6. Sodium carbonate.
7. Cyclone sample mill.
8. Centrifuge.
9. Hot air oven.

3 Methods

3.1 Butyl Etherification [9]

1. Heat the material (starch, butylene oxide, sodium hydroxide, and sodium sulfate) at 140 °C with continuous stirring at 250 rpm for 100 min in Parr stirred reactor (*see* **Note 1**).
2. Precipitate the starches with a high quantity of acetone.
3. Freeze dry the precipitates to obtain the butyl-etherified starch material.
4. Determine the hydrophobicity of the modified starch by testing the wettability in acetone/water solutions.
5. Confirm the grafting of the hydroxybutyl-group onto the starch molecules with ^{1}H-NMR (*see* **Note 2**).

3.2 Hydroxy-propylation of Corn Starch and Cassava Starch [5]

1. Mix chloropropilene glycol (CPG) and cassava starch in a static reactor.
2. Add NaOH (50% (w/v)) to the mixture in the reactor.
3. Dehydrate-modified dihydroxypropyl cassava starch (DHPS) in a flash dryer.
4. Determine the molar substitution degree, retrogradation, and microstructural characteristics of the modified starches (*see* **Note 3**).

3.3 Double Etherification with CHPTMA and CHPS-Na [56]

1. Prepare starch samples by suspending food-grade corn starch in methanol and distilled water mixture with continuous stirring at 40 °C (*see* **Note 4**).
2. Acid-convert the starch with hydrochloric acid (*see* **Note 5**).
3. Prepare starch slurry by adding 324 g of dried acid converted starch to distilled water which contains 32.4 g of sodium sulfate (*see* **Note 6**).
4. Transfer the dispersion in a four-necked round-bottom flask (1000 mL) and immerse it in a water bath.
5. To increase the pH of the dispersion to 10–11, add 3% (w/w) NaOH solution.
6. Stir the dispersion constantly and heat to 45 °C in the water bath.
7. Immediately infuse the dispersion with a mixture of CHPTMA, CHPS-Na, and NaOH solution (*see* **Note 7**).
8. Using a 3% NaOH solution, adjust the dispersion's pH to 11–12.5.
9. Run the reaction for 8 h at 45 °C along with mechanical agitation.
10. Neutralize rough product with a 2 M HCl standard solution to a pH range of 6.5–7.
11. Filter and wash three times with an ethanol solution.
12. Grind the modified starch and sift it through a 100-mesh screen.
13. Dry the esterified sample in an oven at 45 °C for 12 h.
14. Pulverize and again sieve the modified starch through a 100-mesh sieve.
15. Determine the degree of substitution (DS) by nitrogen analysis using the Kjeldahl method and also the properties of the modified starch (*see* **Note 8**).

3.4 Benzyl Chloride in Ethanol (Solvent-Based Etherification) [11]

1. Add 0.5 moles of sodium sulfate per anhydroglucose unit (AGU) in 600 mL of water to the reactor.
2. Next, add 1 mol of sago starch powder per AGU and 0.08 mol of aqueous sodium hydroxide solution per AGU.
3. Heat the dispersion at 60 °C in the reactor and add benzyl chloride to the dispersion with stirring for 24 h.
4. Cool the solution at room temperature for 30 min and adjust the pH to neutral by adding acetic acid.
5. Precipitate the modified starch with 80% (v/v) methanol solution.
6. Filter with a vacuum suction filtration system after being washed with 90% ethanol.
7. Dry the modified starch in a convection oven at 45 °C, approximately for 96 h.
8. Sieve the starch through a wire mesh with 0.425 mm aperture diameter.
9. For solvent-based etherification, carry out the process in 70% ethanol solution (*see* **Note 9**).
10. Determine the degree of substitution (elemental analysis) and other functional properties (*see* **Note 10**).

3.5 Etherification with Dimethyl Sulfoxide (DMSO) [12]

1. Dissolve 50 g of high amylose corn starch (HACS) in 1500 mL of dimethyl sulfoxide (DMSO) (*see* **Note 11**).
2. Add sodium hydroxide powder to the mixture (*see* **Note 12**), and stir for one hour at 60 °C in a nitrogen atmosphere.
3. Add 1-bromopropane drop by drop for 3 h at 60 °C in an oil bath (*see* **Notes 13** and **14**).
4. Add 1-bromopropane very slowly with care (*see* **Note 15**).
5. Put the solution in excess water, neutralize it with HCl (6 N), and stir for a few days.
6. Filter the solution through Whatman filter paper and collect the insoluble part or filtrate on filter paper.
7. Wash multiple times the insoluble part or filtrate with distilled water.
8. Dry it at 80 °C in an oven overnight.
9. Test the degree of substitution by ^{1}H-NMR and other functional properties of the modified starch (*see* **Note 16**).

3.6 1-Allyloxy-2-Hydroxy-Propyl-Starches [13]

1. Mix the starch with distilled water to a 200–400 g/kg slurry.
2. Add NaOH (1.5% or 4% mol/mol dry starch) and Na_2SO_4 (5% or 35% g/g dry starch) to the starch slurry.
3. Increase the temperature of the slurry to 48 °C.

4. After that, add allyl glycidyl ether to the mixture (0.1625 or 1.66 g/g dried starch).
5. Isolate the products by chemical characterization method (*see* **Note 17**).
6. Determine the degree of substitution of modified starches by using ^{1}H-NMR (*see* **Note 18**).

3.7 Etherification with Methyl Methacrylate [14]

1. Make the slurry of starch with methyl methacrylate and isopropyl alcohol in a 250 mL stoppered bottle (*see* **Note 19**).
2. Add the freshly prepared sodium hydroxide (catalyst solution) to the solution.
3. Stir the mixture continuously before heating for 20 h at 55 °C.
4. Grind the reaction mass and neutralize it using alcoholic acetic acid.
5. Filter the mixture and dry the precipitates in an oven (60 °C for 10 h).
6. Then, cool the modified starch to room temperature.
7. Measure the ether % by hydrolysis in terms of molar substitution (*see* **Notes 20** and **21**).

3.8 Hydroxypropyl Banana Starch [57]

1. To prepare the starch, immediately rinse banana fruit slices in a 2:1 (v/w) sodium bisulfite solution at 40 °C.
2. Macerate and combine at low speed (100 rpm) for 5 min, filter the resulting slurry, wash at least three times, and refrigerate overnight at 6 °C.
3. Washed with tap water and centrifuged for 15 min at 3000 rpm.
4. Dry the white sediment at 40 °C for up to 48 h in a convection oven.
5. Take 100 g (db) of starch sample and mix with 120 mL of 0.1% NaOH and 15 g Na_2SO_4.
6. Add 12 mL of propylene oxide after 10 min of reaction.
7. Continue the reaction for 24 h at 40 °C with shaking.
8. Wash the starch suspension three times with distilled water.
9. Use 2 M HCl to maintain the pH of the solution at 5.5.
10. Use distilled water to clean the starch cake.
11. Dry the modified starch at 40 °C in a vacuum oven.
12. Determine the hydroxypropyl content by spectrophotometric method and express in terms of molar substitution (*see* **Note 22**).

3.9 Hydroxy-propylation of Maize Starch [15]

1. Prepare ionic liquid and native starch (IL + NS) mixture of 5% (w/w) starch solution by mixing maize starch with [AMIM] Cl atmosphere (*see* **Note 23**).
2. Stir the mixture at 90 °C for 1.5 h in gaseous N_2.
3. *Preparation of etherified starch (ES)*: Add NaOH to the mixture (IL + NS) and stir properly (*see* **Note 24**).
4. Then, add sodium monochloroacetate at 90 °C and run the reaction for 180 min.
5. Cool the solution to room temperature.
6. Precipitate the modified starch by using ethanol with continuous stirring for 1 h.
7. Next, centrifuge for 40 min at 3000 rpm.
8. To get rid of excess IL, wash the precipitate twice with ethanol.
9. In a vacuum oven, dry the product for 1 hat 45 °C.
10. Modified starch with improved functional properties achieved (*see* **Note 25**).

3.10 Esterification with Citric Acid [18]

1. Weigh 20 g of cassava starch.
2. In 12 mL of distilled water, dissolve (10%, 20%, 30%, or 40% of the starch dry weight) of citric acid.
3. Use 10 M NaOH to bring the pH of the solution to 3.0.
4. Boost the dilution to a final amount of 24 mL.
5. In a stainless-steel tray, combine the citric acid solution (24 mL) and weighed cassava starch (20 g).
6. Condition the solution for 8 h at room temperature.
7. Place the tray in a forced-air oven and dry the solution at 50 °C for 24 h (*see* **Note 26**).
8. Grind the mixture and again, dry the powder at 130 °C for 5 h in a forced air oven.
9. Wash the dried powder three times with distilled water.
10. Centrifuge the solution to remove unreacted citric acid and wash again with distilled water.
11. Air-dry the modified starch at 45 °C and grind it again.
12. Modified starch is thus obtained after procedure with improved functional properties (*see* **Note 27**).

3.11 Citrate Esterification [22]

It proceeds in two steps.

3.11.1 Debranching of Waxy Maize Starch

1. Combine waxy maize starch (5 g, db) with distilled water to get a 10% (m/v) solids content.

2. While stirring continuously, boil the aqueous starch slurry for 30 min.
3. Autoclave the mixture for 20 min at 121 °C (*see* **Note 28**).
4. Cool the mixture to 50 °C.
5. Hydrolyze the starch slurry by adding 200 μL of pullulanase enzyme at 50 °C in a shaking water bath for 24 h.
6. Again, autoclave the sample at 121 °C for 20 min (*see* **Note 29**).
7. Centrifuge the sample at 3000 rpm for 10 min.
8. Store the sample at 4 °C for 8 h for further use (*see* **Note 30**).

3.11.2 Preparation of Ester-Modified Starch

1. Combine 5 g of debranched product and citric acid in a ratio of 5:2 (w/v).
2. Use distilled water to dissolve the mixture.
3. Bring the pH of the solution to 3.5 using a 10 M NaOH solution.
4. Condition the solution for 1 day at room temperature.
5. Dry the samples at 130–150 °C for 3–5 h in the oven, so the esterification reaction takes place.
6. Wash the dried powder three times with absolute ethanol to remove unreacted citric acid.
7. Again, dry it at 30 °C and finely grind the powder (*see* **Note 31**).
8. Determine the degree of substitution and other functional properties of the modified starch.

3.12 Esterification with Octenyl Succinic Anhydride [58]

1. Weigh 20 g potato starch (db) and mix with 20–40% distilled water (w/w) with continuous stirring while mixing.
2. Add 3% NaOH solution to the suspension to change the pH using a pH meter.
3. Add OSA that has been diluted five times with 100% alcohol gradually over the course of 2 h.
4. Wait for appropriate period of time to pass, so that the reaction takes place properly.
5. Use a 3% HCl solution to adjust the pH to 6.5 after the reaction process.
6. After rinsing the mixture twice with distilled water and twice with 70% aqueous alcohol, centrifuge it twice.
7. Dry the sample in oven dryer at 45 °C for 24 h.
8. Sieve the powder by passing through nylon sieve (120 mesh).
9. Finally, the modified starch with improved functional properties can be obtained (*see* **Note 32**).

3.13 Esterification with Saturated Fatty Acid Chlorides [16]

1. Combine 50 mL of NaOH solution (0.25 M) and 1 g of native starch.
2. To ensure that the combination completely dissolves, keep the solution at room temperature on a magnetic stirrer for one hat at 300 rpm.
3. Stir the starch slurry at 500 rpm while adding the precise amount of fatty acid chlorides in dropwise manner at room temperature.
4. Stir the solution for more than 10 min to complete the esterification process.
5. Add 50 mL methanol and 150 m ethanol to the solution to end the reaction.
6. Discard any leftover reactants, such as unreacted FACs and any salts formed during the reaction.
7. Centrifuge the mixture for 15 min at 9000 rpm to obtain the precipitate.
8. Freeze-dry the precipitate for 48 h and keep it in a desiccator containing phosphorus pentoxide (P_2O_5) (*see* **Note 33**).
9. Determine the molar degree of substitution and functional properties of the modified starch (*see* **Note 34**).

3.14 Octenyl Succinic Anhydride Modification [19]

1. Suspend 10 g of the starch in distilled water with agitation.
2. Dilute 2-octen-1-ylsuccinic anhydride (OSA) with isopropanol in the multiples of 5.
3. Add 0.5–3% of diluted OSA over 2 h into the solution.
4. Maintain the pH of the solution to 8.5 by using 0.3% NaOH solution.
5. Rest the solution for 8 h to complete the esterification process at 35 °C, while adjusting the pH to 6.5 with HCl (2 M) solution.
6. Centrifuge the mixture (3000 g) and with the help of distilled water wash the contents for three times and, thereafter, with 70% ethanol for three more times.
7. Oven dry the residue at 40 °C for 24 h.
8. Grind the dried contents to fine powder and sift through 100-mesh sieve.
9. Determine the degree of substitution and the properties of modified starch (*see* **Note 35**).

3.15 Maleic Anhydride Esterified Starch by the Dry Method [59]

1. Grind 5 g of maleic anhydride in a pestle and mortar to make fine powder.
2. Add 20 g corn starch (after removing moisture) to the grounded maleic anhydride and evenly grind the content to fine powder.

3. Transfer the contents to a conical flask measuring 250 mL.
4. Place the flask in water-bath for 1–4 h with intermittent stirring at a controlled temperature of 50–90 °C.
5. Stop heating the content of the flask when the reaction is completed after the fixed time interval.
6. Cool down the content to room temperature.
7. Add the calculated amount of acetone to the flask.
8. Stir the mixture for some time and perform suction filtration.
9. Wash the mixture with acetone for three times.
10. Dry out the contents in oven at 50 °C till the constant weight is achieved.
11. Determine the degree of substitution and other functional properties (*see* **Note 36**).

3.16 Esterification with Ferulic Acid Chloride [57]

The process is completed in two steps.

3.16.1 Preparation of Ferulic Acid Chloride

1. Mix ferulic acid and thionyl chloride.
2. Reflux the contents for 3 h at 80 °C.
3. Remove thionyl chloride with repeated distillation with chloroform after the reaction has been completed.
4. Dry the contents after distillation in a nitrogen atmosphere.

3.16.2 Preparation of Potato Starch Ester

1. Add acid chloride to extract the starch of potato which is dispersed in DMSO.
2. Thereafter, add pyridine to the solution in the ratio 2:1.
3. Reflux the contents for one hat at 110 °C in nitrogen atmosphere with continuous stirring.
4. Precipitate the contents with 95% alcohol (300 mL).
5. Centrifuge the precipitates and wash with 70% alcohol.
6. Dry the residue in vacuum dryer at 50 °C.
7. Determine the degree of substitution and functional properties of the modified potato starch (*see* **Note 37**).

3.17 Esterification with Glutaric Acid [60]

1. To prepare esterified adlay starch (*see* **Note 38**) dissolve glutaric acid in distilled water
2. Use 10 M NaOH to get the pH down to 3.5.
3. Increase the volume with distilled water until it reaches 10 mL.
4. Put 10 g of starch and 10 mL of glutaric acid solution in a stainless-steel bowl.
5. Let the sample condition for 16 h at room temperature.

6. Dry the sample by placing the bowl in a forced-air oven at 50 ° C at for 24 h.
7. Grind the mixture and again dry the powder in a forced-air oven for 3–9 h at 70–130 °C.
8. Rinse the dried powder in 1.5 L of water to eliminate any glutaric acid that has not reacted.
9. Air-dry the sample once more, this time at room temperature, and then powder it
10. Determine the degree of substitution and functional properties of the modified starch (*see* **Note 39**).

3.18 Starch Cross-Linked Phosphate [23]

1. Disperse the weighed quantity of starch in distilled water to prepare a 45% solution of starch.
2. Weigh 4 and 10 g of trimetaphosphate and sodium sulfate, respectively.
3. Mix both chemicals with 200 mL of starch solution.
4. Raise the pH of the mixture to 9.5 with the help of sodium hydroxide or hydrochloric acid.
5. Agitate the slurry at room temperature for about 1 h.
6. Wash the slurry with distilled water for three or more times.
7. Dry the solution in an oven at 48 °C, where the moisture content of the solution is reduced to 10–15%.
8. Heat the sample for 2 h at 180 °C for phosphorylation.
9. Cool the modified starch at room temperature.
10. Recover the starch by centrifugation at 1500 g for 10 min.
11. Wash the starch cake numerous times in distilled water until unreacted phosphate is not found.
12. Bring the suspension to a pH of about 6.5.
13. Dry the recovered starch in a vacuum oven at 40 °C.
14. Calculate the amount of phosphate in terms of molar substitution (*see* **Note 40**) and also the improved functional property of the modified starch (*see* **Note 41**).

3.19 Cross-Linked Oat Starch with $POCl_3$ [61]

1. Take $POCl_3$ in two different concentrations (0.5 and 1 g/kg, dry weight of starch) for cross-linking of oat starch.
2. Dissolve 15 g of starch (db) with 24 mL distilled water with gentle stirring.
3. Add 0.3 g of sodium sulfate (Na_2SO_4).
4. Raise the pH to 11.5 with the help of 0.5 M sodium hydroxide solution.
5. Hold the temperature of the slurry to 25 °C throughout the reaction.

6. With the help of a micropipette, add phosphoryl chloride to the slurry.
7. Seal the reaction jar.
8. Stir continuously the $POCl_3$ and starch solution during their 1 h reaction.
9. After the allotted cross-linking period is passed, use diluted HCl (0.1 M) to bring the starch slurry pH level to 5.5.
10. Following sedimentation, wash the starch three times with distilled water to make the starch free from acid.
11. Dry the modified starch in an oven at 40 °C.
12. Determine the modifications that occur in the native starch (*see* **Note 42**).

3.20 Cross-Linked Starch with Sodium Trimetaphosphate (STMP) [24]

1. Weigh 50 g of banana starch and mix with 70 mL distilled water.
2. Mix 5.96, 0.6, and 5 g of STMP, STPP, and sodium sulfate, respectively, in a round-bottom flask.
3. Using 25 mL of 1.0 mol sodium hydroxide or around 2.0% sodium hydroxide on starch basis to raise the mixture pH to 11.5.
4. Continuously stir the starch slurry.
5. Warm the slurry to 45 °C and maintain the temperature for about 3 h.
6. Neutralize the slurry by adding 1.0 M hydrochloric acid drop by drop to adjust the pH at 6.5.
7. Centrifuge the slurry and collect the modified starch.
8. Dry the starch in an oven at 40 °C after multiple washings with distilled water.
9. Determine the properties modified in the starch (*see* **Note 43**).

3.21 Cross-Linked Starch with Epichlorohydrin [24]

1. Weigh 100 g of banana starch.
2. To the starch, add 150 mL of distilled water, 1.5 g of sodium chloride, and 5.5 mL of epichlorohydrin.
3. Thereafter, add solution of potassium hydroxide which is 6 g of potassium hydroxide in 40 mL of distilled water.
4. Add slowly over 30 min.
5. Stir continuously starch dispersion.
6. Add 50 mL of distilled water and 2 mL of epichlorohydrin.
7. Stir the slurry for about 18 h at room temperature.
8. Filter the slurry through Whatman No. 41 filter paper.
9. Wash the residue with distilled water.

10. Dry the washed content in an oven dried at 40 °C.
11. Determine the cross-linking in the modified powder and the functional properties modified in the powder (*see* **Note 44**).

3.22 Cross-Linking with STMP/STPP of Corn Starch [28]

1. Weigh 50 g of corn starch and mix with different amounts of solution of STMP/STPP which is 99/1% weight by weight based on the weight of starch taken (*see* **Note 45**).
2. Dissolve the solution in 70 mL of distilled water and name with different labels.
3. Adjust the pH of the solutions to 11.0 by using 0.1 N NaOH solution.
4. Keep the slurry in shaking water bath at 45 °C for 3 h.
5. Neutralize the suspensions to pH 6.0 by using 0.1 N HCl solution.
6. Wash the suspension four times or more times with distilled water.
7. Dry the slurry in an electric oven at 40 °C for 24 h.
8. Ground finely the oven-dried samples by using mortar and pestle.
9. Sift the fine powder through a 100-mesh sieve.
10. Determine the functional properties of the cross-linked modified corn starch (*see* **Note 46**).

3.23 Cross-Linked Sago Starch with POCl3 [25]

1. Dry the starch of sago plant in an oven at 45 °C and sieve through a 120-mesh sieve.
2. Cross-linking of native sago starch is done with phosphorous oxychloride ($POCl_3$) in an alkali medium which contain sodium hydroxide.
3. For the procedure, take 50 g of sago starch on dry basis and disperse in 200 mL distilled water.
4. Adjust the pH of the starch slurry to 9.0 by using 0.5 N NaOH solution.
5. Add cross-linking reagent ($POCl_3$) drop by drop at a concentration of 0.5–2.5%.
6. Stir the starch dispersion continuously for 1 h.
7. To complete the cross-linking reaction, store the dispersion for 12 h at room temperature.
8. Stop the reaction by bringing the pH of the starch suspension to 6.5 with the help of 1 N HCl.
9. To assure the elimination of unreacted salt, carry out extensive washing.

10. The cross-linked starch was crushed and sieved (60 mesh) after drying overnight at 40 °C in a vacuum oven.
11. Determine the properties of modified starch (*see* **Note 47**).

3.24 Cross-Linked Tapioca Starch with STMP/STPP [62]

1. Weigh 200 g of tapioca starch and dissolve in 280 mL of 0.1 N sodium hydroxide solution (*see* **Note 48**).
2. Mix the above solution with different concentrations of a mixture of STMP/STPP (*see* **Note 49**).
3. Adjust the pH of solution to 11 with the help of sodium hydroxide.
4. Incubate the starch slurry at 45 °C for 3 h.
5. Treat the suspension with 0.1 N hydrochloric acid to neutralize the pH of the suspension to 6.5.
6. Rinse the solution with distilled water four times properly.
7. Dry the sample in oven at 40 °C overnight.
8. Grind the dried samples and sieve in 100-mesh sifter.
9. Determine the properties of the cross-linked starch (*see* **Note 50**).

3.25 Cross-Linking with Phosphoryl Chloride ($POCl_3$) of Taro Starch [63]

1. Cross-linking of taro starch is done using cross-linking agent phosphoryl chloride ($POCl_3$).
2. With gentle stirring, 0.6 g of Na_2SO_4 is dissolved in 48 mL of distilled water together with 30 g of dried taro starch.
3. Next, 11.5 pH is achieved by adding 0.5 M NaOH.
4. The starch slurry's temperature needs to be held constant at 25 °C.
5. With the use of a micropipette, add $POCl_3$ to the reaction vessel at a concentration of 0.5 or 1.0 g/kg (0.05 or 0.10% based on the dry weight of starch).
6. Wait for 1 h so that $POCl_3$ and starch solution react properly.
7. By adding diluted HCl (0.1 M), further bring down the pH of the starch slurry to 5.5.
8. Recover the starch by centrifuging for 10 min at 5000 rpm.
9. Wash the cake several times with distilled water.
10. Then, dry the modified starch in an oven for 24 h at 45 °C.
11. Determine the properties of modified starch (*see* **Note 51**).

3.26 Oxidation of Hybrid Maize Starch [64]

1. Prepare the paste of starch by dissolving 100 g maize starch in 500 mL of distilled water.
2. Adjust the pH to 9.5 by using 2 M sodium hydroxide solution.
3. Add 10 g of NaOCl to the starch slurry slowly for 30 min and maintain the pH at 9 to 9.5 over the period.

4. Constantly stir the solution at 30 °C.
5. Control the pH by adjusting with 0.5 M hydrochloric acid or sodium hydroxide.
6. After the addition of NaOCl the reaction will proceed for about 10 min.
7. Once the reaction is completed, adjust the pH of solution to 7 with 1 M sulfuric acid solution.
8. Filter the modified maize starch.
9. Wash the modified oxidized starch with distilled water four times.
10. Air-dry the starch at room temperature of about 30 ± 2 °C for the time of 48 h.
11. Determine the properties modified after modification (*see* **Note 52**).

3.27 Oxidation of Sweet Potato Starch [65]

1. Take 100 g starch of sweet potato and disperse in 500 mL of distilled water.
2. Adjust the pH of the solution to 9.5 with the help of sodium hydroxide (2.0 M).
3. Add 10 g of NaOCl to the starch slurry for 30 min over the mixture.
4. Constantly stir the mixture.
5. Maintain the pH of mixture at 9.0–9.5.
6. After the addition of NaOCl, allow the mixture to rest for 10 min so that the reaction is completed.
7. Adjust the pH to 7.0 with the help of 1 M solution of sulfuric acid.
8. Filter the oxidized starch of sweet potato to remove impurities.
9. By using distilled water wash the modified oxidized starch four times.
10. Air-dry the starch at room temperature at 30 ± 2 °C for 48 h.
11. Determine the functional properties of the modified starch (*see* **Note 53**).

3.28 Oxidation of Mucuna Seed Beans Starch [66]

1. Weigh 100 g of Mucuna beans starch.
2. Disperse the starch in 500 mL of distilled water.
3. Adjust the pH of the solution to 9.5 with the help of 0.3 M sodium hydroxide solution.
4. Add 10 g of NaOCl drop by drop to the starch solution slowly over 2 h.
5. Constantly stir the solution while mixing NaOCl.

6. Maintain the pH of the solution to 9.0–9.5 with the help of sodium hydroxide.
7. Cool the vessel in which reaction is taking place in crushed ice and NaCl solution.
8. Allow the solution to rest for 5 h to complete the reaction after the addition of NaOCl.
9. Bring the pH to 7.0 with the help of 0.3 M hydrochloric acid.
10. Filter the slurry through Whatman No. 4 filter paper.
11. Wash the modified starch with distilled water for four times.
12. Air-dry the modified starch at 30 ± 2 °C for 48 h.
13. Determine the functional properties of the modified starch (*see* **Note 54**).

3.29 Oxidation of Cassava Starch [32]

1. Prepare slurry of cassava starch which contain about 40% dry solids by mixing in distilled water.
2. Adjust the pH of the starch slurry to 8–11 with an aqueous sodium hydroxide solution.
3. Adjust the temperature of the cassava starch slurry to 30 °C.
4. Add sodium hypochlorite (3% active chlorine based on starch) drop by drop to the solution over 15 min.
5. Continuously stir the solution to properly mix the contents.
6. Maintain the pH of the solution with proper calculated amount of NaOH or HCl solution when the reagent is added to the solution and after the addition during resting time for completion of reaction.
7. Stir the mixture regularly.
8. Collect the samples at different time intervals to check the completion of reaction.
9. Add sodium bisulfite to stop the process and adjust the pH to 6.5–7.0.
10. Filter the samples and wash them completely with distilled water.
11. Check the outcome of the silver nitrate test and cease washing.
12. Dry the modified starch in an oven set to 50 °C until the sample's weight stabilizes.
13. Determine the functional properties altered in the modified oxidized starch (*see* **Note 55**).

3.30 Oxidation of Breadfruit Starch [20]

1. Extract the starch from breadfruit.
2. Weigh 100 g of the native breadfruit starch.
3. Combine the starch with 500 mL of distilled water to make slurry.

4. Use 0.3 M sodium hydroxide to get the mixture's pH value to 9.5.
5. Weigh 10 g of NaOCl containing 6% of active chlorine, i.e., 0.6 g of Cl per 100 g of starch.
6. Add NaOCl dropwise to the solution over 2 h with constant stirring.
7. Maintain the pH of the solution to 9.0–9.5.
8. Cool the reaction vessel in crushed ice and sodium chloride solution.
9. Allow the reaction to continue for 5 h after the complete addition of NaOCl.
10. Bring the pH of the solution to 7.0 with the help of 0.3 M hydrochloric acid.
11. Filter the slurry with Whatman No. 4 filter paper properly.
12. Wash the obtained residue of modified starch four times with the help of distilled water.
13. Dry the modified starch at room temperature (30 ± 2 °C) for 48 h until the weight of the starch becomes constant.
14. Determine the improved functional properties of the bread-fruit starch (*see* **Note 56**).

3.31 Oxidation of Fermented Cassava Starch [31]

1. Remove the water-soluble fraction from the cassava starch during fermentation.
2. Treat the obtained starch after fermentation with an oxidant in a glass reactor.
3. Resuspend the weighed starch of about 200 g (db) in 500 mL of distilled water.
4. Heat the suspension to different temperatures.
5. Add sodium hypochlorite to the heated solution.
6. Using sodium hydroxide (0.1 N) and hydrochloric acid (0.1 N), adjust the pH of the heated solution.
7. Let the solution rest for 50 min.
8. Remove 65 g of starch (db) from the reactor.
9. Filter the starch using medium-porosity filter paper in a Buchner funnel.
10. Wash the residue with distilled water by using approximately 1200 mL of water.
11. Resuspend the washed filtrate in 600 mL of distilled water.
12. Re-filter the solution.
13. Include the water-soluble component with the oxidized starch (*see* **Note 57**).

14. Reduce the modified starch's moisture content to less than 12% by drying it in a forced-air oven at 42 °C.
15. Use fermented cassava starches that are oven and sun-dried for 8 h as control.
16. Test the functional properties modified in the starch (*see* **Note 58**).

3.32 Oxidation by Ozone [30]

3.32.1 Set Up for the Treatment

1. Ozone (O_3) oxidation is performed in a rotating vessel connected to an ozone generator and oxygen cylinder.
2. Cells in a corona discharge reactor that stream diatomic oxygen into a high-voltage electric field between conductive and dielectric surfaces produce ozone.
3. Seal firmly all the connections like reaction container, generator, and oxygen cylinder which are connected by Teflon tubing.
4. Maintain 8 ± 0.2 mL/s flow rate from the cylinder (*see* **Note 59**).
5. Position two potassium iodide (KI) traps such that they catch the unreacted ozone at the end of the reaction vessel.

3.32.2 Method for Treatment of Starch

6. Take 4% of potassium iodide solution and place about 1 L (total) of the solution to each trap which react with O_3 to form O_2, I_2, and KOH.
7. Fill the reaction vessel with the powdered starch sample to start the reaction.
8. O_3 will produce at various ozone-generating timings.
9. The reactor rotates at 150 revolutions per min (*see* **Note 60**).
10. Continue for an additional 10 min so that O_3 and starch will react properly.
11. Close inlet and outlet of the gas during the time when reaction is taking place.
12. Finally, flush the O_2 from the vessel for 20 min (*see* **Note 61**).
13. Once the procedure is completed, remove the modified oxidized starch from the vessel.
14. Titrate the KI solutions in the traps to determine the extra O_3, which is not reacted to the starch.
15. Perform a blank test using the same steps but without starch before the starch and O_3 reaction.
16. Draw a calibration curve against the amounts of O_3 produced (A) and OGT (t).
17. Estimate the amount of ozone that is reacted with starch by deducting the unreacted amount from the produced amount.
18. Identify the functional characteristics of the starch modified by ozone (*see* **Note 62**).

3.33 Acid Hydrolysis of Corn Starch [34]

1. Prepare 40% slurry of starch by mixing 400 g starch (db) and 0.06 N hydrochloric acid solution (HCl) to make final weight of 1000 g in a heating water bath at about 50 °C.
2. Terminate the reaction when the peak viscosity falls between 200 and 300 BU of the modified starch which is acid-thinned starch (*see* **Note 63**).
3. Thereafter, neutralize the slurry to pH of about 5.5 by using 1 N sodium hydroxide solution.
4. Wash the slurry three times by deionized or distilled water with double the amount of slurry.
5. Filter the slurry.
6. Dry the residue at 45 °C overnight in a convection oven.
7. Determine properties of modified starch (*see* **Note 64**).

3.34 Starch Nanocrystals with Acid Hydrolysis [39]

1. Prepare 40% slurry of sago starch.
2. For this, mix 400 g starch on dry basis with 0.14 N hydrochloric acid solution at about 50 °C and make 1000 g final weight.
3. Incubate the suspension at 6 to 24 h at 6 h interval at 50 °C (*see* **Notes 65** and **66**).
4. Stir the suspensions at 200 rpm in an orbital incubator shaker (*see* **Note 67**).
5. After 6, 12, 18, and 24 h, neutralize the slurry with 1% solution of NaOH to adjust the pH at 5.5.
6. Wash the slurry three times with double the volume of distilled water.
7. Filter the suspension using a Whatman No. 4 filter paper.
8. Dry the starches in an oven at 40 °C overnight.
9. Determine the properties of the modified starch (*see* **Note 68**).

3.35 Acid Hydrolysis of Sweet Potato Starch [65]

1. Weigh 100 g starch of native sweet potato.
2. Mix the weighed starch with 500 mL solution of 0.15 M hydrochloric acid to make starch slurry.
3. Magnetically stir the mixture for 8 h and maintain a temperature of 50 °C.
4. Filter the solution to obtain the modified starch.
5. Wash the residue four times by using distilled water.
6. Air-dry the sample at 30 ± 2 °C for 48 h.
7. Determine the properties of acid-modified starch (*see* **Note 69**).

3.36 Alkali Hydrolysis of Rice Starch [37, 38]

1. Weigh 2 g dry weight of rice starch.
2. Suspend the starch in 50 mL of 0.1% or 0.4% weight by volume of sodium hydroxide which contain 0.1% weight by volume sodium azide (*see* **Note 70**).
3. Keep the starch slurries for rest at 35 °C for 7 and 14 d.
4. For the time period, shake the starch slurry three times a day by hand so that sedimentation of the starches does not occur.
5. Thereafter, centrifuge the solution at 3000 g for 10 min to obtain undissolved residues (*see* **Note 71**).
6. Wash the residues four times with double distilled water and two times with anhydrous ethanol.
7. Dry the modified starch in an oven at 40 °C for 2 d.
8. Grind the dried modified starches into fine powders in a pestle and mortar.
9. Sift the modified starch by passing through a 100-mesh sieve.
10. Do the structural analysis of the obtained starch (*see* **Note 72**).

3.37 Alkali Hydrolysis of Pea Seeds Starch [68]

1. The pea seed starch hydrolysis is done by alkali treatment.
2. For the procedure, weigh 10 g of pea starch dry weight.
3. Suspend the starch in 100 mL sodium hydroxide solution with 0.1 molarity and the solution contains 0.1% sodium azide.
4. The starch slurry is restored for 15 and 30 d at 35 °C with continuous shaking to resuspend the settled or sedimented starch granules.
5. Filter the slurry through Whatman No. 1 filter paper with suction.
6. Wash the residue with the help of distilled water so that the filtrate is neutralized to a neutral pH.
7. Wash it with 95% C_2H_5OH two times.
8. Dry the treated starch at room temperature overnight in gentle air stream.
9. Use the dry weight of starch after alkali treatment divided by the dry weight of starch before treatment to compute the yield of alkali-treated starch.
10. Determine the properties of the modified starch (*see* **Note 73**).

3.38 Dual Modification Using Cross-Linking/ Acetylation/ Hydroxypropylation Methods of Rice Starch [69]

1. Prepare the starch slurry by dissolving 500 g (db) of Japonica and Indica rice varieties in 750 mL distilled water.
2. *Acetylation*: Add $C_4H_6O_3$ (5% solid) dropwise into the slurry meanwhile and stir at ambient temperature while maintaining pH between 7.8 and 8.2 using 4% NaOH solution.
3. After addition of $C_4H_6O_3$, the reaction proceeds for at least 5 min to ensure ample time for completion of reaction.

4. *Hydroxypropylation*: Add propylene oxide (5% of solid) and adjust the pH of starch slurry to 11.5 by adding 1 N NaOH.
5. Incubate the slurry in water bath at 45 °C for 24 h.
6. *Cross-linking*: Add 0.02% of phosphorous oxychloride to starch mixture with a pH of 11.5 maintained by using 1 N NaOH.
7. Agitate the mixture is for 2 h in a water bath at 45 °C.
8. *Dual modification*: First, cross-link the starch by using phosphorous oxychloride (0.02%).
9. Thereafter, add 5% of either acetic anhydride or propylene oxide to the cross-linked starch.
10. Wash all starch slurries with five times volumes of distilled water after the reactions.
11. Neutralize the modified starch slurry with 1 N HCl and dry at 40 °C for 12 h in a convection oven to moisture the content below 10–12%.
12. Determine the properties of modified starch (*see* **Note 74**).

3.39 Dual Modification Using Cross-Linking/ Hydroxypropylation (HPT) Methods of Sweet Potato Starch [70]

1. *Cross-linking*: Weigh 1 kg of sweet potato starch and dissolve in tap water to prepare 39% (w/w) starch slurry.
2. At room temperature, add 0.70 M of sodium sulfate in the starch slurry.
3. By gradually adding 4.4% (w/w) NaOH, bring the pH of the solution to 10.5.
4. Then, for a further 6-h reaction, add 0.002 mol of sodium trimetaphosphate at 35 °C.
5. Neutralize the suspension by using 10 N sulfuric acid.
6. Dewater the suspension over a Buchner funnel.
7. Clean the cross-linked starch with five times the volume of tap water.
8. Dry the modified starch in hot air oven at 40 °C.
9. *Hydroxypropylation*: Take starch slurry (follow **steps 1–3**).
10. At room temperature, add 0.2 mol of propylene oxide with the help of dropping funnel to the bolted container.
11. Leave the suspension for 24 h at 35 °C to complete the reaction.
12. Further, follow **steps 5–8**.
13. *Dual modification*: Cross-link the sweet potato starch and, thereafter, hydroxypropylation of the cross-linked starch is done.
14. Fractionate the starches by sifting on an analytical sieve shake with a mesh size of 20 μm.

15. The vibrational amplitude set at the time of sieving is 1 mm.
16. Rinse the dual modified starch with tap water.
17. Terminate the sieving when the washing water is clear and starch particles are not visually seen in the water.
18. Precipitate the fractionated starch overnight and collect after decantation.
19. Dry out the starch in an oven at 40 °C overnight.
20. Determine the properties of the modified starch (*see* **Note 75**).

3.40 Dual Modification of Wheat Starch Oxidation/Cross-Linking [71]

1. *Cross-linking*: Weigh 30 g of wheat starch and mix with 70 mL distilled water to prepare starch slurry.
2. Mix 2% of sodium phytate on dry basis to the starch slurry.
3. Stir the slurry at 200 rpm for 6 h at a constant temperature 50 ° C in an incubator.
4. Adjust the pH of the starch slurry to 7.4 by adding 0.1 mol/L hydrochloric acid.
5. Isolate the modified starch by placing the starch slurry in a centrifuge that rotates at 3500 revolutions per min and 960 g for 15 min.
6. Wash the sediment obtained after centrifugation with distilled water.
7. Stop the process when the supernatant does not show any response to silver nitrate solution.
8. Dry the modified starch sediment for 24 h at −50 °C in vacuum freeze-drying oven.
9. *Oxidation*: Prepare 40% wheat starch solution by dispersing 40 g starch in 100 mL of distilled water.
10. Raise the pH of the mixture to 8.0 by using 1 M NaOH.
11. Add 0.07% ferrous sulfate (catalyst, db) into the starch slurry maintaining the temperature of 45 °C.
12. Thereafter, add H_2O_2 solution to the starch solution drop by drop within 15 min so that terminal concentration reaches to 12%.
13. Continue the process for 3 h while maintaining the same pH and temperature in an incubator.
14. Stop the process by adjusting the pH to 6.5 using 1 M HCl.
15. Centrifuge the starch slurry at 3500 rpm, 960 g for 20 min to produce oxidized wheat starch.
16. Properly clean the sediment by ultrapure water.
17. Stop the washing of starch when the supernatant showed no response to silver nitrate solution.

18. Dry the modified starch at −50 °C for 24 h in a vacuum freeze-drying oven.
19. *Dual modification*: Dual-modified starch can be prepared by using the methods as described above alternatively.
20. Determine the properties of modified starch (*see* **Note 76**).

3.41 Dual Modification of Tapioca Starch Using Succinylation/Cross-Linking Methods [72]

1. For dual modification, first acid hydrolyze the starch to prepare acid thin starch and then, etherify the acid thin starch with propylene oxide to form hydroxypropylated starch.
2. *Acid hydrolysis*: Add 1.08 N HCl to the tapioca starch under controlled temperature for 24 h at 40 °C.
3. Strain the solution and adjust the pH to 6.0 to 6.5 with the help of 3% NaOH solution.
4. Filter the solution and wash the sample with distilled water.
5. Use conductivity meter to determine whether the filtrate is free from traces of salt.
6. Dry the acid thinned starches for 15 h in hot air oven at 50 °C.
7. *Hydroxypropylation*: Take the 35% acid thinned starch (w/v) from the above procedure.
8. With the help of 15% Na_2SO_4 solution make slight slurry of the starch.
9. Adjust the pH of starch slurry to 11.2 by using 3% NaOH solution at 40 °C.
10. With continuous stirring, add 27% propylene oxide (w/v) to the starch slurry for 24 h.
11. Adjust the pH of starch slurry to 6.5 with the help of 0.1 M HCl solution.
12. Wash the starch solution with deionized water.
13. Dry the modified starch solution in a hot air oven at 40 °C for 12 h.
14. Grind the dried modified starch to powder and sieve through wire mesh.
15. Determine the properties of the modified starch (*see* **Note 77**).

3.42 Dual Modification of Waxy Cassava Starch Using Acid Hydrolysis/Succinylation Method [73]

1. *Acid hydrolysis*: Take 500 mL of conical flask and dissolve 100 g of waxy cassava starch in 300 mL of aqueous HCl solution.
2. Stir the solution and keep at room temperature at about 25 °C.
3. Neutralize the solution drop by drop with 20 g/100 mL solution of sodium hydroxide.
4. Then, wash the acid-hydrolyzed starch three times by using distilled water.

5. Centrifuge the starch slurry at 8000 g for 15 min.
6. Dry the sediment in a hot air oven at 40 °C for 24 h.
7. Grind the starch to fine powder in a mortar and sieve through 250 μm mesh.
8. Store the powdered modified starch plastic sheeting bags at room temperature.
9. *Esterification*: Prepare the acid-treated starch suspension by suspending 5 g of starch in 50 mL of distilled water.
10. Add 2-octen-1-yl succinic anhydride (3% db) to the starch suspension.
11. Add 25 mL of a 0.5 mol/L solution of aqueous sodium hydroxide to the acid-treated starch suspension.
12. Shake the slurry for 24 h.
13. Using phenolphthalein indicator, titrate excess alkali with 0.5 mol/L of hydrochloric acid.
14. Wash the starch solution three times using distilled water and centrifuge at 8000 g for 15 min.
15. Further determine the properties of modified starch (*see* **Note 78**).

3.43 Acetylated Distarch Adipate [74]

1. Prepare native potato starch and retrograded potato starch by freezing and defrosting method.
2. Mix 10 g of starch per 100 g solution.
3. Freeze the solution at −20 °C for 3 d.
4. Defrost the frozen sample at 20 °C for 2 d (*see* **Note 79**).
5. Rinse the starch with distilled water and dry at 30 °C for 24 h in air dryer.
6. Grind and sieve the powder.
7. *Acetylation*: Add acetic acid anhydride 13 mL per 100 g starch to form acetylated starch under reaction conditions.
8. *Cross-linking*: Inject 1 mL of the cross-linking agent and hot-dissolve 20 g of adipic acid in 80 g of acetic anhydride.
9. *Dual modification*: Rinse the acetylated distarch adipate produced by acetylation and, thereafter, cross-linking with de-ionized water.
10. Dry the modified starch in an air dryer for 24 h at 30 °C.
11. Grind and filter through a screen with a mesh diameter of 400 mm.
12. Determine the properties of modified starch (*see* **Note 80**).

3.44 Acetic Acid and Ultrasound [41]

3.44.1 Ultrasonic Treatment

1. Mix one part of isolated starch with four parts of distilled water (w/v).
2. Then, ultrasonicate the starch slurry in probe-type ultrasonic processor for 15 min at 20 kHz with an amplitude of 24 μm.
3. In the ultrasonic treatment of slurry, provide intensity of 140 W.
4. Maintain the temperature of ultrasonic treatment at 40 ± 1 °C with a jacketed container which is connected to circulating water bath.
5. Thereafter, centrifuge the starch slurry at 6000× *g* for 15 min.
6. Drain the supernatant and the residue left is ultrasonicated starch sediment.
7. Then, it is dried in a hot air oven at 45 °C which reduces the moisture content to below 10%.
8. Pulverize the ultrasonicated starch and pass through 100-mesh sieve to obtain a fine powder.

3.44.2 Acetylation

1. Prepare acetylated starch by reacting starch with acetic acid.

3.44.3 Dual Modification

1. It can be done by either first performing ultrasonication followed by acetylation or vice versa.
2. Determine the functional properties of the modified starch (*see* **Note 83**).

3.45 Microwave and Other Chemical Modifications [75]

1. *Microwave treatment*: Weigh native talipot starch having moisture content of 30% maintained by dispersing in DH_2O and keep it for 24 h to set equilibrium at 4 ± 2 °C in an air-tight container.
2. Then, the starch which has gained equilibrium is transferred to glass container. The container should be compatible with microwave.
3. Irradiated the slurry at 2450 MHz and 700 W for 90 s in microwave oven.
4. Dry the irradiated starch of talipot and store for further use.
5. *Oxidation* is done by reacting native starch with hydrogen peroxide (H_2O_2).
6. *Esterification* is done by reacting native starch with octenyl succinic anhydride.
7. *Cross-linking* is done by reacting native starch with 10% STMP/STPP.

8. *Dual modification*: It is done by subjecting microwave-treated starch to further modification by either oxidation, esterification, or cross-linking treatment.
9. Determine the properties improved by the dual modification (*see* **Note 84**).

3.46 Succinylation and Annealing [10]

1. Prepare succinylated starch by reacting 50 g of native corn starch slurry (diluted in 125 mL of distilled water containing 1 g of sodium carbonate) with 1 g of succinic anhydride.
2. Filter the slurry and wash with 95% ethanol.
3. Again, wash with distilled water and dry in an incubator at 25 ° C for 24 h.
4. Grind and sieved in 100 mesh by using a cyclone sample mill.
5. For *annealing process*, weigh 62.5 g of starch and add 125 mL of distilled water to it.
6. Heat the solution in a glass beaker using water bath at 55 °C for 24 h.
7. Centrifuge the suspension at 1665× g (*see* **Note 81**).
8. Dry the wet starch in an incubator at 25 °C for 24 h.
9. Grind the dried starch and sieve in a 100-mesh sieve by using cyclone sample mill
10. *Dual modification*: Take succinylated starch obtained from the above procedure, except drying of the succinylated starch is not done.
11. Anneal the wet starch having a moisture content of 40% and dry in a hot air oven at 55 °C for 24 h.
12. Grind the dried starch and sieve in a 100 mesh by using cyclone sample mill.
13. Determine the properties of modified starch (*see* **Note 82**).

4 Notes

1. Prepare the non-butyl-etherified waxy and high amylose starch as controls without the addition of butylene oxide.
2. Starch becomes compatible with polylactide after modification and the mechanical properties also improve, but there is no effect on stability on modification with this method.
3. Modification improves swelling power, freezing–thaw stability, and viscosity, whereas it lowers down the retrogradation.
4. It removes protein components from the starch.

5. Hydrochloric acid hydrolyses glucosidic bonds between anhydroglucose units of starch, thus reducing the apparent viscosity of the starch.
6. Na_2SO_4 prevents the swelling of starch granules.
7. NaOH and CHPTMA should have the same total mole quality.
8. A degree of substitution value of 0.034 in the modified starch improves the adhesion of starch to cotton and polyester fibers.
9. For solvent-based etherification, pre-disperse the sago starch powder in water and before adding benzyl chloride add C_2H_5OH into the starch slurry once the reactor achieves the temperature of 60 °C.
10. Modified starch shows higher thermal stability and better flow properties.
11. The resulting solution becomes nearly clear.
12. The viscosity of the solution increases by dispersing powdered NaOH to the solution.
13. The molar amounts of 1-bromopropane and powdered NaOH must be equal. The range of 1-bromopropane's molar ratio to hydroxyl groups is 1.0–5.0.
14. The reaction mixture's color shifts from yellow to a yellowish-brown color.
15. Adding 1-bromopropane results in an exothermic reaction, and it has a rather low boiling point (71 °C).
16. Most of the modified etherified starch is amorphous, and the rate of biodegradation decreases as etherification increases. The clear glass transition temperature of modified starch drops as the degree of substitution increases.
17. Chemical analysis can reveal the differences between the modified and native starch's molecular structures.
18. After modification, the starch's granule crystallinity changes. As the degree of substitution increases, the incorporation of allyl glycidyl ether into the starch lowers the gelatinization temperatures, modifies the pasting characteristics, enhances swelling power, and boosts solubility index.
19. The reaction proceeds smoothly when the water-to-solvent medium ratio is 4 mL/4 mL. Since the starch will swell at higher ratios, etherification will be prevented if used at higher ratio.
20. Boil modified starch (1 g) under reflux for 2 h by combining it with 25 mL of 1 N alcoholic (isopropanol) KOH. Use phenolphthalein indicator to titrate with 0.1 N HCl after cooling the substance to room temperature.

21. Molecular substitution is defined as the ratio of 1 mole of a substituent to 1 mole of anhydroglucose.
22. The ability to bind water (water-binding capacity) of modified banana starch increases by 50% than that of native starch.
23. Ionic liquid has outstanding catalytic and dissolving properties.
24. It produces a homogeneous transparent solution.
25. Modification improves swelling power and water solubility index of native starch, while amylose content reduces significantly.
26. For better results, moisture level should be below 5–10% (w/w).
27. Citrate starch gel exhibits better freeze–thaw stability, reduced digestibility, and thus, can be used as resistant starch with greater thermal stability.
28. Autoclaving completely gelatinizes the starch granules.
29. Autoclaving completely inactivates the pullulanase enzyme.
30. Cooling and resting provide time for molecule reassociation.
31. Following the same process, starch without glutaric acid is utilized as a control.
32. The modified starch gelatinizes within a shorter duration and achieves higher viscosity.
33. It avoids moisture absorption.
34. New bands of carbonyl and carboxyl groups in the ester bond suggest successful esterification. It improves emulsifying properties and hydrophobicity, and destroys the crystalline structure of amylopectin of the native starch.
35. Modification increases emulsion stabilization property and, thus, can have applications in high value-added products in food and pharmaceutical industries.
36. Modification destroys the crystalline structure, resulting in improvement of thermoplasticity and decrease in gelatinization temperature.
37. Modification tends to harm the ordered crystalline structure mainly B-type, which leads to increased thermal stability of the modified starch.
38. Isolate the adlay starch by using alkaline steeping method.
39. Adlay starch modification increases the resistant starch content.
40. Molar substitution is defined as the amount of substitution per mole of anhydrous glucose unit.
41. Modified starch shows improvement in freeze–thaw stability.
42. Modification of oat starch by cross-linking decreases the swelling power but increases the syneresis.

43. Modification improves swelling power at 60 °C, lowers down solubility, increases transition temperatures, and decreases enthalpy value.
44. Modification increases the starch's ability to expand and, thus, decreases its solubility.
45. The starch that is not treated to the condition of cross-linking with a combination of STMP/STPP is utilized as a control.
46. Modification by cross-linking reduces the solubility, swelling power, and paste clarity of native starch of maize.
47. Modification improves swelling power and pasting properties of the sago starch.
48. Solid concentration of 41.67% (w/w) is achieved and dissolve with concentrations from 0.25% to 6% (w/w).
49. With the increase in concentration of STMP/STPP, the swelling power and solubility of starch increases.
50. This improves paste clarity, swelling power, solubility, pasting, gelatinization behavior of the modified starch.
51. Chemical modification using $POCl_3$ could improve pasting property, swelling, freeze–thaw stability, texture, and color.
52. Modification of native starch reduces moisture, ash, protein, fat, fiber, pH, amylose content, and swelling power, and increases solubility and water and oil absorption capacity of the starch.
53. Modification of sweet potato starch reduces moisture, ash, and protein content, and improves solubility of swelling power, gel-forming capacity, and paste clarity.
54. Modification reduces the non-starch components' value; increases swelling power, solubility, water and oil absorption capacity, paste clarity; and reduces the viscosity of the starch.
55. Modification decreases viscosity gelatinization temperature, enthalpy of the modified starch, and retrogradation of amylopectin increases slightly, and also prevents retrogradation.
56. Modified starch retains higher moisture content; the modification decreases crude fiber, protein, and fat contents, improves the swelling power, reduces the solubility, pasting temperature, and viscosity of the native starch.
57. Add equally as much oxidized starch as extracted starch is added to the water-soluble fraction.
58. Sun-dried fermented cassava starch exhibits more carboxyl and carbonyl content, greater expansion, and less biscuit hardness. Biscuit hardness is decreased and starch expansion is increased when sodium hypochlorite is used as an oxidizing agent.

59. Use a bubble-type flow meter to calculate the flow rate. It is expected that the ozone flow rate to the reaction vessel equals this flow rate.
60. It makes sure that starch and ozone are in constant contact throughout the reaction.
61. It will flush out O_3 which does not react with starch.
62. Oxidation decreases the swelling power of sago and tapioca starches whereas increases the solubility of sago starch. It increases the swelling power of corn starch and decreases the solubility of tapioca starch. Pasting properties also increases with oxidation.
63. BU is measured by Brabender Viscograph.
64. Modified starches have slightly higher crystallinity and lower amylose content. Modification decreases gelling temperature and gel strength of the starch.
65. Sago starch with different molecular weights can be prepared.
66. Incubation time of less than 6 h has no significant effects and incubation time of 24 h decreases starch molecular weight.
67. Shaking prevent sedimentation of the starch.
68. Acid hydrolysis decreases the molecular weight of the starch, specifically reduces amylopectin.
69. Modification improves the gelling capacity and paste clarity, whereas water and oil absorption capacity, pasting temperature, and viscosity of the modified are decreased by acid-thinning of starch as compared to the native starch.
70. Suspend control starches in 0.1% sodium azide solution without addition of NaOH.
71. Use the supernatant for measuring solubilized carbohydrates to quantify the hydrolysis degree by the anthrone-sulfuric acid method.
72. Alkali-treated rice starches contain less amylose and have higher start and peak gelatinization temperatures. The modified starch hydrolyzes more quickly with HCl and amylolytic enzymes.
73. Modification decreases the amylose content of starch to about 20–25%, decreases crystallinity, reduces pasting viscosities, smaller decreases in swelling power, and increases the rate of in vitro enzymatic breakdown.
74. Various applications are possible by the dual modification of cross-linked rice starches with acetylation and hydroxypropylation. The modified starch shows higher gelatinization temperatures, gel hardness, chewiness, swelling, and lower solubility.
75. Granules of larger sizes are more susceptible to modification processes than those of smaller sizes.

76. Dual modification with cross-linking and oxidation of wheat starch results in higher peak temperature, lower retrogradation, better paste clarity, good water holding capacity, and better freeze–thaw stability, thus, having good efficiency to be used in food, paper, laundry, and textile industries.
77. The paste viscosity and granule size of tapioca starch are comparable to those of maize starch. Cassava or tapioca starch was found to be easier to hydrolyze than maize starch.
78. Hydroxypropyl acid-thinned tapioca starch (HPS) can be successfully used as plasma extender. The dual-modified starch increases the turbidity, colloid osmotic pressure, and viscosity of the plasma extender as compared to native tapioca starch.
79. Precipitated starch has a spongy structure.
80. The dual-modified starch produces retrograded starch; its resistance is increased due to amylolysis; as a result, viscosity of the paste is also increased. It can be used as a thickening agent in foods.
81. It separates the water from the wet starch.
82. Dual modification increases water-binding capacity, swelling power, viscosity, paste clarity, gelatinization temperature, and gelatinization enthalpy,
83. As the swelling power decreased, the paste viscosity also reduced but resistant starch content and slowly digestible starch content increased in the modified starch.
84. In all types of dual modifications, the relative crystallinity and set-back viscosity of the native talipot starch decreases. Modification also lowers the retrogradation tendency, which can be used in frozen foods. The resistant starch content increases as more functional groups are incorporated, which destroys the crystalline structure in starch granules after microwave treatment.

5 Conclusion

Chemical modification of starch through esterification, etherification, cross-linking, oxidation, hydrolysis, and dual modification offers exciting possibilities for expanding native starch applications. The modifications generally affect the structural, physical, and functional properties, including crystallinity, surface properties, solubility, viscosity, swelling ability, pasting, gelatinization properties, thermal, and freeze–thaw stability. The combination of these modifications, along with other physical or enzymatic modifications, has great potential for further development in the properties of the starch.

References

1. Pérez S, Bertoft E (2010) The molecular structures of starch components and their contribution to the architecture of starch granules: a comprehensive review. Starch/Staerke 62: 389–420. https://doi.org/10.1002/star.201000013
2. Sinhmar A, Pathera AK, Sharma S et al (2023) Impact of various modification methods on physicochemical and functional properties of starch: a review. Starch/Staerke 75:2200117. https://doi.org/10.1002/star.202200117
3. Mir SA, Bosco SJD, Shah MA et al (2015) Effect of gamma irradiation on physicochemical properties of brown rice. Int J Food Eng 11: 563–571. https://doi.org/10.1515/ijfe-2015-0038
4. Xu J, Andrews TD, Shi YC (2020) Recent advances in the preparation and characterization of intermediately to highly esterified and etherified starches: a review. Starch/Staerke 72: 1900238. https://doi.org/10.1002/star.201900238
5. Schmitz CS, De Simas KN, Santos K et al (2006) Cassava starch functional properties by etherification – hydroxypropylation. Int J Food Sci Technol 41:681–687. https://doi.org/10.1111/j.1365-2621.2005.01136.x
6. Haroon M, Wang L, Yu H et al (2016) Chemical modification of starch and its application as an adsorbent material. RSC Adv 6:78264–78285. https://doi.org/10.1039/c6ra16795k
7. Chen YF, Kaur L, Singh J (2017) Chemical modification of starch. In: Starch in food: structure, function and applications. Woodhead Publishing, pp 283–321
8. Omoregie Egharevba H (2020) Chemical properties of starch and its application in the food industry. Chem Prop Starch 9. https://doi.org/10.5772/intechopen.87777
9. Wokadala OC, Emmambux NM, Ray SS (2014) Inducing PLA/starch compatibility through butyl-etherification of waxy and high amylose starch. Carbohydr Polym 112:216–224. https://doi.org/10.1016/j.carbpol.2014.05.095
10. Ariyantoro AR, Katsuno N, Nishizu T (2018) Effects of dual modification with succinylation and annealing on physicochemical, thermal and morphological properties of corn starch. Foods 7:133. https://doi.org/10.3390/foods7090133
11. Misman MA, Azura AR, Hamid ZAA (2015) Physico-chemical properties of solvent based etherification of sago starch. Ind Crops Prod 65:397–405. https://doi.org/10.1016/j.indcrop.2014.11.009
12. Teramoto N, Motoyama T, Yosomiya R, Shibata M (2003) Synthesis, thermal properties, and biodegradability of propyl-etherified starch. Eur Polym J 39:255–261
13. Huijbrechts AML, Desse M, Budtova T et al (2008) Physicochemical properties of etherified maize starches. Carbohydr Polym 74: 170–184. https://doi.org/10.1016/j.carbpol.2008.02.001
14. Bayazeed A, Farag S, Shaarawy S, Hebeish A (1998) Chemical modification of starch via etherification with methyl methacrylate. Starch/Staerke 50:89–93. https://doi.org/10.1002/(SICI)1521-379X(199803)50:2/3<89::AID-STAR89>3.0.CO;2-B
15. Bakouri H, Guemra K (2019) Etherification and cross-linking effect on physicochemical properties of Zea mays starch executed at different sequences in 1-butyl-3-methylimidazolium chloride [BMIM]Cl ionic liquid media. Int J Biol Macromol 125:1118–1127. https://doi.org/10.1016/j.ijbiomac.2018.12.140
16. Mirzaaghaei M, Nasirpour A, Keramat J et al (2022) Chemical modification of waxy maize starch by esterification with saturated fatty acid chlorides: synthesis, physicochemical and emulsifying properties. Food Chem 393: 133293. https://doi.org/10.1016/j.foodchem.2022.133293
17. Tian S, Chen Y, Chen Z et al (2018) Preparation and characteristics of starch esters and its effects on dough physicochemical properties. J Food Qual 2018:1–7. https://doi.org/10.1155/2018/1395978
18. Mei JQ, Zhou DN, Jin ZY et al (2015) Effects of citric acid esterification on digestibility, structural and physicochemical properties of cassava starch. Food Chem 187:378–384. https://doi.org/10.1016/j.foodchem.2015.04.076
19. Miao M, Li R, Jiang B et al (2014) Structure and physicochemical properties of octenyl succinic esters of sugary maize soluble starch and waxy maize starch. Food Chem 151:154–160. https://doi.org/10.1016/j.foodchem.2013.11.043
20. Adebowale KO, Olu-Owolabi BI, Olawumi EK, Lawal OS (2005) Functional properties of native, physically and chemically modified breadfruit (Artocarpus artilis) starch. Ind Crops Prod 21:343–351. https://doi.org/10.1016/j.indcrop.2004.05.002

21. Luo T, Yang C, Zhang K et al (2021) Synthesis and characterization of methyltetrahydrophthalic anhydride esterified corn starch by wet method. Mater Express 11:1223–1230. https://doi.org/10.1166/mex.2021.2026
22. Liu Y, Liu J, Kong J et al (2020) Citrate esterification of debranched waxy maize starch: structural, physicochemical and amylolysis properties. Food Hydrocoll 104:105704. https://doi.org/10.1016/j.foodhyd.2020.105704
23. Waliszewski KN, Aparicio MA, Bello LA, Monroy JA (2003) Changes of banana starch by chemical and physical modification. Carbohydr Polym 52:237–242. https://doi.org/10.1016/S0144-8617(02)00270-9
24. Carmona-Garcia R, Sanchez-Rivera MM, Méndez-Montealvo G et al (2009) Effect of the cross-linked reagent type on some morphological, physicochemical and functional characteristics of banana starch (Musa paradisiaca). Carbohydr Polym 76:117–122. https://doi.org/10.1016/j.carbpol.2008.09.029
25. Singh AV, Nath LK (2012) Synthesis and evaluation of physicochemical properties of cross-linked sago starch. Int J Biol Macromol 50:14–18. https://doi.org/10.1016/j.ijbiomac.2011.09.003
26. Woo KS, Seib PA (2002) Cross-linked resistant starch: preparation and properties:1
27. Colivet J, Carvalho RA (2017) Hydrophilicity and physicochemical properties of chemically modified cassava starch films. Ind Crops Prod 95:599–607. https://doi.org/10.1016/j.indcrop.2016.11.018
28. Wang Y, He M, Wu Y et al (2021) Effect of crosslinking agents on the physicochemical and digestive properties of corn starch aerogel. Starch/Staerke 73:619–625. https://doi.org/10.1002/star.202000161
29. Amorim EOC, Doval VC, Cristianini M (2011) Effect of ozonation on the sensory characteristics and pasting properties of cassava starch. Procedia Food Sci 1:914–919. https://doi.org/10.1016/j.profoo.2011.09.138
30. Chan HT, Bhat R, Karim AA (2009) Physicochemical and functional properties of ozone-oxidized starch. J Agric Food Chem 57:5965–5970. https://doi.org/10.1021/jf9008789
31. Dias ARG, Zavareze EDR, Elias MC et al (2011) Pasting, expansion and textural properties of fermented cassava starch oxidised with sodium hypochlorite. Carbohydr Polym 84:268–275. https://doi.org/10.1016/j.carbpol.2010.11.033
32. Sangseethong K, Lertphanich S, Sriroth K (2009) Physicochemical properties of oxidized cassava starch prepared under various alkalinity levels. Starch/Staerke 61:92–100. https://doi.org/10.1002/star.200800048
33. Jiang S, Liu C, Han Z et al (2016) Evaluation of rheological behavior of starch nanocrystals by acid hydrolysis and starch nanoparticles by self-assembly: a comparative study. Food Hydrocoll 52:914–922. https://doi.org/10.1016/j.foodhyd.2015.09.010
34. Wang YJ, Den TV, Wang L (2003) Structures and rheological properties of corn starch as affected by acid hydrolysis. Carbohydr Polym 52:327–333. https://doi.org/10.1016/S0144-8617(02)00323-5
35. Kim HY, Lee JH, Kim JY et al (2012) Characterization of nanoparticles prepared by acid hydrolysis of various starches. Starch/Staerke 64:367–373. https://doi.org/10.1002/star.201100105
36. Nagamori M, Funazukuri T (2004) Glucose production by hydrolysis of starch under hydrothermal conditions. J Chem Technol Biotechnol 79:229–233. https://doi.org/10.1002/jctb.976
37. Nadiha MN, Fazilah A, Bhat R, Karim AA (2010) Comparative susceptibilities of sago, potato and corn starches to alkali treatment. Food Chem 121:1053–1059
38. Su N, Wu Q, Liu Y et al (2014) Hydrogen-rich water reestablishes ROS homeostasis but exerts differential effects on anthocyanin synthesis in two varieties of radish sprouts under UV-A irradiation. J Agric Food Chem 62:6454–6462. https://doi.org/10.1021/jf5019593
39. Fouladi E, Mohammadi Nafchi A (2014) Effects of acid-hydrolysis and hydroxypropylation on functional properties of sago starch. Int J Biol Macromol 68:251–257. https://doi.org/10.1016/j.ijbiomac.2014.05.013
40. Čížová A, Sroková I, Sasinková V et al (2008) Carboxymethyl starch octenylsuccinate: Microwave- and ultrasound-assisted synthesis and properties. Starch/Staerke 60:389–397. https://doi.org/10.1002/star.200800221
41. Khurshida S, Das MJ, Deka SC, Sit N (2021) Effect of dual modification sequence on physicochemical, pasting, rheological and digestibility properties of cassava starch modified by acetic acid and ultrasound. Int J Biol Macromol 188:649–656. https://doi.org/10.1016/j.ijbiomac.2021.08.062
42. Bhargava A, Shelke S, Dilkash M et al (2022) A comprehensive review on catalytic etherification of glycerol to value-added products. Rev Chem Eng. https://doi.org/10.1515/revce-2021-0074

43. Korma SA (2016) Chemically modified starch and utilization in food stuffs. Int J Nutr Food Sci 5:264. https://doi.org/10.11648/j.ijnfs.20160504.15
44. Ačkar D, Babić J, Jozinović A et al (2015) Starch modification by organic acids and their derivatives: a review. Molecules 20:19554–19570. https://doi.org/10.3390/molecules201019554
45. Aadil RM, Zeng XA, Sun DW et al (2015) Combined effects of sonication and pulsed electric field on selected quality parameters of grapefruit juice. LWT 62:890–893. https://doi.org/10.1016/j.lwt.2014.10.025
46. Chen Q, Yu H, Wang L et al (2015) Recent progress in chemical modification of starch and its applications. RSC Adv 5:67459–67474. https://doi.org/10.1039/c5ra10849g
47. Maitra J, Shukla VK (2014) Cross-linking in hydrogels – a review. Am J Polym Sci 4:25–31
48. Masina N, Choonara YE, Kumar P et al (2017) A review of the chemical modification techniques of starch. Carbohydr Polym 157:1226–1236. https://doi.org/10.1016/j.carbpol.2016.09.094
49. Neelam K, Vijay S, Lalit S (2012) Various techniques for the modification of starch and the applications of its derivatives. Int Res J Pharm 3:25–31
50. Palacios-Fonseca AJ, Castro-Rosas J, Gómez-Aldapa CA et al (2013) Effect of the alkaline and acid treatments on the physicochemical properties of corn starch. CYTA J Food 11:67–74. https://doi.org/10.1080/19476337.2012.761651
51. Azmi AS, Malek MIA, Puad NIM (2017) A review on acid and enzymatic hydrolyses of sago starch. Int Food Res J 24:265–273
52. Vargas-León EA, Falfan-Cortes RN, Navarro-Cortez RO et al (2019) Double chemical modification in rice starch: acid hydrolysis optimization process and phosphating. CYTA J Food 17:632–639. https://doi.org/10.1080/19476337.2019.1624623
53. Ashogbon AO, Akintayo ET (2014) Recent trend in the physical and chemical modification of starches from different botanical sources: a review. Starch/Staerke 66:41–57. https://doi.org/10.1002/star.201300106
54. Zia-ud-Din XH, Fei P (2017) Physical and chemical modification of starches: a review. Crit Rev Food Sci Nutr 57:2691–2705. https://doi.org/10.1080/10408398.2015.1087379
55. Haq F, Yu H, Wang L et al (2019) Advances in chemical modifications of starches and their applications. Carbohydr Res 476:12–35. https://doi.org/10.1016/j.carres.2019.02.007
56. Li W, Xu Z, Wang Z et al (2018) Double etherification of corn starch to improve its adhesion to cotton and polyester fibers. Int J Adhes Adhes 84:101–107. https://doi.org/10.1016/j.ijadhadh.2018.03.001
57. Mathew S, Abraham TE (2007) Physico-chemical characterization of starch ferulates of different degrees of substitution. Food Chem 105:579–589. https://doi.org/10.1016/j.foodchem.2007.04.032
58. Hui R, Qi-he C, Ming-liang F et al (2009) Preparation and properties of octenyl succinic anhydride modified potato starch. Food Chem 114:81–86. https://doi.org/10.1016/j.foodchem.2008.09.019
59. Zuo Y, Gu J, Yang L et al (2013) Synthesis and characterization of maleic anhydride esterified corn starch by the dry method. Int J Biol Macromol 62:241–247
60. Kim MJ, Choi SJ, Shin SI et al (2008) Resistant glutarate starch from adlay: preparation and properties. Carbohydr Polym 74:787–796. https://doi.org/10.1016/j.carbpol.2008.04.043
61. Mirmoghtadaie L, Kadivar M, Shahedi M (2009) Effects of cross-linking and acetylation on oat starch properties. Food Chem 116:709–713. https://doi.org/10.1016/j.foodchem.2009.03.019
62. Wongsagonsup R, Pujchakarn T, Jitrakbumrung S et al (2014) Effect of cross-linking on physicochemical properties of tapioca starch and its application in soup product. Carbohydr Polym 101:656–665. https://doi.org/10.1016/j.carbpol.2013.09.100
63. Hazarika BJ, Sit N (2016) Effect of dual modification with hydroxypropylation and cross-linking on physicochemical properties of taro starch. Carbohydr Polym 140:269–278. https://doi.org/10.1016/j.carbpol.2015.12.055
64. Lawal OS, Adebowale KO, Ogunsanwo BM et al (2005) Oxidized and acid thinned starch derivatives of hybrid maize: functional characteristics, wide-angle X-ray diffractometry and thermal properties. Int J Biol Macromol 35:71–79. https://doi.org/10.1016/j.ijbiomac.2004.12.004
65. Chibuzo IM (2012) Physicochemical and retrogradation characteristics of modified sweet potato (*Ipomoea batatas* L. (Lam)) starch. J Agric Food Tech 2:49–55
66. Adebowale KO, Lawal OS (2003) Functional properties and retrogradation behaviour of native and chemically modified starch of

mucuna bean (*Mucuna pruriens*). J Sci Food Agric 83:1541–1546. https://doi.org/10.1002/jsfa.1569

67. Cai J, Yang Y, Man J et al (2014) Structural and functional properties of alkali-treated high-amylose rice starch. Food Chem 145:245–253. https://doi.org/10.1016/j.foodchem.2013.08.059
68. Wang S, Copeland L (2012) Effect of alkali treatment on structure and function of pea starch granules. Food Chem 135:1635–1642. https://doi.org/10.1016/j.foodchem.2012.06.003
69. Lee SJ, Hong JY, Lee EJ et al (2015) Impact of single and dual modifications on physicochemical properties of japonica and indica rice starches. Carbohydr Polym 122:77–83. https://doi.org/10.1016/j.carbpol.2015.01.009
70. Zhao J, Chen Z, Jin Z et al (2015) Effects of granule size of cross-linked and hydroxypropylated sweet potato starches on their physicochemical properties. J Agric Food Chem 63:4646–4654. https://doi.org/10.1021/jf506349w
71. Sun F, Liu J, Liu X et al (2017) Effect of the phytate and hydrogen peroxide chemical modifications on the physicochemical and functional properties of wheat starch. Food Res Int 100:180–192. https://doi.org/10.1016/j.foodres.2017.07.001
72. Sookkasem A, Luang-ni P, Wansuksri R et al (2020) Physical and rheological characteristics of hydroxypropyl, acid-thinned tapioca starch-based plasma expander for blood volume replacement. Starch/Staerke 72:1900224. https://doi.org/10.1002/star.201900224
73. Fonseca-Florido HA, Vázquez-García HG, Méndez-Montealvo G et al (2018) Effect of acid hydrolysis and OSA esterification of waxy cassava starch on emulsifying properties in Pickering-type emulsions. LWT 91:258–264. https://doi.org/10.1016/j.lwt.2018.01.057
74. Zięba T, Gryszkin A, Kapelko M (2014) Selected properties of acetylated adipate of retrograded starch. Carbohydr Polym 99:687–691. https://doi.org/10.1016/j.carbpol.2013.08.064
75. Aaliya B, Sunooj KV, John NE et al (2022) Impact of microwave irradiation on chemically modified talipot starches: a characterization study on heterogeneous dual modifications. Int J Biol Macromol 209:1943–1955. https://doi.org/10.1016/j.ijbiomac.2022.04.172

Chapter 11

Enzymatically Modified Starch

Mitali Madhumita, Puja Priyadarshini Nayak, and Sujosh Nandi

Abstract

Starch is a semicrystalline polymer that is abundantly present in nature as a source of energy storage. It has widespread applications in food and non-food industries, such as pharmaceutical, textile, and paper, due to its low cost, bioavailability, biodegradability, and easy accessibility. The demand for enzymatically modified starch is increasing day by day due to their role in nutritional enhancement and retaining better physico-chemical properties. In addition, the functional properties of native starch, such as thickeners, gelling agents, binders, encapsulants, and swelling agents, make it a foremost choice for all the stakeholders. However, inherent properties like quick retrogradation, thermal and shear instability, and sensitivity to acidic and high-temperature environments restrict its prevalent industrial applications. In this regard, nutritional, functional, physico-chemical, and structural modifications have been made to native starch by chemical, physical, and enzymatic methods. Among them, enzymatic modifications of native starch have been extensively explored as a green technology compared to toxic chemical approaches and energy-intensive physical approaches, resulting in modified starches with altered gelation, viscosity, and solubility characteristics. This chapter provides a comprehensive discussion on numerous protocols and methods of enzymatic modifications of starch.

Key words Enzyme, Starch, Modification, Protocols, Functionalization

1 Introduction

Starch has garnered significant attention from various industries including paper, textiles, pharmaceuticals, cosmetics, food, and packaging, as well as researchers and consumers alike, owing to its plentiful supply, cost-effectiveness, compatibility with biological systems, and sustainable nature. Starch is mostly a plant polysaccharide that is stored as semicrystalline granules with lipids, proteins, and water molecules in the amyloplasts and chloroplasts of plant cells [1]. The properties of starch are influenced by the amylose and amylopectin fractions in the starch granules, as both have different physicochemical properties. Therefore, changes in the fractions of these two polymers can alter the overall properties of starch. The fraction of these two polymers is related to their botanical origin. In

Sneh Punia Bangar (ed.), *Standardized Procedures and Protocols for Starch*, Methods and Protocols in Food Science, https://doi.org/10.1007/978-1-0716-3866-8_11,

food processing industries, starch is used for its various functional and physicochemical properties, including swelling, pasting, encapsulating, thickening, gelatinization, and retrogradation [2]. Behavioral characteristics of native starches, such as low thermal stability, low resistance to shear stress, temperature, and acids, and high hot-paste viscosity, hinder their widespread applications in various sectors of industries, including the food sector. These shortcomings of native starch have created enormous research interest in modifying the functional, physicochemical, and textural properties of starch granules. The modifications of starch molecules are accomplished based on their specific uses, like thickener, stabilizer, encapsulating and coating materials, fat replacer, resistance starch, and so on [3]. Broadly, four different processes, namely physical, chemical, enzymatic, and genetic, are used in the modification process of starch granules.

Physical modifications of starch are achieved by hydrothermal processes (annealing and high moisture treatment), pre-gelatinized processes, and nonthermal treatments (pulse electric field, micronization, high-pressure processing, and ultrasonication), resulting in modified starch with improved thermal stability, textural properties, swelling power, solubility, heat tolerance capacity, viscosity, and crystallinity [4]. On the other hand, in chemical modifications, native starch is treated with hypochlorides, acids, phosphates, etc. by introducing new functional groups at C-2, C-3, and C-6 of anhydroglucose units or by acid hydrolysis, co-polymerization, cross-linking, dextrinization, and oxidation [5]. Although these chemical modification processes have been practiced for several years and are well accepted due to their low cost, they produce harmful effluent, making them environmentally unfriendly. Therefore, these days, a comparatively new and eco-friendly modification technique, namely, enzymatic modification, has been substantially adopted by the research communities and various industries [6]. Enzymatic modification of starch, in particular, stands out as a highly effective and preferable method for reducing the intrinsic viscosity of starch slurry compared to chemical modification.

Enzymatic modification shows enormous advantages over other methods, such as high yield, low energy consumption, mild reaction conditions, lower production of undesirable byproducts and co-products, and the use of substrate (starch) specific enzymes for target modifications. The enzymatically modified starch meets all the requirements of the food products with the advantage of having high yield and reaction efficiency, better utilization capacity, and simple operative activity. For example, α-amylase enzyme, which is commonly used in modified starch, like potato and pea, improves the structural and physicochemical properties, water solubility, swelling capacity, greater thermal stability, branching degree, amylopectin chain length distribution, and light transmittance, making it ideal for industrial applications [7].

Over the years, starch from several sources has been utilized for enzymatic modification purposes, by considering several enzymes, successfully modifying the chain length of starch, introducing new branch points, debranching, and substituting phosphates [8, 9]. Hence, it can be corroborated that the researchers have adopted several methods to accomplish the necessary modifications in starch, which are summarized in this chapter to provide comprehensive knowledge about all the methods.

1.1 Enzymes and Their Principles for Starch Modification

Enzymes (hydratases) are used in enzymatic modification to speed up the formation of highly functional derivatives of starch. Syrups made from glucose, high-fructose corn, or cassava are the most common products of enzymatic modification. The enzymes modify starch by transferring glycosidic bonds and residues. This is called starch depolymerization resulting in oligosaccharides. In this modification process, the enzymes cleave the β-1,4 link between the glucose molecules. The principles of enzymatic modification of starch (EMS) revolve around the use of specific enzymes to target and modify starch molecules.

1.2 Role of Enzymes in Starch Modification

Enzymatic modification of starch is a controlled and sustainable process that can lead to a wide range of starch-derived products. The structure of starch can be directly altered by enzymatic treatment, resulting in modified branch chain length distribution, amylose-to-amylopectin ratio, and molecular weight [10].

1.3 Types of Enzymes Used in Starch Modification

Starch-modifying enzymes are classified into four groups, namely endoamylases, exoamylases, debranching enzymes, and transferases, according to the reactions they catalyze and the products they generate when they interact with starch. These groups are diverse and show a wide range of substrate and product specificities [11, 12]. The basic working principles of these enzymes are described as follows.

1.3.1 Endoamylases

Endoamylases catalyze the hydrolysis of the internal glycosidic bonds within starch molecules randomly [13]. These enzymes cleave the α-1,4-glycosidic bonds that link individual glucose units within the starch polymer, forming shorter starch fragments. Endoamylases are employed in various applications, including the production of sweeteners, syrups, and other starch-derived products in the food, beverage, and biofuel industries.

1.3.2 Exoamylases

Exoamylases catalyze the hydrolysis of the external glycosidic bonds at the ends of polysaccharide chains. These enzymes work on the nonreducing ends of starch molecules on both the α-1,4 and α-1,6 bonds, progressively releasing individual sugar units [13]. Exoamylases are essential in the final stages of starch hydrolysis, breaking down larger oligosaccharides into individual sugar units. They find applications in industries such as food and beverage (for sugar production), biofuel production (as part of starch-to-ethanol processes), and various enzymatic processes where controlled hydrolysis of starch is required.

1.3.3 Debranching Enzymes

Debranching enzymes are the most common type of starch-converting enzyme used today that cleave α-1,6-glycosidic bonds in amylopectin, breaking down highly branched regions of starch. Debranching enzymes, like pullulanase and isoamylase, break down α-1,6-glycosidic bonds at the polysaccharide branching site of starch. This creates amylose, maltotriose, or maltooligosaccharides [14]. Debranching enzymes are essential in the starch industry and other applications where the controlled breakdown of starch is required.

1.3.4 Transferases

Transferase enzymes are important for modifying starch because they make new glycosidic links or rearrange existing ones. The principle of transferase enzymes in the modification of starch lies in their ability to facilitate the transfer of glycosyl groups from one molecule to another. Amylomaltase is a transferase enzyme that breaks down the α-1,4-glycosidic linkages on the donor molecule, moves some of the donors to the glycosidic acceptor, and then builds new glycosidic links [12, 14].

Some types of enzymes used in the food processing sector are listed in Table 1.

2 Protocols for Different Enzymatic Modifications of Starch

2.1 Materials

2.1.1 Alpha-Amylase

Alpha-amylase (EC 3.2.1.1) works on an exo-action pattern and hydrolyzes the α-1,4-glucosidic linkages from the nonreducing chain ends, producing maltose. In addition, this enzyme can have an endo-action pattern, hydrolyzing internal glucosidic linkages to liberate maltooligosaccharides. α-amylase has been considered for modification of starch in cereals, root, and tuber crops. Several works have been reported for modifying starch from tapioca, red rice, wheat, and waxy maize using α-amylase [28–30]. The specific methods and protocols for starch modification using α-amylase may vary depending on the source and type of starch, the source and type of α-amylase, the reaction conditions, and the desired degree of hydrolysis. However, a general protocol for modification of starch using α-amylase can be followed as described below.

Starch Sample Preparation

1. Weigh a known amount of the selected starch sample using an analytical balance.

Table 1
Different enzymes used for modification of starch

Sl. no.	Starch source	Enzymes used	Reference
1	Corn	Amyloglucosidase and maltogenic amylase α-amylase and glucoamylase	[15] [16]
2	Wheat	α-amylase Maltogenic α-amylase, pullulanase, and isoamylase	[17] [18]
3	Rice	α-amylase and glucoamylase β-Amylase, transglucosidase, and pullulanase	[19] [20]
4	Barley	Isoamylase and pullulanase	[21]
5	Oats	β-amylase and transglucosidase	[22]
6	Sorghum	Pullulanase	[23]
7	Pigeon pea	α-amylase and glucoamylase	[24]
8	Potato	4-α-glucanotransferases Amylosucrase	[25] [17]
9	Sweet potato	Branching enzyme, β-amylase, and transglucosidase	[7]
10	Cassava	α-amylase and amyloglucosidase	[26]
11	Quinoa	α-amylase and saccharifying enzyme	[17]
12	Loquat	α-amylase and glycosylase	[27]

2. Dissolve the starch in distilled water or a buffer solution at a mass-to-volume ratio of 2:4 to prepare a starch solution.
3. Heat the starch solution to a temperature above its gelatinization point (e.g., 80–100 °C).
4. Maintain the temperature for a sufficient time (e.g., 15–30 min) while stirring.
5. Continue heating and stirring until the solution becomes clear and homogeneous.

Preparation of α-Amylase Solution

1. Calculate the required amount of α-amylase enzyme for the desired enzymatic activity.
2. Prepare the α-amylase solution by dissolving the calculated amount of enzyme in a suitable buffer solution such as sodium acetate or sodium phosphate.
3. Maintain the pH of the buffer solution within the optimal range for α-amylase activity, typically pH 5–7.

2.1.2 Beta-Amylase

The exo-splitting enzyme β-amylase (EC 3.2.1.2) shortens the exterior side chains of amylopectin by releasing maltose from the nonreducing end of polyglucan chains and producing β-maltose and other maltodextrins. The α-1,4-glycosidic bond of D-anhydroglucose monomers can be broken by β-amylase from the nonreducing end of the glucan chain, but it will cease when α-1,6 linkage or an additional group remains. To produce β-maltose and β-limit dextrin, the starch is partially hydrolyzed, which shortens the starch branches and trims their outer branches [31]. Over the years, several researchers have utilized β-amylase in the process of modifying starch sourced from waxy maize, rice, octenylsuccinic anhydride (OSA), etc. [32, 33]. However, the protocol for the modifications of starch may vary depending on the source and type of starch, the source and type of β-amylase, the reaction conditions, and the desired degree of hydrolysis. However, a general protocol for modification of starch using β-amylase can be followed as described below.

Starch Sample Preparation

1. Measure a predetermined quantity of the chosen starch sample using an analytical balance.
2. Dissolve the starch in either distilled water or a buffer solution with a mass-to-volume ratio of 2:4 to create a starch solution.
3. Heat the starch solution to a temperature exceeding its gelatinization point (e.g., 80–100 °C).
4. Maintain the temperature for an adequate duration (e.g., 15–30 min) while stirring continuously.
5. Continue heating and stirring until the solution attains a clear and uniform consistency.

Preparation of β-Amylase Solution

1. Determine the necessary quantity of β-amylase enzyme for the desired enzymatic activity.
2. Dissolve the β-amylase in an appropriate buffer solution, such as sodium acetate, to produce an enzyme solution.
3. Maintain the pH of the buffer solution within the ideal range for β-amylase activity, typically between pH 4 and 5.5.

2.1.3 Pullulanase

Pullulanase enzymes (EC 3.2.1.41) are debranching enzymes that catalyze the hydrolysis of α-1,6-glycosidic linkages in starch. They are used in starch modification to increase solubility, reduce viscosity, improve digestibility, and produce modified starches with specific functional properties, such as high clarity, low gelatinization temperature, and resistance to retrogradation.

Starch Solution Preparation

1. Weigh 10 g of starch and dissolve it in 100 mL of distilled or deionized water in a beaker.
2. Heat the starch solution to 50 °C on a magnetic stirrer, stirring continuously to ensure complete dissolution.
3. Maintain the starch solution at 50 °C for 30 min.
4. After 30 min, adjust the pH of the starch solution to 6 using 1 N HCl or 1 N NaOH, if necessary.
5. Allow the starch solution to cool to room temperature before use.

Preparation of *Pullulanase* Solution

1. Determine the necessary quantity of pullulanase enzyme to achieve the desired level of enzymatic activity.
2. Dissolve the pullulanase in an appropriate buffer solution, such as sodium acetate, to generate an enzyme solution.
3. Ensure that the pH of the buffer solution remains within the ideal range conducive to pullulanase activity, typically ranging from pH 5.5–6.5.

2.1.4 Branching Enzyme

Branching enzyme (BE; EC 2.4.1.18), also known as starch branching enzyme (SBE), and 1,4-α-glucan branching enzyme belong to the transferase group. BE breaks down the 1,4-α bonds on the linear chains of amylopectin and amylose. This is followed by a transglycosil process that forms a new link with the 1,6-α -glycosidic bonds and introduces branching in the starch molecule [34–37]. Starch modification using BE can be performed as follows.

Starch Solution Preparation

1. Prepare a starch solution as described in the protocol for starch modification using pullulanase.

Preparation of *BE* Solution

1. Dissolve 10 mg of branching enzyme in 10 mL sodium acetate buffer.
2. Ensure the pH of the buffer is within the range of 6–7.
3. Mix the branching enzyme thoroughly with the buffer solution until completely dissolved.

2.1.5 Protocol for the Determination of Starch with Glucoamylase

The determination of the starch depends on the dispersion of starch that occurs by the mode of action of enzyme, swelling, action of acids and salt used. Starch can be determined by two methods: direct method by optical rotation of the solution and indirect method by saccharifying action of enzyme. Based on the disadvantages of the direct method or acid hydrolysis, enzymatic hydrolysis gains more attention for starch determination. The enzyme catalyzes the hydrolysis of α-D glycosidic linkages liberating D-glucose by which the substrates are increased. Therefore, the starch is transferred to D-glucose using a selected enzyme. One example of the determination of starch using glucoamylase and its procedure is discussed in detail.

Sample Preparation

The selected sample of the enriched starch must be dried or treated with ethanol. Then, the sample was ground to obtain a particle size <0.5 mm and hydrothermally treated with glucoamylase to destroy the starch granules. A sample of <2 g containing <0.5 g of starch can be taken for the sample analysis.

1. Dry or treat the selected sample of enriched starch, possibly with ethanol.
2. Grind the treated sample to obtain a particle size <0.5 mm.
3. Hydrothermally treat the sample using glucoamylase to destroy the starch granules.
4. Take a sample weighing less than 2 g, containing less than 0.5 g of starch, for further analysis.

Hydrolysis of Starch

1. Take 20–25 mL of distilled water with the sample for hydrolysis, maintaining the pH between 5 and 7 to form a suspension.
2. Heat the suspension material at 135 °C for 1 h with continuous stirring, then allow it to cool.
3. Add 2.5 mL of acetate buffer to the cooled suspension.
4. Immerse the suspension in a water bath at a specific temperature corresponding to the glucoamylase used.
5. Add 5 mL of glucoamylase solution to initiate enzymatic hydrolysis, and incubate for 2 h.
6. Filter the dispersion after enzymatic hydrolysis and collect the residue to determine liberated glucose.

2.2 Method for Different Enzymatic Modifications of Starch

2.2.1 Procedure for Alpha-Amylase Enzymatic Hydrolysis of Starch

1. Introduce the prepared enzyme solution into the starch suspension, ensuring thorough mixing for dispersion.
2. Initiate a timer to monitor the reaction time closely, crucial for enzymatic modification.
3. Maintain the reaction mixture at the optimal temperature, typically within a range of 30–70 °C, using a water bath or incubator.
4. Continuously stir or agitate the mixture at regular intervals to promote uniformity and efficacy of the reaction.
5. Ensure consistent contact between the α-amylase enzyme and the starch substrate for precise and controlled hydrolysis.
6. Terminate enzymatic activity by placing the reaction mixture in an ice bath, followed by filtration.
7. Obtain modified starch by washing the sediment with distilled/deionized water, followed by centrifugation and drying.

The effect of modification can be obtained by calculating the degree of hydrolysis (DH) of the modified starch using the following equation:

$$DH\ (\%) = \frac{T_c \times 0.9}{M_s} \times 100$$

where

T_c = amount of total carbohydrate in the supernatant

0.9 = the conversion coefficient

M_s = weight of native starch (on dry basis).

2.2.2 Procedure for Beta-Amylase Enzymatic Hydrolysis of Starch

1. Add the prepared enzyme solution of β-amylase to a starch suspension and mix thoroughly for the enzymatic hydrolysis.
2. Set a timer to monitor the critical reaction time.
3. Maintain the reaction mixture at the optimal temperature (30–70 °C) for β-amylase activity.
4. Stir the mixture regularly to ensure even contact between the enzymes and starch.
5. Stop the reaction by cooling and filtering the mixture.
6. Obtain modified starch by washing and drying the sediment after filtration.

2.2.3 Procedure for Alpha-Glucosidase Enzymatic Hydrolysis of Starch

α-Glucosidase (EC 3.2.1.20), also known as maltase or α-D-glycoside glucohydrolase, is a member of the glycosyl hydrolase (GH) family. The primary function of α-glucosidases is to liberate α-D-glucose by the exohydrolysis of α-1,4-glycosidic bonds at the nonreducing end of polysaccharides, like maltooligosaccharides

[30]. The steps for starch modification using α-glucosidase are similar to that of α-amylase with slight variations in hydrolysis time. However, few researchers modified the procedures based on sources of α-glucosidase to enhance the properties of modified starch. α-glucosidase produced from *Bacillus* species works well at pH ranges of 4.5–6 and temperature of 60–70 °C. It is advised to utilize this enzyme in conjunction with β-amylase to make maltose syrup and glucoamylase to make glucose syrup. The bacterium *Thermus thermophiles* generates thermostable α-glucosidase, which is ideal for one-step starch processing and operates best at pH 6.2 and temperature of 85 °C [34].

2.2.4 Procedure for Isoamylase Enzymatic Hydrolysis of Starch

Isoamylase, also known as glycogen-6-glucanohydrolase, is a debranching enzyme with the enzyme classification EC 3.2.1.68. Isoamylase is another endo-mechanism enzyme that specifically acts on α-1,6-glycosidic linkages of amylopectin and dextrins [34]. The standard protocol for the modification of starch using isoamylase can overlap with the steps described for starch modifications using α-amylase. However, pH and temperature conditions may vary based on the specific enzyme source. Isoamylases typically work well within a temperature range of 50–70 °C. Hydrolysis of starch using isoamylase occurs at a comparatively lower pH (4.5–3.5) of the starch solution, which is maintained by adding an appropriate amount of dilute HCl [21, 35]. Ulbrich et al. in 2021 [36] used isoamylase to perform the enzymatic hydrolysis of potato starch. The modification involved making a 40% w/w starch slurry, heating it to 35 °C, and adjusting the suspension‘s pH to 4.5 before adding the required amount of isoamylase.

2.2.5 Procedure for Pullulanase Enzymatic Hydrolysis of Starch

The procedure for hydrolyzing starch using pullulanase requires the use of several essential pieces of equipment and materials. Equipment such as a magnetic stirrer, a heating bath, a pH meter, measuring cylinders, beakers, pipettes, filter paper, and a vacuum filtration apparatus are necessary. The materials needed include the previously prepared starch solution, the enzyme solution as per the earlier instructions, and distilled or deionized water.

1. Add the enzyme solution to the starch solution in a beaker and mix thoroughly.
2. Incubate the enzyme-starch mixture at 50 °C on a magnetic stirrer with continuous stirring.
3. Maintain the reaction temperature at 50 °C for 1–2 h, depending on the desired extent of starch modification.
4. Halt the enzymatic reaction by heating the mixture to 80 °C on a heating bath and maintaining this temperature for 10 min.
5. Allow the mixture to cool to room temperature.

6. Filter the reaction mixture through filter paper using a vacuum filtration apparatus.
7. Remove any residual enzyme by washing the filter cake with distilled or deionized water.
8. Dry the modified starch in an oven at 50 °C until it reaches a constant weight.

This comprehensive procedure ensures successful starch hydrolysis using pullulanase. Although the abovementioned protocol is the general steps for enzymatic hydrolysis of starch using pullulanase, over the years, researchers have slightly modified the process conditions according to the sources of starch to obtain higher yields and improved quality modified starch [37–44].

2.2.6 Procedure for Branching Enzyme for Starch Hydrolysis

1. Add the branching enzyme solution to the starch solution and mix thoroughly.
2. Incubate the reaction mixture at 30 °C on a magnetic stirrer with continuous stirring for 1–2 h, depending on the desired level of starch modification.
3. Terminate the reaction by heating the mixture to 80 °C using a heating bath and maintaining this temperature for 10 min.
4. Allow the reaction mixture to cool down to room temperature.
5. Remove insoluble material by centrifugation at 10,000 × g for 10 min.
6. Transfer the supernatant carefully to dialysis tubing and dialyze against a dialysis buffer at 4 °C for 24 h.
7. Collect the modified starch from the dialysis tubing and dry it in an oven at 50 °C until a constant weight is achieved.
8. These precise steps ensure successful starch modification using the branching enzyme.

2.2.7 Procedure for Cyclodextrin Glucosyltransferase Enzymatic Hydrolysis of Starch

The major function of cyclodextrin glucosyltransferases (CGTases; EC 2.4.1.19) is to act as catalysts in the manufacture of cyclodextrins from starch. These enzymes are produced by a broad variety of bacteria, especially *Bacillus* strains. There are a few reactions of practical interest that these enzymes can catalyze, such as cyclization, disproportionation, and coupling [45]. Catalyzation of starch with CGTases can be accomplished by several methods. However, in general, the following steps can be considered:

1. First, liquefy the selected starch using CGTase at a rate of 2 units/g of starch.
2. Incubate the reaction mixture at 90 °C on a magnetic stirrer, stirring continuously for a maximum of up to 2 h depending on the desired degree of starch modification.

3. To stop the reaction, cool the system down to 45 °C.
4. Add CGTase again at a rate of 2 units/g of starch and allow reaction for a period of 12 h to obtain CGTase-catalyzed starch products.

2.2.8 Procedure for Alpha-Glucanotransferases/Amylomaltase Enzymatic Hydrolysis of Starch

Glucanotransferases are classified into distinct enzyme families due to the unique mechanisms by which they catalyze the transfer segment of α-glucan. They are capable of conserving high molecular products compared to hydrolases. Amylomaltase (AM; $(1 \rightarrow 4)$-α-d-glucan:$(1 \rightarrow 4)$-α-d-glucan 4-α-d-glycosyltransferase; E.C 2.4.1.25) is an intracellular enzyme responsible for the intra- and intermolecular attachment of new α-1,4-glucosidic links within and between α-glucan molecules. This catalytic reaction is also known as disproportionation since it involves the transfer of similar links. In addition to its limited hydrolytic activity, AM can catalyze intramolecular α-glucan transfer events, resulting in cyclic compounds [46]. Recently, it has acquired much attention in the food industry due to its unique ability to create thermos-reversible starch, resistant starch, cyclodextrins, and highly branched dextrin [47]. α-glucanotransferases have been used for modifications of rice starch, corn starch, tapioca starch, potato starch, and rice flour [48]. For modifications of these starches using α-glucanotransferases, researchers have followed a similar method.

In this method, starch has to be gelatinized first, followed by reducing the starch slurry temperature to room temperature. Then the enzyme has to be added to the slurry and kept for incubation for several hours to days with continuous stirring. To stop the enzyme reaction, the slurry is then required to be heated. Washing and drying are employed after heating to obtain the modified starch [25, 47, 49, 50]. However, a slight modification in the method was carried out by Hassanein et al. in 2023 [51] who maintained the pH of the potato starch slurry at 6 with sodium phosphate buffer during the gelatinization process. In addition, they provided a 10-d incubation time at 30 °C, which is a comparatively longer time of incubation than the incubation time adopted by other researchers.

2.2.9 Determination of Liberated Glucose

The aliquot is again diluted with distilled water to make up a volume of 1 mL and to this mixture 5 mL of enzyme buffer chromogen is added. Along with this, 0.25 mL of 5 N HCl is added and then this solution is incubated at 20 °C for 45 min. The absorbance is measured.

$$\text{Calculation of\%starch} : \%\text{starch} = 2.25 \times \frac{(M \times V_1)}{(V_0 \times E \times M_s)}$$

where, M is the weight of D-glucose determined from the standard curve in µg, V_1 is the volume after dilution in mL, V_0 is the volume of the aliquot in mL, E is the weight of the sample in g, and M_s is the percentage of the dry weight of the sample.

3 Effect of Starch Composition Analysis on EMS

Starch consists of amylose and amylopectin that is bonded by 1,4-α -glycosidic linkage and 1,4-α along with 1,6-α -glycosidic linkage, respectively. The starch granules in both amylose and amylopectin vary in different parameters like size, shape, crystalline, etc. depending on the length of chain, branched chain, and branching density. Starch composition maintains the quality and nutritional value of the food products. Starch is widely used as functional agent in several food industries due to its several functional properties like swelling, emulsifying, gelling, and gelatinization. Because of high cost of enzyme, the use of enzymatically modified starch in industry is limited. Still, there is a huge interest in academia and industry to use enzymatically modified starch to achieve desirable properties, such as physicochemical, functional, and textural, to extend its sustainability. Wang et al. in 2022 [52] reported on the performance and properties of the paper that is made by the enzymatically modified starch by using α-amylase, β-amylase, pullulanase, glucoamylase, and compound enzymes. The enzymolysis mechanism and the performance of the sizing agent of the enzymatically modified starch-based paper were analyzed. The tensile index and the bending stiffness of the developed paper are increased by 93.9% and 251%, respectively. Moreover, chemical composition, thermal stability, and morphology of the developed paper were also analyzed. The contribution of starch in food products significantly impacts several properties such as water absorption capacity, gelatinization, crystallization, rheological, and retrogradation behavior. Gaenssle et al. in 2022 [53] reported the effect of different glycogen branching enzymes and their mode of action on the amylose content of potato, tulip, wrinkled pea, and waxy corn starch. Glycoside hydrolase family 57 glycogen branching enzymes (GH57GBEs) were found to be significant with the degree of branching than other branched enzymes. Among all the branch enzymes, GH57 GBEs generated more branched enzymes than other enzymes. Xu et al. in 2023 [33] worked on ultrasonicated and enzymatic treated adlay seed starch to analyze the emulsification performance. The esterification efficiency and the degree of substitution were found to be improved in both enzymatic- and ultrasound-treated starch. Moreover, the emulsion stability and the emulsifying activity were found to be better in this treated starch. Gao et al. in 2023 [54] reported the effect of structural and physicochemical properties on the enzyme of the treated pea starch. The results indicated that the pea starch treated with α-amylase showed a lower amylose content but a higher branching degree with a relative crystallinity nature. Another study by Almeida et al. in 2022 [55] reported the effect of digestibility and morphological characteristics of enzymatic hydrolysis (α-amylase and amyloglucosidase) on treated red rice starch.

The parameters of the treated starch (including surface area, hydrolysis index, glycemic index, SEM, FT-IR, and XRD.) were analyzed to study the characteristics behavior. Zheng et al. in 2023 [56] worked on enzymatic hydrolysis of different starch, including maize, pea, and potato, to study the effect of annealing treatment. The results showed that the morphology of all treated starch was not changed by the annealing treatment. Moreover, it was reported that as annealing time increased, a peak crystallinity of the starch was achieved in maize and potato starch while a relative crystallinity was achieved in pea starch. Compared to other modifications, enzymatic modification minimizes the formation of undesirable products and reduces side reactions. Because of the advantages like lower energy cost and high yield, enzymatic modifications are mostly preferred.

4 Conclusion

It is reported that the enzymatic modification and its potential application are much effective for starch modification. The starch treated by enzymes provides a promising enzymatic modification method to prepare different products with good digestibility and is highly effective. The relationship between the different properties of the starch and the enzymatic hydrolysis was also well discussed. The functional properties of the starch can be enhanced through different enzymatic modifications. The method of starch modification with protocol is also addressed, which is accurate, reproducible, rapid, and inexpensive. The application of the branching enzyme in food preservation is limited and needs to be investigated in future studies.

References

1. Perin D, Murano E (2017) Starch polysaccharides in the human diet: effect of the different source and processing on its absorption. Nat Prod Commun 12:837–853. https://doi.org/10.1177/1934578x1701200606
2. Cornejo-Ramírez YI, Martínez-Cruz O, Del Toro-Sánchez CL, Wong-Corral FJ, Borboa-Flores J, Cinco-Moroyoqui FJ (2018) The structural characteristics of starches and their functional properties. CYTA – J Food 16:1003–1017. https://doi.org/10.1080/19476337.2018.1518343
3. Bangar SP, Ashogbon AO, Singh A, Chaudhary V, Whiteside WS (2022) Enzymatic modification of starch: a green approach for starch applications. Carbohydr Polym 287:119265. https://doi.org/10.1016/j.carbpol.2022.119265
4. Zia-ud-Din, Xiong H, Fei P (2017) Physical and chemical modification of starches: a review. Crit Rev Food Sci Nutr 57:2691–2705. https://doi.org/10.1080/10408398.2015.1087379
5. Onyango C (2016) Starch and modified starch in bread making: a review. Afr J Food Sci 10:344–351. https://doi.org/10.5897/ajfs2016.1481
6. Punia S (2020) Barley starch modifications: physical, chemical and enzymatic – a review. Int J Biol Macromol 144:578–585. https://doi.org/10.1016/j.ijbiomac.2019.12.088
7. Guo L (2018) Sweet potato starch modified by branching enzyme, β-amylase and transglucosidase. Food Hydrocoll 83:182–189. https://doi.org/10.1016/j.foodhyd.2018.05.005

8. Park K-H, Park JH, Lee S, Yoo SH, Kim JW (2008) Enzymatic modification of starch for food industry, carbohydrate-active enzymes: structure, function and applications. Woodhead Publishing Limited. https://doi.org/10.1533/9781845695750.2.157
9. Wang S, Wang J, Liu Y, Liu X (2020) Starch modification and application. Starch Structure, Functionality and Application in Foods, pp 131–149
10. Park SH, Na Y, Kim J, Kang SD, Park KH (2018) Properties and applications of starch modifying enzymes for use in the baking industry. Food Sci Biotechnol 27:299–312. https://doi.org/10.1007/s10068-017-0261-5
11. Amaraweera SM, Gunathilake C, Gunawardene OHP, Manamperi A, Fernando CAN, Kulatunga AK, Manipura A (2021) Development of starch-based materials using current modification techniques and their applications: a review. Molecules 26:6880
12. Mamo G, Mattiasson B (2020) Alkaliphiles in biotechnology. Springer
13. Gopinath SCB, Anbu P, Arshad MKM, Lakshmipriya T, Voon CH, Hashim U, Chinni SV (2017) Biotechnological processes in microbial amylase production. Biomed Res Int 2017:9. https://doi.org/10.1155/2017/1272193
14. Prongjit D, Lekakarn H, Bunterngsook B, Aiewviriyasakul K, Sritusnee W, Arunrattanamook N, Champreda V (2022) In-depth characterization of debranching type I pullulanase from Priestia Koreensis Hl12 as potential biocatalyst for starch Saccharification and modification. Catalysts. https://doi.org/10.3390/catal12091014
15. Davoudi Z, Azizi MH, Barzegar M (2022) Porous corn starch obtained from combined cold plasma and enzymatic hydrolysis: microstructure and physicochemical properties. Int J Biol Macromol 223:790–797. https://doi.org/10.1016/j.ijbiomac.2022.11.058
16. Leyva-lόpez R, Palma-rodríguez HM, Lόpez-torres A, Capataz-tafur J, Bello-pérez LA, Vargas-torres A (2019) Use of enzymatically modi fi ed starch in the microencapsulation of ascorbic acid: microcapsule characterization, release behavior and in vitro digestion. Food Hydrocoll 96:259–266. https://doi.org/10.1016/j.foodhyd.2019.04.056
17. Jung D, Park C, Kim H, Gyu T, Lee B, Baik M, Yoo S, Seo D (2022) Enzymatic modification of potato starch by amylosucrase according to reaction temperature: effect of branch-chain length on structural, physicochemical, and digestive properties. Food Hydrocoll 122:107086. https://doi.org/10.1016/j.foodhyd.2021.107086
18. Zhai Y, Li X, Bai Y, Jin Z, Svensson B (2022) Maltogenic α-amylase hydrolysis of wheat starch granules: mechanism and relation to starch retrogradation. Food Hydrocoll 124:107256. https://doi.org/10.1016/j.foodhyd.2021.107256
19. Xiao W, He H, Dong Q, Huang Q, An F, Song H (2023) Effects of high-speed shear and double-enzymatic hydrolysis on the structural and physicochemical properties of rice porous starch. Int J Biol Macromol 234:123692. https://doi.org/10.1016/j.ijbiomac.2023.123692
20. Li H, Gui Y, Li J, Zhu Y, Cui B, Guo L (2020) Modification of rice starch using a combination of autoclaving and triple enzyme treatment: structural, physicochemical and digestibility properties. Int J Biol Macromol 144:500–508. https://doi.org/10.1016/j.ijbiomac.2019.12.112
21. Wang C, Zhang X, Tian X, Zhang Z, Zhang X (2023) Physical and enzymatic modifications of starch from blue highland barley and their characterizations, digestibility, and lipolysis inhibitory activities. LWT 180:114722. https://doi.org/10.1016/j.lwt.2023.114722
22. Shah A, Masoodi FA, Gani A, Ashwar B (2018) Dual enzyme modified oat starch: structural characterisation, rheological properties, and digestibility in simulated GI tract. Int J Biol Macromol 106:140–147. https://doi.org/10.1016/j.ijbiomac.2017.08.013
23. Semwal J, Meera MS (2023) Modification of sorghum starch as a function of pullulanase hydrolysis and infrared treatment. Food Chem 416:135815. https://doi.org/10.1016/j.foodchem.2023.135815
24. Benítez RB, Elvira Tabares WF, Lenis Velásquez LA, Hurtado Sánchez CI, Salinas Cruel OA (2021) Enzymatic hydrolysis as a tool to improve total digestibility and techno-functional properties of pigeon pea (*Cajanus cajan*) starch. Heliyon 7:e07817. https://doi.org/10.1016/j.heliyon.2021.e07817
25. Christensen SJ, Madsen MS, Zinck SS, Hedberg C, Sørensen OB, Svensson B, Meyer AS (2023) Enzymatic potato starch modification and structure-function analysis of six diverse GH77 4-alpha-glucanotransferases. Int J Biol Macromol 224:105–114. https://doi.org/10.1016/j.ijbiomac.2022.10.107
26. Jorge F-F, Jairo C-CER-SES-M, H´ector C-V (2023) Hydrothermal processes and simultaneous enzymatic hydrolysis in the production of modified cassava starches with porous-

surfaces. Heliyon 9:e17742. https://doi.org/10.1016/j.heliyon.2023.e17742

27. Li J, Fan J, Hu F (2023) Ultrasound-assisted acid/enzymatic hydrolysis preparation of loquat kernel porous starch: a carrier with efficient palladium loading capacity. Int J Biol Macromol 247:125676. https://doi.org/10.1016/j.ijbiomac.2023.125676
28. Almeida RLJ, dos Santos Pereira T, de Andrade Freire V, Santiago ÂM, Oliveira HML, de Sousa Conrado L, de Gusmão RP (2019) Influence of enzymatic hydrolysis on the properties of red rice starch. Int J Biol Macromol 141:1210–1219. https://doi.org/10.1016/j.ijbiomac.2019.09.072
29. Grewal N, Faubion J, Feng G, Kaufman RC, Wilson JD, Shi YC (2015) Structure of waxy maize starch hydrolyzed by maltogenic α-amylase in relation to its retrogradation. J Agric Food Chem 63:4196. https://doi.org/10.1021/jf506215s
30. Zhai X, Wu K, Ji R, Zhao Y, Lu J, Yu Z, Xu X, Huang J (2022) Structure and function insight of the α-glucosidase QsGH13 from *Qipengyuania seohaensis* sp. SW-135. Front Microbiol 13. https://doi.org/10.3389/fmicb.2022.849585
31. Cheng W, Sun Y, Xia X, Yang L, Fan M, Li Y, Wang L, Qian H (2022) Effects of β-amylase treatment conditions on the gelatinization and retrogradation characteristics of wheat starch Wen. Food Hydrocoll 124:107286
32. Li H, Li J, Xiao Y, Cui B, Fang Y, Guo L (2019) In vitro digestibility of rice starch granules modified by β-amylase, transglucosidase and pullulanase. Int J Biol Macromol 136:1228–1236
33. Xu Y, Sun L, Gu Y, Cheng G, Fan X, Ding Y, Zhuang Y (2023) Improving the emulsification performance of adlay seed starch by esterification combined with ultrasonication and enzymatic treatment. Int J Biol Macromol 242:124839. https://doi.org/10.1016/j.ijbiomac.2023.124839
34. Hutabarat DJC, Stevensen J (2023) Physicochemical properties of enzymatically modified starch: a review. IOP Conference Series: Earth and Environmental Science. https://doi.org/10.1088/1755-1315/1169/1/012093
35. Sorndech W, Sagnelli D, Meier S, Jansson AM, Lee B, Hamaker BR, Rolland-sabaté A, Hebelstrup KH, Tongta S, Blennow A (2016) Structure of branching enzyme- and amylomaltase modified starch produced from well-defined amylose to amylopectin substrates. Carbohydr Polym 152:51–61. https://doi.org/10.1016/j.carbpol.2016.06.097
36. Ulbrich M, Asiri SA, Bussert R, Flöter E (2021) Enzymatic modification of granular potato starch using Isoamylase—investigation of morphological, physicochemical, molecular, and techno-functional properties. Starch 73:1–9. https://doi.org/10.1002/star.202000080
37. Asiri SA, Flöter E, Ulbrich M (2019) Enzymatic modification of granular potato starch—effect of debranching on morphological, molecular, and functional properties. Starch/Staerke 71. https://doi.org/10.1002/star.201900060
38. Demirkesen-Bicak H, Tacer-Caba Z, Nilufer-Erdil D (2018) Pullulanase treatments to increase resistant starch content of black chickpea (Cicer arietinum L.) starch and the effects on starch properties. Int J Biol Macromol 111:505–513. https://doi.org/10.1016/j.ijbiomac.2018.01.026
39. Hu A, Chen X, Wang J, Wang X, Zheng J, Wang L (2021) Effects on the structure and properties of native corn starch modified by enzymatic debranching (ED), microwave assisted esterification with citric acid (MCAE) and by the dual ED/MCAE treatment. Int J Biol Macromol 171:123–129. https://doi.org/10.1016/j.ijbiomac.2021.01.012
40. Huang TT, Zhou DN, Jin ZY, Xu XM, Chen HQ (2015) Effect of debranching and heat-moisture treatments on structural characteristics and digestibility of sweet potato starch. Food Chem 187:218–224. https://doi.org/10.1016/j.foodchem.2015.04.050
41. Li X, Wang Y, Lee BH, Li D (2018) Reducing digestibility and viscoelasticity of oat starch after hydrolysis by pullulanase from *Bacillus acidopullulyticus*. Food Hydrocoll 75:88–94. https://doi.org/10.1016/j.foodhyd.2017.09.009
42. Precha-Atsawanan S, Puncha-arnon S, Wandee Y, Uttapap D, Puttanlek C, Rungsardthong V (2018) Physicochemical properties of partially debranched waxy rice starch. Food Hydrocoll 79:71–80. https://doi.org/10.1016/j.foodhyd.2017.12.014
43. Shi J, Sweedman MC, Shi YC (2018) Structural changes and digestibility of waxy maize starch debranched by different levels of pullulanase. Carbohydr Polym 194:350–356. https://doi.org/10.1016/j.carbpol.2018.04.053
44. Zeng F, Ma F, Gao Q, Yu S, Kong F, Zhu S (2014) Debranching and temperature-cycled crystallization of waxy rice starch and their digestibility. Carbohydr Polym 113:91–96. https://doi.org/10.1016/j.carbpol.2014.06.057

45. Li Y, Li C, Gu Z, Cheng L, Hong Y, Li Z (2019) Digestion properties of corn starch modified by α-D-glucan branching enzyme and cyclodextrin glycosyltransferase. Food Hydrocoll 89:534. https://doi.org/10.1016/j.foodhyd.2018.11.025
46. Sorndech W, Meier S, Jansson AM, Sagnelli D, Hindsgaul O, Tongta S, Blennow A (2015) Synergistic amylomaltase and branching enzyme catalysis to suppress cassava starch digestibility. Carbohydr Polym 132:409–418. https://doi.org/10.1016/j.carbpol.2015.05.084
47. Leoni C, Gattulli BAR, Pesole G, Ceci LR, Volpicella M (2021) Amylomaltases in extremophilic microorganisms. Biomolecules 11:1335. https://doi.org/10.3390/biom11091335
48. Chen Y, McClements DJ, Peng X, Chen L, Xu Z, Meng M, Ji H, Long J, Qiu C, Zhao J, Jin Z (2023) Research progresses on enzymatic modification of starch with 4-α-glucanotransferase. Trends Food Sci Technol 131:164–174. https://doi.org/10.1016/j.tifs.2022.11.025
49. Cho KH, Auh JH, Ryu JH, Kim JH, Park KH, Park CS, Yoo SH (2009) Structural modification and characterization of rice starch treated by Thermus aquaticus 4-α-glucanotransferase. Food Hydrocoll 23:2403–2409. https://doi.org/10.1016/j.foodhyd.2009.06.019
50. Park HR, Kang J, Rho SJ, Kim YR (2020) Structural and physicochemical properties of enzymatically modified rice starch as influenced by the degree of enzyme treatment. J Carbohydr Chem 39:250–266. https://doi.org/10.1080/07328303.2020.1788574
51. Hassanein WS, İspirli H, Dertli E, Yilmaz MT (2023) Structural characterization of potato starch modified by a 4, 6-α-glucanotransferase B from *Lactobacillus reuteri* E81. Int J Biol Macromol 242:124988. https://doi.org/10.1016/j.ijbiomac.2023.124988
52. Wang T, Wang F, Ma R, Tian Y (2022) Enzymatically modified starch for paper surface sizing: enzymes with different action modes and sites. Carbohydr Polym 291:119636. https://doi.org/10.1016/j.carbpol.2022.119636
53. Gaenssle ALO, van der Maarel MJEC, Jurak E (2022) The influence of amylose content on the modification of starches by glycogen branching enzymes. Food Chem 393:133294. https://doi.org/10.1016/j.foodchem.2022.133294
54. Gao L, Wan C, Wang H, Wang P, Yang P, Eeckhout M, Gao J (2023) Changes in the structural and physicochemical characterization of pea starch modified by Bacillus-produced α-amylase. Innov Food Sci Emerg Technol 86:103376. https://doi.org/10.1016/j.ifset.2023.103376
55. Almeida RLJ, Rios NS, dos Santos ES (2022) Modification of red rice starch by a combination of hydrothermal pretreatments and α-amylase hydrolysis. Carbohydr Polym 296:119963. https://doi.org/10.1016/j.carbpol.2022.119963
56. Zheng Y, Chai Z, Kong X, Chen S, Ye X, Tian J (2023) Effect of annealing treatment on the physicochemical properties and enzymatic hydrolysis of different types of starch. Food Chem 403:134153. https://doi.org/10.1016/j.foodchem.2022.134153

Index

A

B

C

Sneh Punia (ed.), *Standardized Procedures and Protocols for Starch*, Methods and Protocols in Food Science, https://doi.org/10.1007/978-1-0716-3866-8,

D

E

F

G

H

I

Q

R

S

T

U

V

Printed in the USA
CPSIA information can be obtained
at www.ICGtesting.com
CBHW061705200524
8775CB00006B/7

9 781071 638651